MOLECULAR MECHANISMS OF ENZYME ACTION

NRI SYMPOSIA ON MODERN BIOLOGY

MOLECULAR MECHANISMS OF ENZYME ACTION

Edited by

Yasuyuki Ogura
Yuji Tonomura
Takao Nakamura

UNIVERSITY PARK PRESS
Baltimore • London • Tokyo

UNIVERSITY PARK PRESS
Baltimore · London · Tokyo

Library of Congress Cataloging in Publication Data

NRI Symposium on Modern Biology, 1st, Tokyo, 1966.
Molecular mechanisms of enzyme action.

Includes bibliographies.
1. Enzymes—Congresses. I. Ogura, Yasuyuki, 1910– ed. II. Tonomura, Yūji, ed. III. Nakamura, Takao, 1929– ed. IV. Nomura Sōgō Kenkyujo. V. Title.
QP601. N18 1966 574.1'925 72-8875
ISBN 0-8391-0738-2

UTP No. 3045-67694-5149
Printed in Japan.

Originally published by
UNIVERSITY OF TOKYO PRESS

PREFACE

In April 1965 the Nomura Securities Company, in celebration of its fortieth anniversary, founded the Nomura Research Institute of Technology and Economics (NRI). The aim of the institute was to provide independent, objective and impartial research covering wide range of disciplines in the human, social and natural sciences to government and industry on a contract basis. The life sciences have been emphasized as an area in great need of laboratory study since the need to solve the practical problems presented by increasing population and rapidly deteriorating environment, the need to achieve better control of diseases, and many other demands are being made of the knowledge of the life processes.

The NRI plans to hold symposia in order to provide better understandings of the life processes particularly in connection with new and progressive comprehensive areas of the biological disciplines. The mechanism of enzyme action was selected as the topic of the first symposium because of its importance in clarifying life processes at molecular level. It is by the catalytic actions of enzymes that living organisms, including human beings, are "alive," performing growth and proliferation and responding in coordination with changes in ambient conditions. Enzymes are, in addition, also being utilized for practical purposes, such as use as drugs for treating diseases, in diagnostics, food processing, and industrial catalysis. Moreover, knowledge of the mechanism of enzyme action is indispensable not only for practical applications but also for solving many other problems directly related to human life.

We were commissioned to organize the first NRI Symposium on Modern Biology held in 1966, and the occasion coincided with the retirement of Professor Yasuyuki Ogura of the University of Tokyo. It was decided that the symposium be held in his honor considering the great contributions he has made to enzymology.

Professor Ogura was born in 1910 and graduated from the Tokyo Imperial University (now the University of Tokyo), Faculty of Science, Department of Botany in 1935. Following graduation he attended the graduate school of the University of Tokyo under Professor Hiroshi Tamiya, and received his Ph. D. in 1951. He spent two years, from 1953 to 1955, in the United States working on enzymes with Professor Briton Chance at the Johnson Foundation Laboratory. He became a professor in the Department of Biochemistry, Faculty of Science, the University of Tokyo in 1960. Professor Ogura has devoted much of his time to studying reaction mechanism of enzymes and has become a leader in the field. Many of his disciples are spread throughout the world and are actively contributing to the field of enzymology.

The symposium was honored by having an opening lecture given by Professor Hiroshi Tamiya on the history of the study of oxidoreductive enzymes in Japan, which was the main object of Professor Ogura's investigation. The program included many topics on oxidation-reduction enzymes, which is the research area specialized in by Professor Ogura, as well as several topics on hydrolyzing enzymes.

It is to our great satisfaction that the symposium, which presented the state of enzyme study in Japan, was such a success. We hope that the symposium results will contribute to both further progress in science and the solving of practical problems.

We would like to extend thanks to all those helped to make the symposium success, in particular, to Professor Tamiya who gave help and advice during all stages of the symposium, to all the speakers who so kindly responded to our requests, and to Dr. Takahisa Ohta and his associate who made arrangements for the symposium to proceed smoothly. We also indebted to Professor Tamiya for his great efforts in reviewing the draft of this book. Finally, we wish to express our appreciation to the University of Tokyo Press for their helpful cooperation in the publication of this volume.

Yuji Tonomura
Takao Nakamura
Hiroyuki Matsumiya

CONTENTS

PREFATORY NOTE

EARLY DAYS (1883–1930) OF THE STUDY OF OXIDOREDUCTIVE ENZYMES IN JAPAN

Hiroshi Tamiya

It is a profound pleasure for me to have been asked to make an opening address at this meeting held in honor of Professor Yasuyuki Ogura, with whom I worked for many years in the field of enzymology especially in relation to biological oxidoreductive processes.

As far as I know, the man who first conducted a study in the relevant field in Japan was Hikorokuro Yoshida (*1*). He published a paper entitled " Chemistry of lacquer (Urushi) " in the Journal of the Chemical Society (London) as early as in 1883. From what I have read, Yoshida's career is not clear except that his work was performed at the then newly established Chemical Laboratory of the Imperial Geological Survey. He pursued—in an astonishingly authentic Western fashion—a detailed chemical survey on the nature and behavior of the tree juice of Urushi (*Rhus vernicifera*) which had traditionally been used in China and the Far East as material for preparing world famous lacquer wares. From samples of raw Urushi juice, he separated a substance that was found to play an essential role in the lacquering process, and, based on the results of elementary analysis, he assigned to it a chemical formula $C_{14}H_{18}O_2$, naming it " urushic acid " (it was, however, revealed by later workers to be a diphenol having the composition $C_{21}H_{34}O_2$ and renamed " urushiol "). The key event occurring in the lacquering process is the gradual hardening or " drying " of the spread Urushi juice accompanied by the darkening of its color. This process was explained by Yoshida as resulting from oxidation of urushic acid to a substance he called " oxyurushic acid " which " polymerized " to form the resinous matter. On performing an elementary analysis of this reaction product, he concluded that in this oxidative process one molecule of urushic acid had taken up one atom of oxygen of air.

The point of primary importance from our point of view is that the reaction in question was proved by Yoshida to be caused by a catalytic agent, which was " albuminoid " in nature, being water-soluble but alcohol-insoluble and losing its activity (by being coagulated) at temperatures higher than 63°C. Moreover, he demonstrated that the optimum temperature for this catalytic agent was at around 20°C, and that it required the presence of air and sufficient moisture for its action. The need of moisture was a remarkable finding at that time, since the phenomenon of hardening of lacquer had been thought to be a process of drying. He proudly wrote that these findings " bear out the practical experience of our lacquer men, *viz.*, that lacquer dries best in the rainy season; it dries better in summer than in winter—a damp atmosphere of about 20–30°C being just the state of air during the rainy season of a year."

Thus, Yoshida was the pioneer in our enzymology and discoverer of one of the most important oxidative enzymes, laccase, as it was later called by G. Bertrand (*2*) of France. It is interesting to note that Yoshida did not use the term " enzyme " or " oxidase " in his paper; instead, he called the agent merely " diastase " or " diastatic matter " which meant the enzymes in general at that time.

The highly laudable, methodical adequacy and logical scrupulousness we see in the work of Yoshida are something that had been lacking in the conventional " science " of old Japan, and we cannot but think that he had some masterly instruction in performing his research.

Soon after the Meiji Restoration, several governmental colleges were established in Tokyo in 1877, and as their teaching staff a number of scholars were invited from Western countries, mostly from Germany and Britain for the fields of fundamental and applied natural sciences. Among these teachers, there was a British scientist named R. W. Atkinson who taught analytical and applied chemistry at the College of Science. Although his name was not mentioned in Yoshida's paper, I presume he might have been Yoshida's instructor, since he was, most probably, the first to introduce knowledge about enzymes into this country. In a paper (*3*) published in 1881, he reported on his discovery of a strong diastatic enzyme in the mycelium of the " Koji "-mold, *Aspergillus oryzae*, which had been traditionally used in the process of fermentation of Japanese rice wine (sake), soy-sauce, and others.

During a period of almost 20 years after the publication of Yoshida's paper, there was no work done in Japan on the oxidoreductive enzymes until a scientist named Oscar Loew came from Germany to teach agricultural chemistry at the College of Agriculture. He was an outstanding plant biochemist whose name we still find in modern text books of plant physiology and biochemistry for his classical works on the mineral nutrition of higher plants. His sojourn in Japan was from 1893 to 1897 and from 1900 to 1907, and in 1901 he published a paper (*4*) on the hydrogen-peroxide decomposing enzyme contained in tobacco leaves, naming it " catalase." Among the students educated by him, there were some brilliant ones who later became professors of agricultural chemistry at the College of Agriculture. One of them, Keijiro Aso, performed a penetrating study on the oxidizing enzymes in higher plants (*5*). Using tissues of various plants (radish, potato, apple, bamboo, *etc.*), he investigated the properties of their oxidase(s)—with guaiac tincture or tetra-methyl-*p*-phenylenediamine as reagents—and peroxidase(s), with guaiacol, *p*-phenylenediamine or tetra-methyl-*p*-phenylenediamine as reagents. He demonstrated that the activities of these enzymes were all inhibited by hydrogen cyanide and phenylhydrazine. Using the juice obtained from a radish root, he attempted the fractionation of the enzymes, and found that both groups of the enzymes were precipitated when the juice was treated with ethylalcohol or saturated with ammonium sulphate. He also observed that on adding a small quantity of acetic acid to the juice the enzymes were freed from some impurities by precipitation of the latter.

Aso (*6*) also performed an interesting experiment making clear the cause of difference in color between black tea of Western countries and green tea of Japan. He showed that the blackening of tea leaves during the so-called " fermentation "-process—which, in fact, has nothing to do with the activities of microorganisms—was a result of the action of an oxidase upon tannin contained in the leaves. In the processing of Japanese green tea, the newly harvested leaves are subjected, before drying, to steaming which kills the oxidase leaving the leaves maintaining their original green color.

Umetaro Suzuki who became later famous for his discovery of oryzanin, the anti-beriberi vitamin now called B_1, was also a student of Loew. He found (*7*) that the mulberry trees which were suffering from " dwarf-trouble " contained abnormally large amounts of oxidase and peroxidase in their leaves.

Toward the end of the 19th century, the Japanese government ceased its policy of inviting teachers from foreign countries, and the job of teaching young students was handed over to those who had been educated by the foreign teachers. At the same time, a number of promising young people were sent abroad to acquire first-hand knowledge of Western sciences. This was the first study-abroad period which lasted until the outbreak of World War I (1914). Among the works done by Japanese scientists during this period, we can find only a few that were concerned with oxido-reductive enzymes. Yamada (*8*) studied, in collaboration with A. Jodlbauer at Munich, the effects of ultraviolet and visible lights upon the action of peroxidase obtained from horseradish. Sano (*9*) studied at Würzburg, together with K. B. Lehmann, the tyrosinase contained in various plant tissues and bacteria. In the literature, their work is cited as being the first to report on the vulnerability of tyrosinase toward hydrogen cyanide.

During the period from 1910 to 1912 Keita Shibata went to Europe and studied plant physiology under W. Pfeffer at Leipzig and organic chemistry under M. Freund at Frankfurt am Main. Although he published no enzymological studies during this period, the work he performed at Pfeffer's laboratory was, as will be described later, of importance in relation to his later work on cytochrome. During his stay in Europe, his brother Yuji Shibata—who later became professor of inorganic chemistry at the University of Tokyo, and to whom we owe the introduction of spectroscopic techniques to chemistry in Japan—studied the chemistry of complex-salts at A. Werner's laboratory at Zürich. After returning to Japan, the brothers started (in 1916), in fruitful collaboration, a unique investigation of using metal-complex salts as enzyme models. They found that some complex-salts, especially those forming labile aquo-complexes, display oxidase-like actions upon various polyphenols. The large number of works done along this line by the Shibata brothers and their co-workers are summarized in a monograph (*10*) they published in 1936.

In 1913, M. Nagayo, then a new returnee from Germany and professor of pathology at the Medical College of the University of Tokyo, published a paper (*11*) reporting that in Germany the study of the " Nadi-reaction " (caused by the " Nadi-oxidase ") in various tissues and cells was attracting the attention of biologists and medical scientists. Stimulated by this paper, many of Nagayo's students began studies on this reaction from a histo-pathological standpoint. One of these students, S. Katsunuma, who performed most extensive investigations on this subject, summarized the results of his and his colleagues' studies in a monograph (*12*) published in Germany in 1924. From the enzymological point of view, the significance of Katsunuma's work lay in the following points.

It had already been generally agreed among scientists that the Nadi-reaction shown by raw tissues (without fixation with formalin) was a reaction caused by an oxidative enzyme, and that the enzyme might contain iron atom, since the reaction had been known to be inhibited by cyanide. To make clear the distribution of iron in various tissues and cells, Katsunuma performed vital staining of them using Quincke's reagent, *i.e.*, a solution of ammonium sulfide which, in the presence of iron, produces easily discernible ferro-sulfide. He demonstrated that the granules stained by this method coincided almost exactly with those showing the enzymic Nadi-reaction. He thus provided strong experimental evidence in favor of the iron-containing nature of the Nadi-oxidase which is now called cytochrome *c* oxidase. Much importance should be attached to the finding made by Katsunuma that the Nadi-positive granules were present in cytoplasm, leucocytes and so forth, but not in nuclei, which is in coincidence with our modern knowledge about the distribution of mitochondria.

Among the studies performed in Japan during the pre-World War I period, I should like to draw attention to that of E. Yamasaki (*13*) who worked at the laboratory of Kikunae Ikeda of the University of Tokyo. Ikeda was a professor of physical chemistry, whose name is now better known as the discoverer of the " essence of deliciousness " (monosodium glutamate). Ikeda motivated Yamasaki to study the kinetics of catalase reaction which, under certain conditions, did not follow the first-order rate law, as it was generally held at that time. With the assumption that, during the process of reaction, catalase underwent gradual inhibition by its substrate (hydrogen peroxide) and product (oxygen), Yamasaki presented a kinetic formula which was shown to coincide well with experimental data.

During World War I (1914–1918), the Japanese scientific world was practically isolated from that of Western countries. With the beginning of the 1920s, a sort of study-abroad rush occurred, in view of the remarkable revival of science, especially in Europe. It was in 1920 that Thunberg in Sweden propounded his dehydrogenase theory, and several years later (1925), there appeared almost simultaneously two most important papers, one by Keilin on cytochrome and the other by Warburg on his Atmungsferment.

Among the countries in Europe, the one most swarmed with Japanese scientists (mostly medical) was Germany. Various kinds of oxidoreductive enzymes were investigated by them under the guidance of: O. Warburg at Berlin-Dahlem (Sakuma, *14*; Tanaka, *15*; Toda, *16*; working mostly on the role of iron in oxidoreductive reactions; Fujita, *17*, on the effect of carbon monoxide upon the respiration of leucocyte, *etc.*), M. Jacoby in Berlin (Tsuchihashi, *18*, on catalase), J. Wohlgemuth in Berlin (Koga, *19*; Maeda, *20*; Yamasaki, *21*; Hizume, *22*; Sugihara, *23*; Tateyama, *24*, on polyphenol oxidase), Th. Brugsch and H. Horsters in Berlin (Harada, *25*, on succino-dehydrogenase), C. Neuberg at Berlin-Dahlem (Kumagawa, *26*, on dehydrogenase activity of yeast), R. Willstätter at Munich (Suminokura, *27*, on laccase), *etc.*

Some went to Sweden or to England, *e.g.*, to H. v. Euler's institute at Stockholm (Nakamura, *28*, on catalase), F. G. Hopkins' institute at Cambridge (Kodama, *29*, *30*, on xanthine oxidase) and C. L. Evans' laboratory in London (Tsubura, *31*, on succino-dehydrogenase). U.S.A. was, at that time, not the country for the study of oxidoreductive enzymes; in the literature I could find only one Japanese, Ishikawa (*32*), who studied formico- and succino-dehydrogenases of bacteria in cooperation with A. J. Kendall at Chicago.

Unfortunately, the majority of these studies, except a few to be mentioned below, were scientifically of minor importance, having been pursued, in subordinate manners, under the guidance of their respective teachers. In fact, many of the scientists mentioned above did, after returning, little to contribute to the study of oxidoreductive enzymes in this country.

One of the exceptions may be the work of Kodama (*29*) performed at

Cambridge dealing with the xanthine dehydrogenase of milk. It was a pioneer work in measuring the redox-potential of the reaction system of oxidoreductive enzymes. Using a gold electrode, he showed that the reaction system of xanthine-dehydrogenase could become reactive to the electrode only in the presence of an oxidoreductive pigment such as methylene blue. By measuring the change of redox-potential of such a system, he followed the process of the electron-transport from the substrate to the pigment mediated by the enzyme.

Working with blood catalase at Jacoby's laboratory in Berlin, Tsuchihashi (*18*) succeeded in making the enzyme free from hemoglobin by shaking their mixture with alcohol and chloroform followed by adsorption to calcium phosphate. This procedure was used afterwards by many investigators working on blood catalase. It may be of interest to note here that Tsuchihashi—having observed that, with the advancement of the purification of the enzyme, there occurred a gradual decrease of nitrogen and protein contents in the preparation—stated that if the enzyme could be purified to a sufficient degree, it would become protein-free. Undoubtedly he was influenced by the concept of R. Willstätter who maintained at that time that the essential principle of enzymes was not necessarily protein. It was only 3 years later that Sumner succeeded in purifying urease in the form of a crystallized protein.

Fujita who studied at Warburg's institute wrote, after returning to Japan, a book (*33*, in Japanese) describing in detail the manometric technique of his German teacher. Although the technique had already been in use in some laboratories at that time, Fujita's book contributed much to the popularization of this method which was, and still is, indispensable in various fields of biochemistry and cell physiology.

Incited by, or independently of, the influences of the returnees from Europe, investigations on oxidoreductive enzymes in Japan became gradually active towards the middle of the 1920s. Kanda (*34*) of the Institute for Physical and Chemical Research performed a series of biochemical studies on the bioluminescence of a marine crustacean, *Cypridina*. He purified luciferin (the substrate of luciferase) to a considerable degree by salting it out from its solution with ammonium sulfate followed by precipitation with phosphotungstic acid.

An interesting finding was reported by Itano and Arakawa (*35*) in 1928, who investigated, at the Oh-hara Institute for Agricultural Research, the catalase contained in cells of a thermophilic cellulose-decomposing bacterium. The enzyme was found to display its activity even at temperatures as high as 80–100°C.

A long series of work (*36*; nearly 40 separate papers) entitled " Report of the peroxidase reaction " was published beginning in 1925 by A. Sato and his staff at the Medical School, Tohoku University. They made comparative studies on the peroxidase activities—using various measuring techniques—of leucocytes, milk, fetuses, *etc.* of rabbits, guinea pigs and men. From an enzymological point of view, however, their studies seemed to be of rather minor interest. Of the findings they reported, I cite here only one, namely that the peroxidase activity of rabbit milk became weaker than normal when the animal was made deficient of vitamin B_1, but restored its normal level on recovering from avitaminosis. Along a similar line, Yaoi (*37*) reported in 1928 that the methylene blue reducing power of pigeon muscle became weaker than normal when the animal was made deficient of vitamin B_1.

This retrospective account would not be regarded as being appropriate to its title unless it deals with the earlier activities of the Shibata school which later produced an array of brilliant works by E. Yakushiji, K. Okunuki, A. Takamiya, Y. Ogura and others on cytochromes, oxidases, dehydrogenases, catalase, *etc.*

The first paper which made Shibata's school widely known in the field of oxidoreductive enzymology was entitled " Untersuchungen über die Bedeutung des Cytochroms in der Physiologie der Zellatmung " by Shibata and Tamiya (*38*). The idea propounded in this paper, which was called the " theory of oxygenation of cytochromes " was, we must now admit, erroneous in some points, but contained certain material worthy of reappraisal from the angle of modern enzymology.

There had been a cogent background which led Shibata and Tamiya to their " oxygenation theory." In his studies on the metabolism of a mold *Aspergillus oryzae*, Tamiya (*39*) found that the mode of respiration of its mycelial mat—growing on the surface of culture solution, and respiring by virtue of its mycelia exposed to open air—was, in many respects, different from those of other aerobic microorganisms, such as yeast and bacteria, whose respiration takes place under conditions of submersion in the solution. It was demonstrated that the strength of the mycelial respiration was markedly dependent on the oxygen partial pressure, showing a maximum in the air containing about 70% O_2, whereas the respiration of submerged microorganisms has been known to be

O_2-saturated at a very low oxygen partial pressure (*e.g.*, at 10^{-5} atm in the case of *Micrococcus candidus* according to Warburg and Kubowitz, *40*). When the mycelial mat was submerged in the culture solution, its aerobic respiration was immediately suppressed, and instead, the mycelia actively performed the process of alcoholic fermentation; namely there occurred the well-known Pasteur-Meyerhof reaction.

On the other hand, the mycelial mats of various *Aspergillus* species contained far smaller amounts of cytochromes (*a*, *b* and *c*) compared with yeast, and their Nadi-reaction (as well as the activity of succino-dehydrogenase) was nil or markedly weaker than those of ordinary aerobic organisms (Tamiya and Hida, *41*). Furthermore, the respiration of mold mycelia was found to be, although somewhat sensitive to cyanide, not at all inhibited by carbon monoxide. It was, however, later found by Ogura and Nagahisa (*42*) that when the mold was subjected to submerged culture ("shaking culture"), it became, unlike the aerial mycelia of the surface culture, rich in cytochromes, concomitantly with the change of its respiration to the CO-sensitive type.

Thus, it was beyond doubt that there was some essential difference in the makeup of respiratory enzyme systems between the cells exposed directly to air and those growing submerged in culture solutions. These findings of Tamiya induced Shibata to infer that the aerobic cells growing in solutions would contain some substance(s) which combines, with strong affinity, with dissolved oxygen and stores it for use by respiratory enzyme systems. This idea recalled to Shibata's mind the work he performed earlier at the institute of W. Pfeffer (*43*). On Pfeffer's suggestion, he performed a study of retesting the work of A. J. Ewart of England (*44*) who reported that some pigment-containing bacteria showed, just like hemoglobin, a capacity for combining, reversibly, with oxygen of air.

To save space, I shall not describe here in detail the experimental technique used in Shibata's work. Using more than 10 species of pigment-forming bacteria (*e.g. Bac. violaceus* and *Bac. brunneus*) as well as a strain of rose-colored yeast and a mold *Monascus purpureus*, he obtained results which were not only affirmative of Ewart's statement, but also indicative of the inhibitory effects of cyanide and carbon monoxide upon the O_2-binding process in question. That the substance(s) responsible for the phenomenon of O_2-binding might have had a close relation to the pigments contained in those organisms was, according to Shibata, evidenced by the observation that no such phenomenon took place in the colorless mutants of *Bac. violaceus* and *Bac. brunneus* or the strains of the same bacteria that had been made colorless by being grown under O_2-deficient conditions. As Shibata admitted later to me, it was regrettable that he did not use ordinary, " colorless " aerobic micro-organisms as controls in his experiments. Having read Keilin's paper on cytochromes, he became convinced that the O_2-binding (or -storing) capacity might not have been limited to the " pigment-forming " microorganisms, but a character in common to all aerobic organisms respiring in liquid media, and that the substance(s) responsible for the phenomenon might, most probably, be cytochromes themselves.

The main experiment, from which Shibata and Tamiya came to believe to have demonstrated the oxygenation and deoxygenation of cytochromes was as follows. In the cells of baker's yeast, which had been subjected to heating at 67°C for 1 hr followed by addition of 5% ethylurethane, all the components of cytochrome were in the " oxidized " forms, and the cells showed no oxygen uptake (in an experiment lasting for 30 min), indicating that they had lost " completely " the activity of dehydrogenase system. On subjecting these cells to anaerobic conditions (evacuation in a Thunberg-type tube), all bands of the " reduced " cytochromes reappeared within about 1 min, and on resupplying air to this cell suspension the " reduced " bands of cytochromes disappeared almost instantaneously.

It might have been quite possible that the yeast cells used in this experiment had still retained a minute part of their dehydrogenase activity, so that the observations described above could not be claimed to be firm evidence for the theory that *all* the cytochrome components performed the processes of oxygenation and deoxygenation. It was, however, demonstrated in 1954 by one of the students of Shibata, K. Okunuki and his staff (*45*), that among the cytochromes, component *a* actually has the property of performing oxygenation and deoxygenation, supplying oxygen to the cytochrome-oxidase system. It should be added that Shibata and Tamiya may have been right in claiming that the agent which oxidizes cytochromes was, contrary to Keilin's contention, not the " Nadi-oxidase," and that the Nadi-system competes with dehydrogenase system for the oxygen supplied by the oxygenating system.

During the early 1930s, we had already become aware of the fallacy in the original, simple " oxygenation theory " of Shibata and Tamiya in view of various experimental data that had accumulated both at home and abroad. In 1937 Ogura and I (*46*) revised the original "oxygenation theory" stating that it was the "Atmungsferment" of Warburg that performs oxygenation and deoxygenation to initiate the electron-transport chain of cellular respiration, in which each of the cytochrome components undergoes, in sequence, the ferri-ferro transformation.

From the early years of the 1930s, the study of oxidoreductive enzymes in this country developed at an ever increasing rate. Indeed, in respect to the rate of advancement in this field, one year in our present time might well be compared to 10 or more years during the period treated in this article. It would give pleasure to me, if this account can give a brief picture of how our predecessors, at first learnt eagerly, albeit passively, Western sciences and later attained scholarly independence of thought in the world of science.

Summary

A review was made of the history of the study of oxidoreductive enzymes in Japan during a period of almost half a century starting from 1883, when H. Yoshida published a paper dealing with laccase. Apparently this investigation was performed under the guidance of some teacher coming from Europe. After a certain period of learning from visiting foreign teachers, there was a period (from about 1900 to 1914), in which a limited number of Japanese scientists went to Europe for the study of the enzymes.

Remarkable progress was then made at home, being directly or indirectly enlightened by the knowledge acquired by those studying abroad. After an interruption caused by World War I, the second period of study-abroad took place from about 1920. Partly incited by the studies performed abroad, but more stimulated, through literature, by the brilliant work of many Western scientists, original work by Japanese scientists began to appear toward the end of the 1920s. The article closed its review with 1930, when Dr. Ogura's teacher, K. Shibata, and his coworkers started their original work on various kinds of oxidoreductive enzymes.

REFERENCES

1 H. Yoshida, *J. Chem. Soc.*, **43**, 472 (1883).
2 G. Bertrand, *Compt. Rend.*, **118**, 1215 (1894); *Ann. Chim. Phys.*, **12**, 115 (1897).
3 R. W. Atkinson, *Mem. Tokyo Imp. Univ. Sci. Dept.*, **6**, 1 (1881); *Proc. Roy. Soc. (London)*, **32**, 299 (1881). See also H. Tamiya, *Proc. Int. Symp. Enzyme Chem., Tokyo and Kyoto*, 21 (1957).
4 O. Loew, *U.S. Dept. Agr. Rept.*, **68**, 47 (1901).
5 K. Aso, *Bull. Coll. Agr. Tokyo Imp. Univ.*, **5**, 207 (1902).
6 K. Aso, *Bull. Coll. Agr. Tokyo Imp. Univ.*, **4**, 255 (1900–1902).
7 U. Suzuki, *Bull. Coll. Agr. Tokyo Imp. Univ.*, **4**, 267, 359 (1900–1902).
8 K. Yamada, *Biochem. Z.*, **8**, 61 (1908).
9 K. B. Lehmann and Sano (initial of first name not given), *Arch. Hyg.*, **67**, 99 (1909).
10 K. Shibata and Y. Shibata, " Katalytische Wirkungen der Metallkomplexverbindungen," Iwata Inst. Plant Biochem., Tokyo, pp. 219 (1936).
11 M. Nagayo, *Nisshin-Igaku*, **2**, 1133 (1913) (in Japanese).
12 S. Katsunuma, " Intrazelluläre Oxydation und Indophenolblausynthese. Histochemische Studie über die ' Oxydasereaktion ' im tierischen Gewebe," Gustav Fischer, Jena, pp. 232 (1924).
13 E. Yamasaki, *Tokyo-Kagakukaishi*, **35**, 1261 (1914) (in Japanese); **36**, 253 (1915); *Sci. Repts. Tohoku Imp. Univ., I Ser.*, **9**, 13 (1920).
14 S. Sakuma, *Biochem. Z.*, **142**, 68 (1923); O. Warburg and S. Sakuma, *Pflügers Arch. Ges. Physiol.*, **200**, 203 (1923).
15 K. Tanaka, *Biochem Z.*, **157**, 425 (1925).
16 S. Toda, *Biochem. Z.*, **172**, 17, 34 (1926).
17 A. Fujita, *Biochem. Z.*, **197**, 189 (1928).
18 M. Tsuchihashi, *Biochem. Z.*, **140**, 63 (1923).
19 T. Koga, *Biochem. Z.*, **141**, 430 (1923).
20 K. Maeda, *Biochem. Z.*, **143**, 347 (1923).
21 Y. Yamasaki, *Biochem. Z.*, **147**, 203 (1924).
22 K. Hizume, *Biochem. Z.*, **147**, 216 (1924).
23 J. Wohlgemuth and N. Sugihara, *Biochem. Z.*, **163**, 260 (1925).
24 R. Tateyama, *Biochem. Z.*, **163**, 297 (1925).
25 Y. Harada, *Biochem. Z.*, **164**, 271 (1925).
26 H. Kumagawa, *Biochem. Z.*, **121**, 150 (1921).
27 K. Suminokura, *Biochem. Z.*, **224**, 292 (1930).
28 K. Nakamura, *Hoppe-Seyler's Z. Physiol. Chem.*, **139**, 140 (1924).
29 K. Kodama, *Biochem. J.*, **20**, 1095 (1926).

30 M. Dixon and K. Kodama, *Biochem. J.*, **20**, 1104 (1926).
31 S. Tsubura, *Biochem. J.*, **19**, 397 (1925).
32 A. J. Kendall and M. Ishikawa, *J. Inf. Dis.*, **44**, 282 (1929).
33 A. Fujita, " The Method and Application of Manometric Techniques in the Field of Medical and Biological Studies," Iwanami-Shoten, Tokyo, pp. 546 (1932) (in Japanese).
34 S. Kanda, *Am. J. Physiol.*, **68**, 435 (1924); *Sci. Papers Inst. Phys. Chem. Res.*, **9**, 265 (1928); **13**, 246 (1930); **18**, Suppl. 1, 15 (1932).
35 A. Itano and S. Arakawa, *Bull. Soc. Agr. Chem.*, **4**, 24 (1928) (in Japanese).
36 A. Sato and his coworkers (S. Yoshimatsu, S. Sekiya, K. Shoji, K. Tokué, T. Arakawa and K. Suzuki), *Am. J. Dis. Child.*, **29**, 301, 313 (1925); *Tohoku J. Exp. Med.*, **7**, 111 (1926); **9**, 642 (1927); **12**, 281, 295, 445, 459 (1929); **16**, 83, 90, 97, 107, 111, 118, 228, 232, 236 (1930).
37 H. Yaoi, *Med. World*, **8**, 85 (1928).
38 K. Shibata and H. Tamiya, *Acta Phytochim.*, **5**, 23 (1930).
39 H. Tamiya, *Acta Phytochim.*, **4**, 227 (1929).
40 O. Warburg and F. Kubowitz, *Biochem Z.*, **214**, 5 (1929).
41 H. Tamiya and T. Hida, *Acta Phytochim.*, **4**, 343 (1929).
42 Y. Ogura and M. Nagahisa, *Bot. Mag.* (*Tokyo*), **51**, 597 (1937).
43 K. Shibata, *Jahrb. Wiss. Bot.*, **51**, 179 (1912).
44 A. J. Ewart, *J. Linn. Soc. Bot.*, **33**, 123 (1897).
45 K. Okunuki and I. Sekuzu, *Seitai-no-Kagaku* (*Science of Living Organisms*), **5**, 265 (1954) (in Japanese); I. Sekuzu, S. Takemori, T. Yonetani and K. Okunuki, *J. Biochem.*, **46**, 43 (1959); Y. Orii and K. Okunuki, *J. Biochem.*, **53**, 489 (1963); **57**, 45 (1965).
46 H. Tamiya and Y. Ogura, *Acta Phytochim.*, **9**, 123 (1937).

REACTION MECHANISM OF AMINE OXIDASE FROM *ASPERGILLUS NIGER*

Yasuyuki Ogura, Haruo Suzuki and Hideaki Yamada
Department of Biophysics and Biochemistry, Faculty of Science, University of Tokyo, Tokyo, and the Research Institute for Food Science, Kyoto University, Kyoto

Amine oxidases are known to catalyze the oxidative deamination of various amines (RCH_2NH_2) in the presence of molecular oxygen, forming stoichiometric amounts of aldehyde (RCHO), ammonia and hydrogen peroxide. The reaction can be expressed as follows:

$$RCH_2NH_2+O_2+H_2O=RCHO+NH_3+H_2O_2\,.$$

At present it is thought that amine oxidases can be divided into two types on the basis of their prosthetic groups: that is, flavin enzymes containing flavin adenine dinucleotide (FAD) and enzymes having both cupric copper and pyridoxal phosphate. The amine oxidases from *Micrococcus rubens* (*1*, *2*) and *Sarcina lutea* (*3–5*) were found to have FAD as the prosthetic group. On the other hand, the enzymes from beef plasma (*6*, *7*), pig plasma (*8*, *9*) and pig kidney (*10–12*) are reported to contain cupric copper and pyridoxal phosphate as essential components. Adachi and Yamada (*13*) prepared amine oxidase in crystalline form from a mycelial extract of *Aspergillus niger* grown in culture medium containing amine as the sole nitrogen source. They reported that this enzyme has a molecular weight of 252,000. By ESR measurements and chemical analysis, Yamada and his coworkers (*13–15*) found that the *Aspergillus* enzyme contained 2–3 atoms of cupric copper per molecule, and that its activity was lost on removal of the bound copper ions and partially recovered on addition of copper ion. The *Aspergillus* enzyme had an absorption spectrum very similar to those of the enzymes from beef plasma (*6*, *7*), pig plasma (*8*, *9*) and pig kidney (*10–12*), and these latter are all markedly inhibited by carbonyl reagents, such as hydrazine, phenylhydrazine and semicarbazide. The reaction of the *Aspergillus* enzyme was also strongly inhibited by these carbonyl reagents. Pyridoxal compounds were demonstrated by bioassay with *Saccharomyces carlsbergensis* in a fraction of a digest of the *Aspergillus* enzyme (*13*). These results suggest that the *Aspergillus* enzyme contains pyridoxal phosphate as a prosthetic group, beside cupric copper. However, the pyridoxal phosphate content per molecule has not yet been determined. The *Aspergillus* enzyme used in this study was prepared by the method of Adachi and Yamada (*13*).

Titration of the Aspergillus Enzyme with Phenylhydrazine

The absorption spectrum of the amine oxidase changed distinctly on addition of phenylhydrazine. That is, an absorption band appeared at

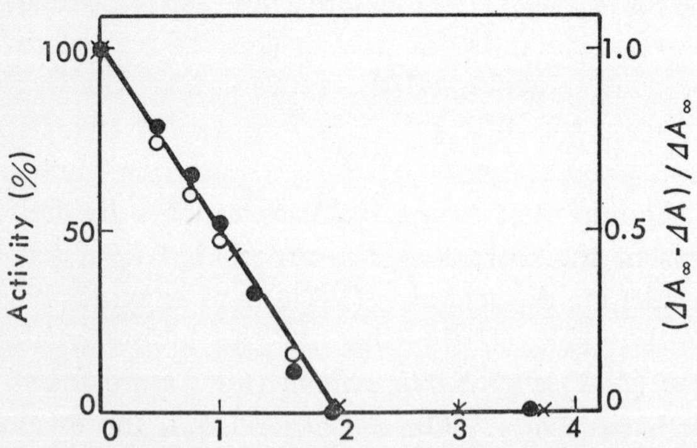

FIG. 1. Titration of the amine oxidase with phenylhydrazine or hydrazine. ● change in the absorbance at 442.5 mμ caused by phenylhydrazine. ○ activity change caused by phenylhydrazine. × activity change caused by hydrazine. Buffer: 0.06 M phosphate (pH 7.1).

442.5 mμ and increased in intensity with increase in the phenylhydrazine concentration (*16*). The peak at 442.5 mμ seems to be due to absorption of an enzyme-phenylhydrazine compound, which was formed by the reaction of this reagent with carbonyl groups in the enzyme. This reaction was found to be irreversible. Making use of these results, the amine oxidase was titrated with phenylhydrazine to determine the number of active carbonyl groups in the enzyme molecule. As shown in Fig. 1, increase in the absorbance at 442.5 mμ with corresponding decrease in enzyme activity was linearly related to the amount of phenylhydrazine added until the molar ratio of phenylhydrazine to amine oxidase reached 2.0. No further change was observed at higher molar ratios. These data suggest that there are two carbonyl groups in the amine oxidase molecule which react with the reagent and that these carbonyl groups are essential for enzyme activity.

Anaerobic Titration of the Amine Oxidase with n-Butylamine

It is known that *n*-butylamine is deaminated oxidatively to *n*-butyraldehyde by the amine oxidase under aerobic conditions, and that the amine oxidase is bleached by the substrate under anaerobic conditions (*17*). As shown in Fig. 2A, the broad absorption band around 420–490 mμ of the native enzyme decreased on anaerobic titration with *n*-butylamine, and this was replaced by two lower maxima at 470 and 440 mμ. The difference spectrum between the native enzyme and the bleached form showed two absorption maxima at 495 and 456 mμ. When the amine oxidase was titrated anaerobically with other amines, such as ethylamine, histamine and benzylamine, the spectra of the bleached forms thus obtained were identical with that of the enzyme bleached with *n*-butylamine (Fig. 2B). This is compatible with the results obtained by Hill and Mann (*18*) for the amine oxidase of pea seedlings.

The amine oxidase was titrated with *n*-butylamine by measuring the change in absorbance at 495 mμ under anaerobic conditions (Fig. 3). The data indicate that on transformation to the bleached form one mole of amine oxidase reacts with two moles of *n*-butylamine. To see whether *n*-butylamine and phenylhydrazine react competitively with the essential carbonyl groups in the amine oxidase, phenylhydrazine was added to the reaction medium after the enzyme had been bleached by addition of a stoichiometric amount of *n*-butylamine under anaerobic conditions.

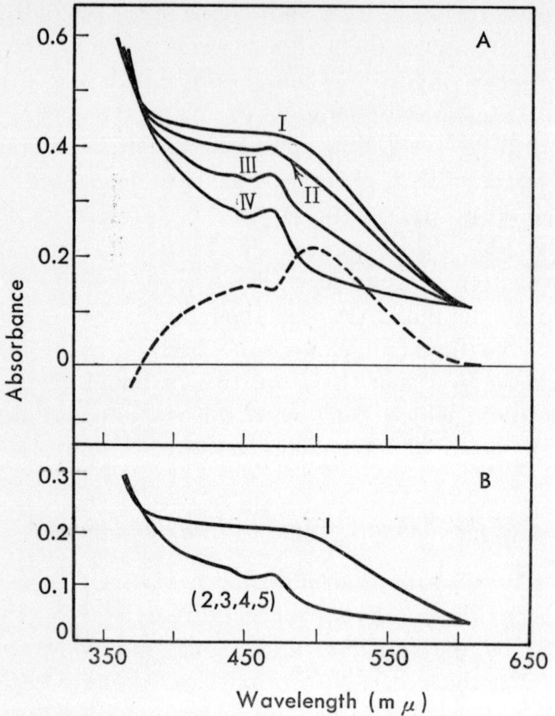

FIG. 2. A) Spectral change of the amine oxidase on treatment with *n*-butylamine (pH 7.1). I, the native enzyme; II and III, partly bleached forms; IV, the fully bleached enzyme. Broken line, difference spectrum of native form minus bleached form. B) Absorption spectra of the amine oxidase bleached anaerobically with different amines. Native enzyme (1), oxidase treated with *n*-butylamine (2), ethylamine (3), histamine (4) and benzylamine (5).

The absorption spectrum of the enzyme-phenylhydrazine compound with a band at 442.5 mμ did not appear. When the substrate in the mixture was exhausted by aeration, the spectrum of this compound appeared. This indicates that the substrate and inhibitor compete for the carbonyl groups in the amine oxidase.

The absorption spectrum of the *Aspergillus* enzyme shows a broad band around 420–490 mμ similar to that of amine oxidases of beef plasma (480 mμ), pig plasma (470 mμ), and pig kidney (470 mμ). A band having a sharp maximum at 440 mμ (beef plasma) (*19*), 420 mμ (pig plasma) (*9*), 440 mμ (pig kidney) (*10*) or 420 mμ (pig kidney) (*11*) ap-

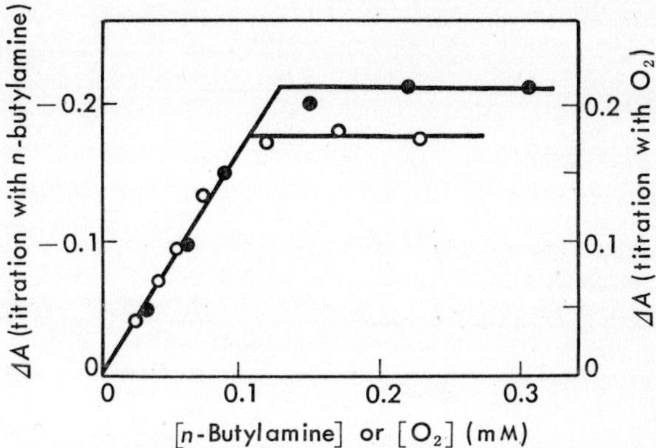

FIG. 3. Spectrophotometric titration of the amine oxidase with *n*-butylamine under anaerobic conditions (●) and of the bleached form with molecular oxygen (○). Concentration of amine oxidase: 65 μM (●), and 54 μM (○). Volume of reaction medium (pH. 7.1): 3.0 ml.

peared on addition of phenylhydrazine to the animal oxidases. Thus, the optical properties of the animal oxidases are similar to those of the *Aspergillus* enzyme, suggesting that the *Aspergillus* enzyme contains copper and pyridoxal phosphate. If this is so, the two active carbonyl groups in the *Aspergillus* enzyme, which react with phenylhydrazine and the substrate, may both be an aldehyde group at position 4 of pyridoxal phosphate. It seems, therefore, possible that the *Aspergillus* enzyme contains two moles of pyridoxal phosphate per mole of enzyme.

To determine the amount of *n*-butyraldehyde produced anaerobically from *n*-butylamine by the amine oxidase reaction, the enzyme was titrated spectrophotometrically with *n*-butylamine in the presence of yeast al-

TABLE I. Stoichiometry of the Formation of *n*-Butyraldehyde from *n*-Butylamine under Anaerobic Conditions

	Concentration (μM)		
	Exp. 1	Exp. 2	Exp. 3
Amine oxidase	26	39	45
n-Butylamine added	41.6	20	40
n-Butyraldehyde formed	43	16	42
Amine oxidase bleached	20.2	9.7	18.3

Buffer: 0.06 M phosphate (pH 7.1). Volume of the reaction medium: 2.5 ml.

TABLE II. Stoichiometry of the Formation of Ammonia in the Amine Oxidase Reaction

	Concentration (μM)	
	Exp. 1	Exp. 2
Amine oxidase	35	35
n-Butylamine added anaerobically	50	37
Amine oxidase bleached	32	18.5
Ammonia formed in absence of oxygen	0	0
Molecular oxygen added	40.6	33
Amine oxidase oxidized	16.7	16.0
Ammonia formed on addition of oxygen	24	33

Volume of reaction medium: 2.3 ml for Exp. 1, and 2.5 ml for Exp. 2. Buffer: 0.06 M phosphate (pH 7.5).

cohol dehydrogenase and NADH under anaerobic conditions. The moles of the bleached form produced were determined from the change in absorbance at 495 mμ and the moles of *n*-butyraldehyde produced were estimated from the decrease in absorbance at 340 mμ, assuming that an equivalent amount of NADH disappeared with formation of *n*-butyraldehyde. Table I shows that the molar ratio of [*n*-butylamine added] : [amine oxidase bleached] : [*n*-butyraldehyde produced] was 2: 1: 2; an equivalent amount of *n*-butyraldehyde was produced from *n*-butylamine even under anaerobic conditions.

To see whether ammonia was produced in the enzymic reaction in the absence of molecular oxygen, a mixture of amine oxidase, L-glutamic acid dehydrogenase, α-ketoglutarate and NADH was titrated anaerobically with *n*-butylamine, taking advantage of the L-glutamic acid dehydrogenase reaction.

$$\alpha\text{-Ketoglutarate} + \text{NADH} + \text{NH}_3 + \text{H}^+ \rightleftharpoons \text{L-glutamate} + \text{NAD}^+ + \text{H}_2\text{O}$$

No ammonia was produced by the enzyme reaction in the absence of oxygen, since no absorbance change at 340 mμ was observed (Table II).

Anaerobic Titration of the Bleached Amine Oxidase with Molecular Oxygen

The *Aspergillus* enzyme which has been bleached with a substrate is known to regain its original color on aeration (*17*), like the animal oxidases described above (*12*). To study the stoichiometry of this backward reaction, a solution of the enzyme was bleached with a stoichiometric amount of *n*-butylamine and then titrated with molecular oxygen. The absorbance at 495 mμ increased almost linearly with increase in the oxygen concentration until the absorption spectrum of the native enzyme was fully restored (Fig. 3). The data indicate that two moles of molecular oxygen are consumed per mole of bleached enzyme. Although ammonia was not detected during formation of *n*-butyraldehyde from *n*-butylamine under anaerobic conditions, a marked decrease in absorbance at 340 mμ together with increase in absorbance at 495 mμ, due to formation of the native enzyme, was observed on titration with molecular oxygen in the presence of the L-glutamic acid dehydrogenase system. The data in Table II show that the approximate molar ratio of [native amine oxidase formed] : [ammonia produced] : [oxygen consumed] was 1: 2: 2. Hydrogen peroxide was not formed in the enzyme reaction under anaerobic conditions, because no oxygen was liberated on addition of catalase (*16*). Under aerobic conditions, however, a stoichiometric amount of hydrogen peroxide was produced with consumption of oxygen, and the molar ratio of [*n*-butylamine oxidized] : [O_2 exhausted] : [H_2O_2 produced] was 1: 1: 1.

Based on these data, the following two steps are proposed as the reaction of the *Aspergillus* enzyme.

$$E\langle{}^{CHO}_{CHO} + 2RCH_2NH_2 = E\langle{}^{CH_2NH_2}_{CH_2NH_2} + 2RCHO \qquad (1)$$

$$E\langle{}^{CH_2NH_2}_{CH_2NH_2} + 2O_2 + 2H_2O = E\langle{}^{CHO}_{CHO} + 2H_2O_2 + 2NH_3 \qquad (2)$$

Reaction 1 is a transamination which takes place anaerobically, and Reaction 2 is an oxidative deamination occurring in the presence of molecular oxygen. It seems likely that the native amine oxidase $\left(E\langle{}^{CHO}_{CHO}\right)$ has two moles of pyridoxal phosphate per mole of the enzyme, and that in Reaction 1 these prosthetic groups change to pyridoxamine phosphate in the bleached form $\left(E\langle{}^{CH_2NH_2}_{CH_2NH_2}\right)$. A similar stoichiometric equation was proposed by other authors (*9*, *20*, *21*) without conclusive evidence to support the reaction scheme described above.

Rate of the Overall Reaction

To obtain further information on the reaction mechanism of the amine oxidase, the reaction was analyzed from the overall reaction rate measured by oxygen consumption (*22*). In our kinetic experiments (*22*), it was assumed that the oxidative deamination of amine is catalyzed by an enzyme unit composed of a single active carbonyl group (presumably PLP); the concentration of the enzyme unit (e) was determined spectrophotometrically, taking the difference between the extinction coefficients at 442.5 mμ of the enzyme unit and its complex with phenylhydrazine as 36.1 $\text{mM}^{-1}\ \text{cm}^{-1}$. Figure 4 shows the rate of the overall reaction, v/e, in the presence of various concentrations of substrate and oxygen. Linear relationships were found between e/v and $1/[O_2]$, which ran parallel with each other in the presence of various given concentrations of *n*-butylamine. The velocities, $V_{ap}{}^{O_2\to\infty}/e$, at a sufficient concentration of oxygen and various given concentrations of the substrate were estimated from the intercepts on the ordinate. The maximum velocity, V_m/e, obtainable at sufficient concentrations of both substrate and

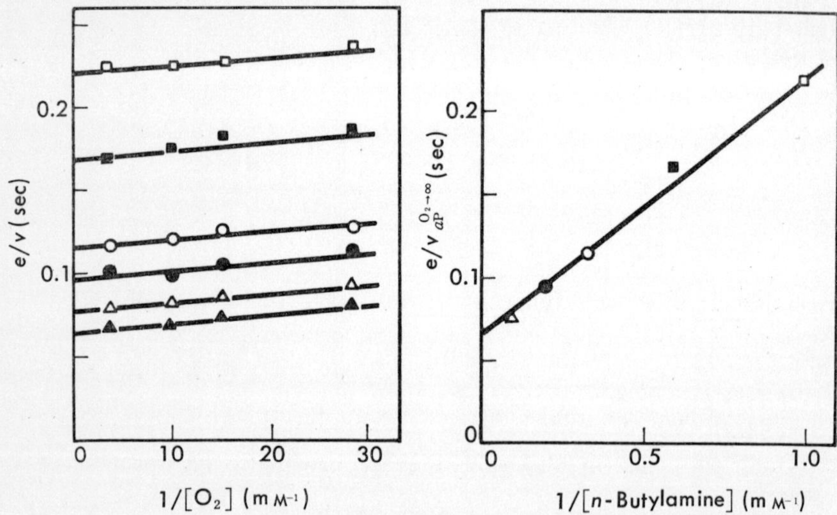

FIG. 4. Plots of e/v *vs.* $1/[O_2]$ and $e/V_{ap}{}^{O_2\to\infty}$ *vs.* $1/[s]$. Enzyme unit concentration: 0.367 μM. *n*-Butylamine concentration: 1 mM (□), 2 mM (■), 3 mM (○), 5 mM (●), 10 mM (△) and ∞ (▲). Buffer: 0.06 M phosphate (pH 7.5). Temperature: 25°C.

TABLE III. Values of V_m/e, K_m and Kinetic Constants, A and B, at pH 7.5 and 25°C as Estimated from the Overall Reaction Kinetics

Substrate	V_m/e (sec^{-1})	K_m (mM)	A (M sec)	B (M sec)
n-Butylamine	15.4	2.36	1.52×10^{-4}	4.3×10^{-7}
Ethylamine	9.5	20	2.0×10^{-3}	5.3×10^{-7}
Benzylamine	8.7	0.56	6.4×10^{-5}	6.0×10^{-7}

oxygen was estimated by plotting the reciprocals of the $V_{ap}^{O_2\to\infty}/e$ values against the reciprocals of the substrate concentrations. The Michaelis constant, K_m, for the substrate at a sufficient concentration of oxygen was obtained from the slope of the linear relationship between $e/V_{ap}^{O_2\to\infty}$ and $1/[s]$. The lines obtained by plotting the e/v values at various oxygen concentrations against 1/[s] were also found to run parallel with each other. Based on these data, the rate of the overall reaction may be expressed as follows:

$$\frac{e}{v}=\frac{e}{V_m}+\frac{A}{[s]}+\frac{B}{[O_2]}, \tag{3}$$

where A and B are kinetic constants. The values of V_m/e, K_m, A and B for *n*-butylamine together with those for ethylamine and benzylamine are summarized in Table III. As can be seen, the value of B is nearly constant with all three substrates.

Absorption Spectrum of the Enzyme in the Steady State

Spectrophotometric measurements were made by the " stopped flow " method to see if a transient intermediate was formed in the steady state of the reaction. A mixture of the enzyme and *n*-butylamine was mixed in a flow apparatus with an equal volume of air-saturated buffer. Figure 5 shows typical oscillograph traces of the absorbance changes at 335 and 495 mμ, which indicate that the bleached form (presumably PMP form) is rapidly oxidized to a steady state level and returns to the PMP form upon exhaustion of the dissolved molecular oxygen. Similar experiments were carried out with various substrates and at different wavelengths, and the maximum absorbance changes on the traces were plotted against the wavelength. The difference absorption spectra thus obtained between the steady state level and the PMP form are shown in

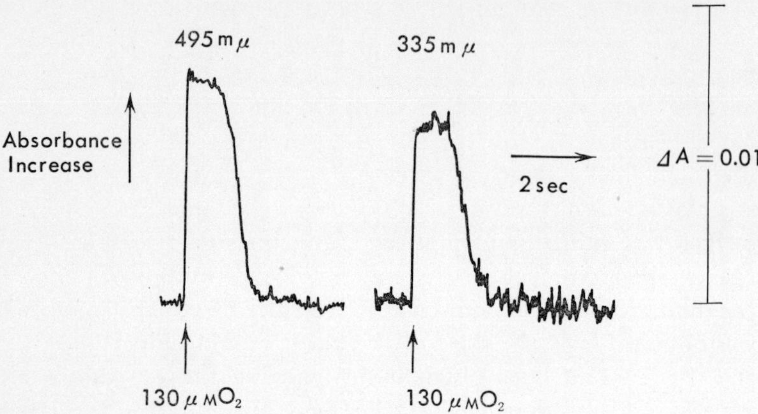

FIG. 5. Oscillograph traces of the absorbance changes at 335 and 495 mμ as obtained by the "stopped flow" method.

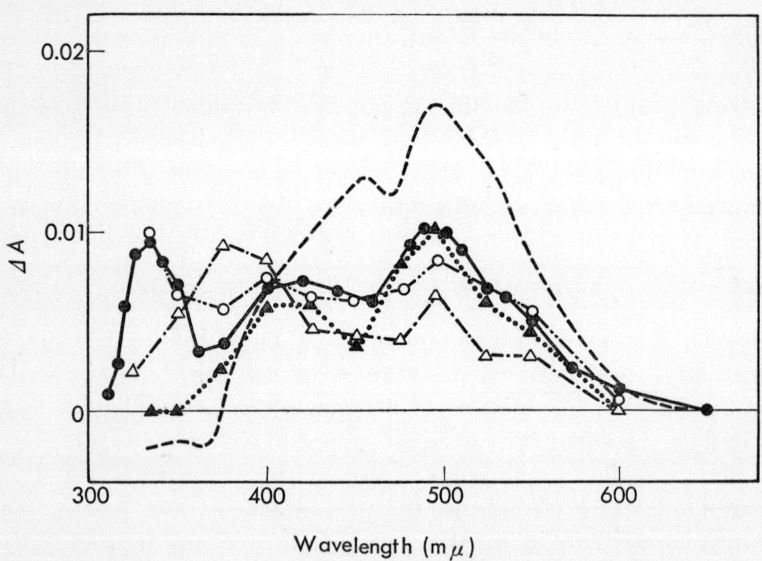

FIG. 6. Difference absorption spectra of the steady state minus PMP level. Substrate used: *n*-butylamine (●), ethylamine (▲), benzylamine (△) and histamine (○). Broken line: difference spectrum of PLP minus PMP form.

Fig. 6; difference maxima were observed at 335, 420 and 495 mμ with *n*-butylamine, at 420 and 495 mμ with ethylamine, at 375 and 495 mμ with benzylamine, and at 335, 400 and 495 mμ with histamine. Thus, the difference spectrum varies with the substrate used, and is distinctly different from the spectra of the PLP and PMP forms. This shows that a transient intermediate(s) is formed in the steady state of the reaction. The absorption spectrum in the steady state may be a composite of the spectrum of the PLP form and that of an intermediate(s). It should be noted that the spectrum in the steady state is very similar to a composite of spectra of the intermediates of glutamate aspartate transaminase, which is known to contain pyridoxal phosphate as a prosthetic group (*23–31*).

Rate of Transamination

The rate of transamination (Reaction 1) was measured by the " stopped flow " method from the absorbance changes at 495 and 335 mμ, on mixing a solution of the PLP form with *n*-butylamine under anaerobic conditions. The oscillograph traces obtained showed that the absorbance values at 335 and 495 mμ decreased gradually after the flow had stopped (Fig. 7). The rapid increase in the absorption at 335 mμ seems to indicate that the intermediate(s) described above is formed very rapidly in the transamination from the PLP form to the PMP form under anaerobic conditions, since no absorption peak was observed at this wavelength in the difference spectrum of the PLP minus PMP form. These data indicate that the equilibrium between the PLP form plus substrate and the intermediate(s) was established in the pre-steady

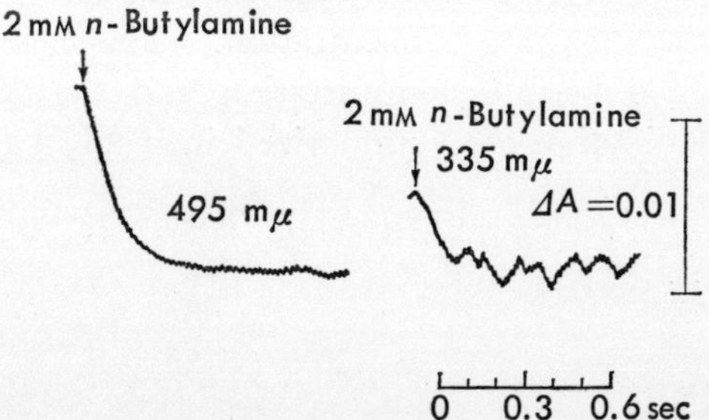

Fig. 7. Oscillograph traces obtained on mixing a solution of the enzyme with *n*-butylamine (pH 7.5, 25°C).

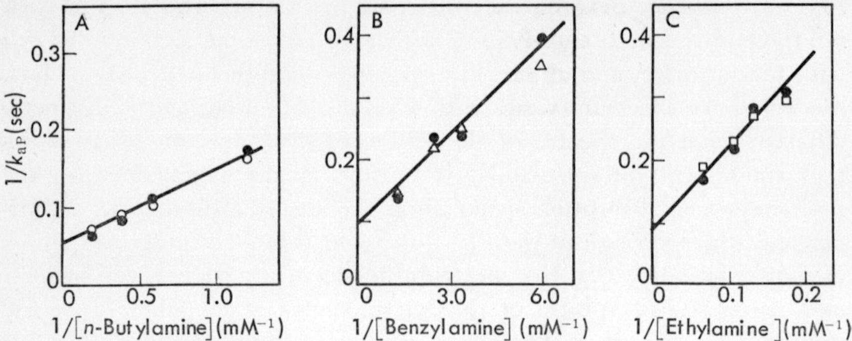

FIG. 8. Relationships between reciprocals of the k_{ap} value and of the substrate concentration. Absorbance changes measured at 495 mμ (●), 335 mμ (○), 375 mμ (△), 420 mμ (□).

state. The relationship obtained by plotting log ΔA_t values against the reaction time was linear (*16*), showing that the decrease in absorbance proceeded according to the first order reaction kinetics with respect to the enzyme concentration (Fig. 7). The value of the apparent rate constant of Reaction 1, k_{ap}, was calculated by the following equation.

$$\log \Delta A_t = -k_{ap}t/2.303+C\,, \tag{4}$$

where t is the time in seconds after the flow has stopped, ΔA_t is the difference in absorbance between a mixture of the PLP form and intermediate(s) and the PMP form at time t as a fixed wavelength, and C is a constant. Figure 8 shows the plots of the reciprocals of k_{ap} values against the reciprocals of the substrate concentration. The relationships obtained with *n*-butylamine, benzylamine and ethylamine were all linear regardless of the wavelength at which they were measured. Thus, k_{ap} was a function of the substrate concentration. The $k_{ap}{}^{max}$ values obtainable at a sufficient substrate concentration and the Michaelis constants, $K_m{}^s$, of transamination are summarized in Table IV.

TABLE IV. Values of $k_{ap}{}^{max}$ and $K_m{}^s$ at pH 7.5 and 25°C Obtained by the "Stopped Flow" Method

Substrate	$k_{ap}{}^{max}$ (sec^{-1})	$K_m{}^s$ (mM)
n-Butylamine	18.0	2.0
Ethylamine	11.1	15.0
Benzylamine	10.3	0.5

Rate of Oxidation of the PMP Form

The rate of oxidation of the PMP form, k_{ox}, with molecular oxygen was measured by the " continuous flow " method, mixing the PMP form prepared anaerobically on addition of *n*-butylamine or ethylamine with a buffer containing oxygen. The concentration of substrate added in this experiment was much lower than the K_m value but higher than the enzyme unit concentration; the PMP form may, therefore, be almost fully oxidized to the PLP form with molecular oxygen under the condition of $e \ll [O_2]$, without appearance of the spectrum of the steady state level. The value of k_{ox} was estimated using the following equation.

$$\log \frac{e}{e-e_t} = \frac{k_{ox}t}{2.303}, \qquad (5)$$

where e_t is the concentration of the PLP form of the enzyme unit at t and t is the time in seconds after mixing the PMP form with molecular oxygen. The relationship between the k_{ox} value and oxygen concentration is shown in Fig. 9 and indicates that the k_{ox} value increases almost

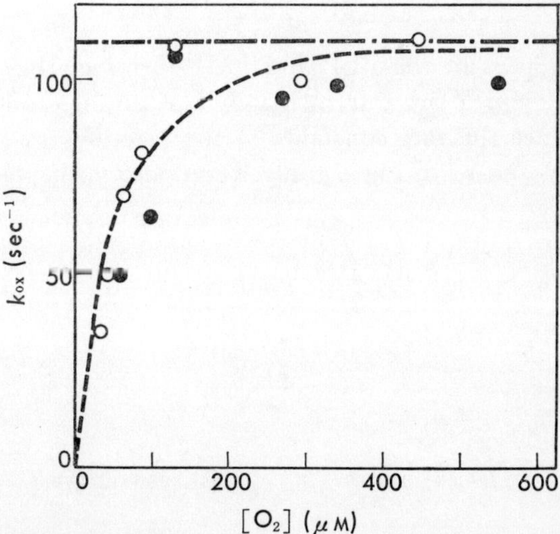

FIG. 9. Relationship between k_{ox} value and oxygen concentration. The PMP form was prepared anaerobically by addition of *n*-butylamine (●) or ethylamine (○).

linearly with increase in oxygen concentration up to 100 μM. It was thus found that the reaction proceeds according to the second order reaction kinetics with respect to both enzyme unit and oxygen concentration. Under conditions of $[O_2]>100$ μM, the k_{ox} value was constant, 110 sec^{-1}, being independent of the oxygen concentration. Namely, the reaction was of the zero order with respect to oxygen concentration. These results indicate that there are at least two reaction steps in the oxidation of the PMP form with oxygen.

Reaction Mechanism of the Amine Oxidase

It was found that 1) an intermediate(s) appeared in the transamination (Reaction 1), and 2) there were at least two reaction steps in the oxidation of the PMP form with oxygen. The following reaction mechanism is considered to account for these data and also those obtained in titration experiments.

$$E_L+S \underset{k_{-1}}{\overset{k_{+1}}{\rightleftharpoons}} X \xrightarrow{k_{+2}} E_M+P \tag{6}$$

$$E_M+O_2 \underset{k_{-3}}{\overset{k_{+3}}{\rightleftharpoons}} Y \xrightarrow{k_{+4}} E_L+NH_3+H_2O_2\,, \tag{7}$$

where E_L and E_M represent the PLP and PMP forms of the enzyme unit, respectively, X and Y are intermediates, S is amine, and P is an aldehyde. The *k's* are the rate constants of the reactions indicated.

In the experiments shown in Fig. 7, only Reaction 6 takes place. In such a case,

$$\frac{d[E_L]}{dt} = -k_{+1}[E_L][s]+k_{-1}[x]$$

$$\frac{d[x]}{dt} = -(k_{-1}+k_{+2})[x]+k_{+1}[E_L][s]$$

$$\frac{d[E_M]}{dt} = k_{+2}[x]\,.$$

under conditions where $k_{+1}[s]$ and $k_{-1} \gg k_{+2}$ (*32*), we obtain

$$[E_L] \doteqdot \frac{e}{k_{+1}[s]+k_{-1}}\left[k_{+1}[s]\exp\left(-k_{+1}[s]-k_{-1}\right)t + k_{-1}\exp\left(-\frac{k_{+1}[s]k_{+2}}{k_{+1}[s]+k_{-1}}\right)t\right]$$

$$[\mathrm{x}] \doteqdot \frac{e}{k_{+1}[\mathrm{s}]+k_{-1}}\Big[-k_{+1}[\mathrm{s}] \exp(-k_{+1}[\mathrm{s}]-k_{-1})t$$

$$+k_{+1}[\mathrm{s}] \exp\Big(-\frac{k_{+1}[\mathrm{s}]k_{+2}}{k_{+1}[\mathrm{s}]+k_{-1}}\Big)t\Big].$$

The first terms of these equations account for a rapid decrease of the PLP form and a rapid formation of the intermediate(s). The second terms indicate the rather slow decrease in the concentrations of the PLP form and intermediate(s). In the experiments described above (Fig. 7), we could only measure the rate of this slow reaction. Then we have

$$[\mathrm{E_L}] \doteqdot \frac{k_{-1}e}{k_{+1}[\mathrm{s}]+k_{-1}} \exp\Big(-\frac{k_{+1}k_{+2}[\mathrm{s}]}{k_{+1}[\mathrm{s}]+k_{-1}}t\Big)$$

$$[\mathrm{x}] \doteqdot \frac{k_{+1}e[\mathrm{s}]}{k_{+1}[\mathrm{s}]+k_{-1}} \exp\Big(-\frac{k_{+1}k_{+2}[\mathrm{s}]}{k_{+1}[\mathrm{s}]+k_{-1}}t\Big).$$

The absorbance difference (ΔA_t) at t can be expressed as follows;

$$\Delta A_t = \Delta\varepsilon_{\mathrm{L-M}}[\mathrm{E_L}]+\Delta\varepsilon_{\mathrm{I-M}}[\mathrm{x}],$$

where $\Delta\varepsilon_{\mathrm{L-M}}$ and $\Delta\varepsilon_{\mathrm{I-M}}$ are the differences between the molar extinction coefficients of the PLP form and PMP form and between those of the intermediate(s) and the PMP form, respectively. The concentration of substrate is almost constant during the reaction, since $e \ll [\mathrm{s}]$. Therefore,

$$\frac{d \ln(\Delta A_t)}{dt} = k_{\mathrm{ap}} = \frac{k_{+2}}{1+\dfrac{k_{-1}}{k_{+1}[\mathrm{s}]}}. \tag{8}$$

Thus, k_{ap} is the rate of Reaction 6 per unit concentration of the enzyme unit. This formula explains the fact that the k_{ap} values measured at a given substrate concentration but at different wavelengths are identical with each other. It is obvious that

$$k_{\mathrm{ap}}{}^{\max} = k_{+2} \text{ and } K_m{}^{\mathrm{s}} = \frac{k_{-1}}{k_{+1}}.$$

The intermediate, Y, postulated from the kinetic standpoint may be either the O_2-PMP complex (E_MO_2) or an imino form of the enzyme unit produced by dehydrogenation of the PMP moiety with oxygen (Scheme 1). The latter is more likely, but the spectral properties of this intermediate are unknown. If this hypothesis is tenable, hydrogen peroxide may be produced on formation of the imino form, and ammonia may be produced by hydration in the reaction from the imino form to the PLP form. Neglecting the reverse reaction, $Y \longrightarrow E_M + O_2$, in Reaction 7, we have (*32*)

$$[E_L] = e\left[1 - \frac{1}{k_{+3}[O_2] - k_{+4}}\{k_{+4}\exp(-k_{+3}[O_2])t - k_{+3}[O_2]\exp(-k_{+4})t\}\right].$$

When the concentration of oxygen is sufficiently low, $k_{+3}[O_2] \ll k_{+4}$. Then,

$$[E_L] = e[1 + \exp(-k_{+3}[O_2]t)]$$

$$\ln\frac{e - [E_L]}{e} = -k_{ox}t$$

$$k_{ox} = k_{+3}[O_2]. \tag{9}$$

When the concentration of oxygen is sufficiently high, $k_{+3}[O_2] \gg k_{+4}$.

$$[E_L] = e[1 - \exp(-k_{+4})t]$$

$$k_{ox} = k_{+4}. \tag{10}$$

From the k_{ox}-$[O_2]$ relationship (Fig. 9), the k_{+4} value was found to be 110 sec^{-1} regardless of the kind of substrate used under conditions where $[O_2] > 100\ \mu M$, and the k_{+3} value was roughly estimated from the tangent of the k_{ox}-$[O_2]$ curve at the origin to be $1.4 \times 10^6\ M^{-1}\ sec^{-1}$.

In the steady state of the overall reaction, the rate, v/e, is expressed by Eq. 11, assuming $k_{-1} \gg k_{+2}$ and $k_{-3} = O$ as described above,

$$\frac{e}{v} = \frac{1}{k_{+2}} + \frac{1}{k_{+4}} + \frac{k_{-1}}{k_{+1}k_{+2}[s]} + \frac{1}{k_{+3}[O_2]}. \tag{11}$$

From Eqs. 3 and 11,

$$\frac{e}{V_m} = \frac{1}{k_{+2}} + \frac{1}{k_{+4}} \tag{12}$$

and

$$B = \frac{1}{k_{+3}}. \tag{13}$$

It is noteworthy that the V_m/e value calculated by applying the values of k_{+2} and k_{+4} to Eq. 12 is in good agreement with the value obtained from the overall reaction kinetics with the same substrate (Tables III

TABLE V. Kinetic Constants of the Amine Oxidase Reaction at pH 7.5 and 25°C

Kinetic constants \ Substrate	*n*-Butylamine	Ethylamine	Benzylamine
k_{+2} (sec^{-1})	18.0	11.1	10.3
k_{+4} (sec^{-1})	110	110	—
V_m/e calculated (sec^{-1})	15.4[a]	9.2[a]	8.5[a]
k_{+3} (M^{-1} sec^{-1})	2.3×10^6[b]	1.9×10^6[b]	1.7×10^6[b]
	1.4×10^6[c]	1.4×10^6[c]	—
$k_{+2}A$ (mM)	2.7	22	0.64

[a] Values were calculated from the values of k_{+2} and k_{+4}. [b] Values were estimated from the data on the overall reaction kinetics (Table III). [c] Values were obtained from the k_{ox}–[O_2] relationship (Fig. 9).

and V). The k_{+3} value obtained by Eq. 13 from the overall reaction kinetics is almost the same as that obtained by the " continuous flow " method. The equation $k_{+2}A=K_m{}^s=k_{-1}/k_{+1}$, obtained from Eqs. 3, 8 and 11, explains the fact that the value of $k_{+2}A$ approximately agrees with the $K_m{}^s$ value for the same substrate (Tables IV and V).

When glutamate aspartate transaminase, which is known to be a pyridoxal phosphate enzyme, was mixed with aspartate or glutamate together with the corresponding product (oxalacetate or α-ketoglutarate), respectively, peaks appeared at 330, 365, 430 and 492 mμ in the absorption spectrum (*23–29*). Fasella (*30*) suggested that the peaks at 365 and 430 mμ were due to the absorption of the aldimine form of the transaminase, and the peak at 330 mμ was due to the absorption of the ketimine form of the enzyme. Jenkins (*25*) suggested that the peak at 490 mμ was due to the absorption of a quinoid-type of the enzyme. The similarity of the steady state spectrum of the amine oxidase to that of the transaminase suggests that the aldimine, ketimine and quinoid forms are formed as intermediates in the steady state of the amine oxidase reaction. If this is true, the intermediate(s), X, formed in the steady state may not be a single complex but a mixture of the above complexes. By analogy to the transaminase reaction, the PLP form of the amine oxidase may change to the aldimine form on reaction with the substrate, and then the aldimine form may change to the ketimine form *via* the quinoid type of the enzyme (Scheme 1). Finally, the PMP form may be produced by hydration of the ketimine form.

It was reported that the amine oxidase molecule contains cupric copper

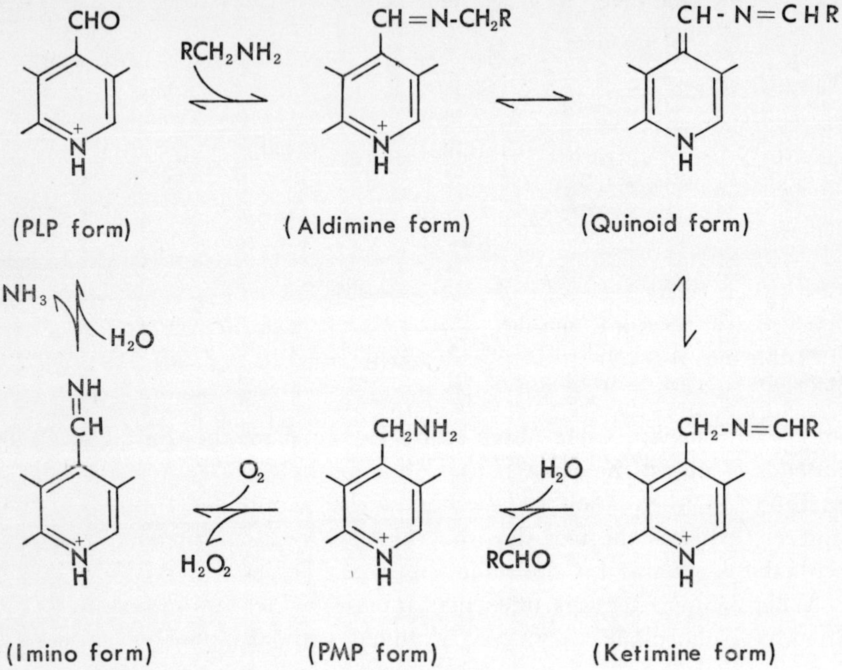

CHO | RCH_2NH_2 | $CH{=}N\text{-}CH_2R$ | $CH\text{-}N{=}CHR$

(PLP form) (Aldimine form) (Quinoid form)

NH_3 H_2O

NH
CH | O_2 H_2O_2 | CH_2NH_2 | H_2O RCHO | $CH_2\text{-}N{=}CHR$

(Imino form) (PMP form) (Ketimine form)

Scheme 1

as an essential component besides the active carbonyl group (pyridoxal phosphate), and that the copper does not change its valence during the reaction (*15*). The latter fact leaves the role of copper ions in the reaction unexplained. By kinetic experiments, Suzuki and Ogura (*33*) confirmed that 8-hydroxyquinoline, which is a Cu^{2+}-chelate reagent, non-competitively inhibits the transamination step (Reaction 6) in the amine oxidase reaction, but not the oxidation step of the PMP form with oxygen. Therefore, the bound copper ions seem to participate in the transamination step.

Summary

The amine oxidase of *A. niger* was titrated with phenylhydrazine (inhibitor) by measurement of the changes in absorbance at 442.5 mμ and its activity. It was found that there are two carbonyl groups essential for activity in the oxidase molecule. Anaerobic titration of the enzyme with *n*-butylamine (substrate) revealed that one mole of the native enzyme reacts with two moles of *n*-butylamine, and that an equivalent amount of *n*-butyraldehyde is produced from *n*-butylamine even under anaerobic conditions. In this reaction, the native enzyme having a broad absorption band around 420–490 mμ was converted to a bleached form whose spectrum showed two lower absorption maxima at 470 mμ and 440 mμ. However, ammonia and hydrogen peroxide were not produced in the absence of molecular oxygen. Anaerobic titration of the bleached enzyme with molecular oxygen showed that one mole of the bleached enzyme reacts with two moles of molecular oxygen, and two moles each of ammonia and hydrogen peroixde and one mole of the native enzyme are formed by the reaction. It was thus confirmed that the amine oxidase reaction is composed of two steps of reaction; the transamination which takes place anaerobically, and the oxidative deamination which occurs in the presence of molecular oxygen.

The steady state spectrum of the enzyme under aerobic conditions varied with the kind of substrate used; maxima were observed at 335, 420 and 495 mμ with *n*-butylamine, 420 and 495 mμ with ethylamine, 375 and 495 mμ with benzylamine, and 335, 400 and 495 mμ with histamine. This shows that transient intermediates are formed in the steady state of the reaction. The amine oxidase reaction was analyzed based on the data of the overall reaction rate and those obtained by the " flow " method. The following scheme was proposed as the reaction mechanism of the amine oxidase to account for the kinetic data.

$$E_L + n\text{-butylamine} \overset{K=2\times10^{-3}\text{M}}{\rightleftharpoons} X \overset{k_{+2}=18\ \text{sec}^{-1}}{\longrightarrow} E_M + n\text{-butyraldehyde}$$

$$E_M + O_2 \overset{k_{+3}=2\times10^{9}\text{M}^{-1}\text{sec}^{-1}}{\longrightarrow} Y + H_2O_2 \overset{k_{+4}=110\ \text{sec}^{-1}}{\longrightarrow} E_L + NH_3 + H_2O_2\,,$$

where E_L and E_M are the native and bleached forms of the enzyme unit (based on the active carbonyl group), respectively, and X and Y are intermediates.

References

1 H. Yamada, O. Adachi and K. Ogata, *Agr. Biol. Chem.*, **29**, 1148 (1965).

2 O. Adachi, H. Yamada and K. Ogata, *Agr. Biol. Chem.*, **30**, 1202 (1966).

3 H. Yamada, T. Uwajima, H. Kumagai, M. Watanabe and K. Ogata, *Biochem. Biophys. Res. Commun.*, **27**, 350 (1967).

4 H. Kamagai, H. Matsui, K. Ogata and H. Yamada, *Biochim. Biophys. Acta*, **171**, 1 (1969).

5 H. Kumagai, H. Yamada, H. Suzuki and Y. Ogura, *J. Biochem.*, **69**, 137 (1971).

6 H. Yamada and K. T. Yasunobu, *J. Biol. Chem.*, **237**, 3077 (1962).

7 H. Yamada and K. T. Yasunobu, *Biochem. Biophys. Res. Commun.*, **8**, 387 (1962).

8 F. Buffoni and H. Blaschko, *Proc. Roy. Soc. London*, **161**, 153 (1964).

9 H. Blaschko and F. Buffoni, *Proc. Roy. Soc. London*, **163**, 45 (1965).

10 B. Mondovi, M. T. Costa, A. Finazzi-Agro and G. Rotilio, *in* " Pyridoxal Catalysis: Enzymes and Model System," ed. by E. E. Snell, A. E. Braunstein, E. S. Severin and Yu. M. Torchinsky, Interscience Publications, New York, p. 403 (1966).

11 E. Y. Goryachenkova, L. I. Shcherbatiuk and C. I. Zemaraev, *in* " Pyridoxal Catalysis: Enzymes and Model Systems," ed. by E. E. Snell, A. E. Braunstein, E. S. Severin and Yu. M. Torchinsky, Interscience Publications, New York, p. 391 (1966).

12 H. Yamada, H. Kumagai, H. Kawasaki, H. Matsui and K. Ogata, *Biochem. Biophys. Res. Commun.*, **29**, 723 (1967).

13 O. Adachi and H. Yamada, *Agr. Biol. Chem.*, **33**, 1707 (1969).

14 H. Yamada, O. Adachi and K. Ogata, *Agr. Biol. Chem.*, **29**, 912 (1965).

15 H. Yamada, O. Adachi and T. Yamano, *Biochim. Biophys. Acta*, **191**, 751 (1969).

16 H. Suzuki, Y. Ogura and H. Yamada, *J. Biochem.*, **69**, 1065 (1971).

17 H. Yamada, O. Adachi and K. Ogata, *Agr. Biol. Chem.*, **29**, 864 (1965).

18 J. M. Hill and P. J. G. Mann, *Biochem. J.*, **91**, 171 (1964).

19 H. Yamada and K. T. Yasunobu, *J. Biol. Chem.*, **238**, 2669 (1963).

20 K. T. Yasunobu and H. Yamada, *in* " Chemical and Biological Aspects of Pyridoxal Catalysis," ed. by E. E. Snell, P. M. Fasella, A. Braunstein and A. Rossi-Fannelli, Pergamon Press, London, p. 453 (1962).

21 C. M. McEwen, Jr., K. T. Cullen and A. J. Sober, *J. Biol. Chem.*, **241**, 4544 (1966).

22 H. Suzuki, Y. Ogura and H. Yamada, *J. Biochem.*, in press.

23 W. T. Jenkins, *J. Biol. Chem.*, **234**, 1121 (1961).

24 G. G. Hammes and P. Fasella, *J. Am. Chem. Soc.*, **84**, 4644 (1962).

25 W. T. Jenkins, *J. Biol. Chem.*, **239**, 1742 (1964).

26 W. T. Jenkins and R. T. Taylor, *J. Biol. Chem.*, **240**, 2907 (1965).

27 P. Fasella, *Biochem. J.*, **95**, 2 (1965).

28 W. T. Jenkins and L. D'Ari, *J. Biol. Chem.*, **234**, 2845 (1966).

29 A. E. Evangelopuolos and I. W. Sizer, *in* "Pyridoxal Catalysis: Enzymes and Model Systems," ed. by E. E. Snell, A. E. Braunstein, E. S. Severin and Yu. M. Torchinsky, Interscience Publications, New York, p. 1 (1966).
30 P. Fassella, *in* "Pyridoxal Catalysis: Enzymes and Model Systems," ed. by E. E. Snell, A. E. Braunstein, E. S. Severin and Yu. M. Torchinsky, Interscience Publications, New York, p. 1 (1966).
31 G. H. Czerlinski and J. Malkwitz, *Biochemistry*, **4**, 1127 (1965).
32 A. A. Frost and R. G. Pearson, "Kinetics and Mechanism," John Wiley and Sons Inc., New York, p. 166 and p. 175 (1961).
33 H. Suzuki and Y. Ogura, *J. Biochem*, in press.

Received for publication December 16, 1971.

PURPLE INTERMEDIATE OF D-AMINO ACID OXIDASE

Kunio Yagi
Institute of Biochemistry, Faculty of Medicine, University of Nagoya, Nagoya

One of the most effective avenues of approach to an understanding of the mechanism of enzyme action may be the elucidation of the mode of interaction between the enzyme and its substrate. For this purpose, the events which occur during the lifetime of the complex between the enzyme and substrate should be clearly delineated. From this point of view our studies have been carried out using D-amino acid oxidase as a material which gave valuable information on forces governing the complex formation and on oxidoreduction reactions traceable by measuring the optical properties of its coenzyme, flavin adenine dinucleotide.

Some years ago, we succeeded in crystallizing a purple complex by mixing the enzyme with its substrate, D-alanine. The crystals were found to contain equimolar amounts of the enzyme and substrate moieties. Thereafter, our effort concentrated on the elucidation of the properties of the complex and the identity of this complex with the swiftly appearing purple intermediate observed by rapid reaction methods. This paper describes the highlights of this series of investiga-

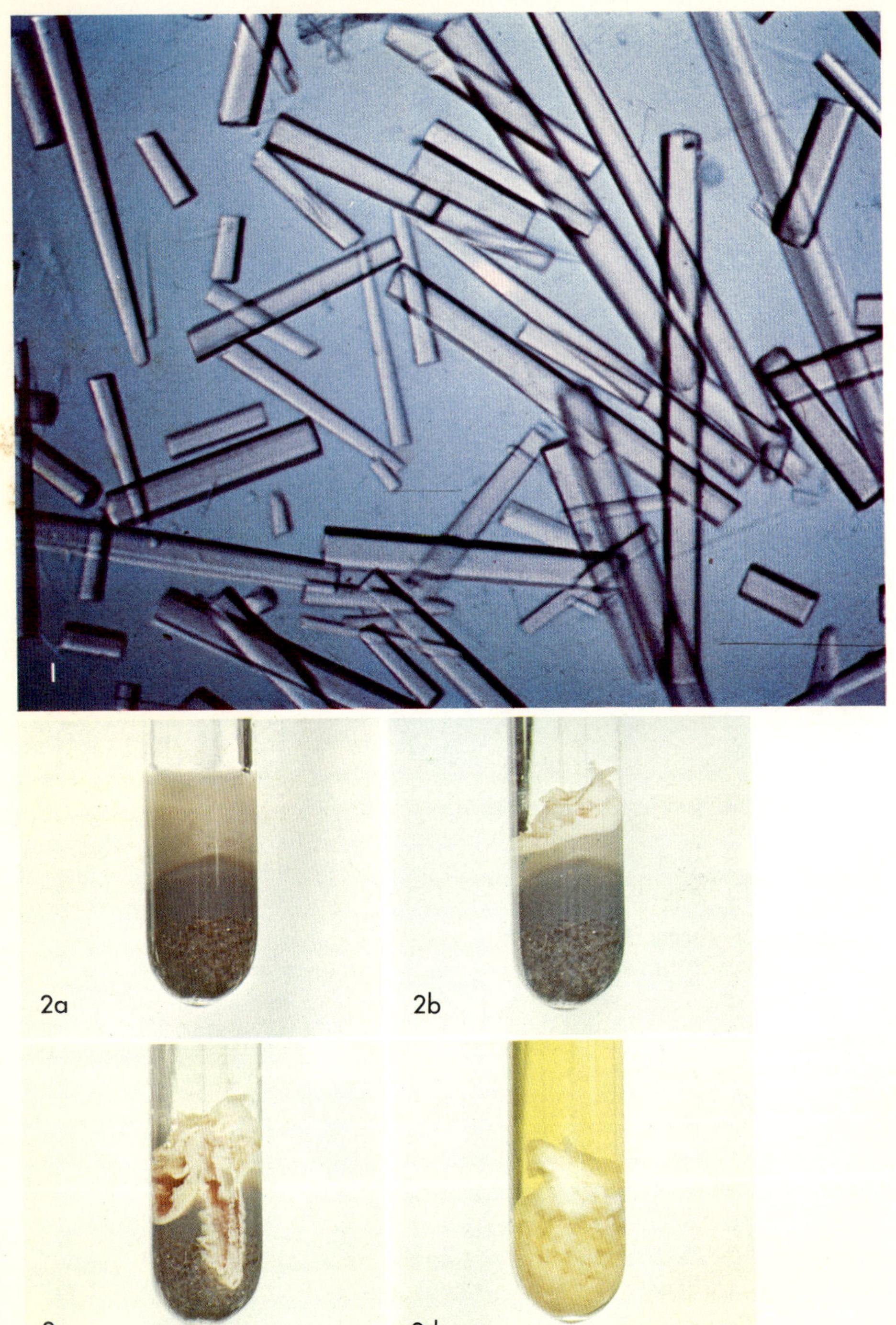
1
2a
2b
2c
2d

tions, especially the studies of a purple colored intermediate which give us valuable insight into the action mechanism of this enzyme.

Crystallization of the Purple Complex

Although the lifetime of an intermediate complex which is formed during the enzymic reaction is extremely short under the optimum conditions, we predicted that the isolation by crystallization of the intermediate complex could be possible if the intermediate was accumulated in the reaction medium under certain specified conditions. For this purpose, we tried to eliminate an electron acceptor (molecular oxygen) and to shift both pH and temperature from their optimum values for catalytic action of D-amino acid oxidase. The original procedure for the crystallization of the intermediate complex was as follows (*1*). One gram of the crystals of the enzyme-benzoate complex was dissolved in 50 ml of O_2-free 0.017 M pyrophosphate buffer, pH 8.3, and 1 g of D-alanine was added by stirring. The pH of the solution was brought to 6.1 by the addition of acetic acid, and then 6 g of ammonium sulfate was added to the solution. The precipitate was collected by centrifugation and dissolved in 50 ml of the same buffer. D-Alanine was added and the above procedure was repeated. The precipitate was dissolved in a minimum volume of the same buffer and the solution (pH 7.2; saturation of ammonium sulfate, 0.05) was transferred to a vessel, which was filled with O_2-free N_2 and kept at 5°C overnight. A crop of purple colored crystals was obtained as shown in Photo 1.

On aeration, the crystals gave equimolar amounts of the oxidized enzyme and the products, indicating that the crystals were composed of equimolar amounts of the enzyme and substrate moieties.

The reflectance spectrum of the crystals was found to show a broad absorption band in the vicinity of 550 mμ. The spectrum was essentially in accord with that of the mother liquor, the purple solution shown in Fig. 1.

← PHOTO 1. Crystals of the purple complex (×500).

PHOTO 2. Changes in color of the purple complex crystals caused by the addition of trichloroacetic acid. a: crystal sediments washed with O_2-free cold distilled water under anaerobic conditions; b: 30 sec after the addition of O_2-free 20% trichloroacetic acid to the washed crystals; c: just after the agitation of the crystals with a glass rod; d: about 20 hr after c.

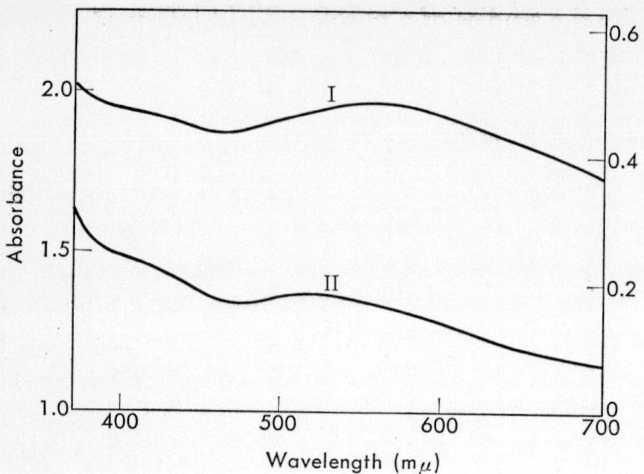

FIG. 1. Reflectance spectrum of the purple complex crystals. I: reflectance spectrum of the crystals; II: absorption spectrum of the mother liquor. The scale on the left of the ordinate is for curve I, and that on the right for curve II.

It was later found that the purple solution could also be prepared by adding pyruvate and ammonium ions to the enzyme preparation fully reduced with excess D-alanine. In the presence of 5.0×10^{-2} M D-alanine and 5.0×10^{-2} M ammonium sulfate, formation of the purple intermediate depended on the concentration of pyruvate and reached the maximum when pyruvate concentration was elevated to 4.0×10^{-1} M. On the other hand, when pyruvate which was present in the purple solution was eliminated by the addition of an appropriate amount of hydrogen peroxide, the absorption band in the vicinity of 550 mμ decreased, indicating that the purple complex was converted to the fully reduced enzyme (*2*) (see the typical spectra shown in Fig. 2).

These results indicate that the purple complex is involved in the equilibrium of the anaerobic reaction as shown by Eq. 1,

$$E_{ox}+S \rightleftharpoons E_{purple} \rightleftharpoons E_{red}+P, \qquad (1)$$

where E_{ox}, E_{red} and E_{purple} represent the oxidized enzyme, the fully reduced enzyme and the purple complex, respectively.

In support of this view, it was found that the anaerobic addition of excess benzoate, which was known to be exchanged for the substrate,

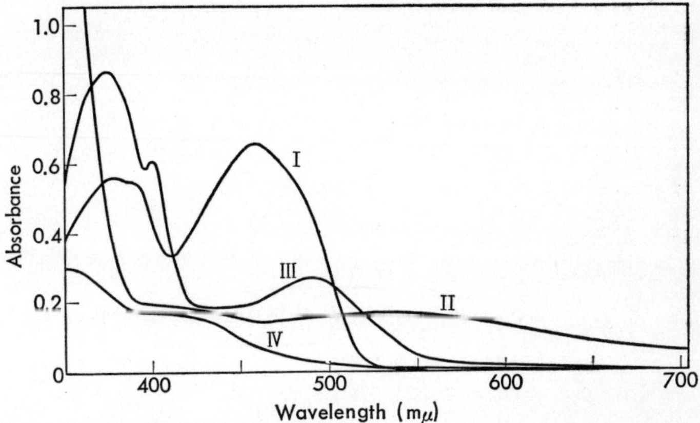

FIG. 2. The absorption spectra of the oxidized enzyme, the purple complex, the semiquinoid enzyme and the fully reduced enzyme. I: oxidized enzyme (5.86×10^{-5} M with respect to the bound FAD); II: the purple complex solution (oxidized enzyme was mixed with 5.0×10^{-2} M D-alanine in the presence of 5.0×10^{-2} M ammonium sulfate and 1.0×10^{-1} M lithium pyruvate); III: the semiquinoid enzyme (oxidized enzyme was irradiated in the presence of 0.02 M EDTA); IV: the fully reduced enzyme (oxidized enzyme was mixed with excess D-alanine (3.3×10^{-2} M)). II–IV, under anaerobic conditions. Measurements were made at pH 8.3.

to the purple solution yielded a characteristic spectrum of the oxidized enzyme-benzoate complex as shown in Fig. 3. This indicates the shift of the equilibrium in Eq. 1 towards the left caused by elimination of free E_{ox} due to formation of the E_{ox}-benzoate complex.

Considering the interconversion between the purple intermediate and the fully reduced enzyme, "the pyruvate-stabilized semiquinoid enzyme" reported by Kubo *et al.* (*3*) is considered to be essentially the same as the entity which was obtained in crystalline form.

Although these results seem to be enough to conclude the occurrence of the purple intermediate in an equilibrium state of the enzyme-substrate interaction, a possible equilibrium represented by Eq. 2 should be considered.

$$\begin{array}{c} \text{purple complex} \\ \downarrow\uparrow \\ E_{ox}+S \rightleftharpoons X \rightleftharpoons E_{red}+P \end{array} \qquad (2)$$

If Eq. 2 is valid, the purple complex may not be involved in the catalytic

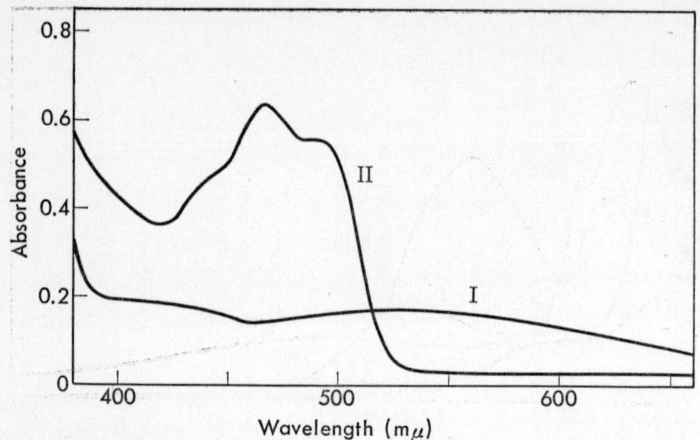

FIG. 3. Reaction of the purple complex with benzoate under anaerobic conditions. I: the purple complex (5.86×10^{-5} M with respect to the bound FAD); II: purple complex was mixed with sodium benzoate (final concentration, 3.0×10^{-1} M). Measurements were made at pH 8.3.

reaction of this enzyme. To check this point, the rapid reaction method was adopted, and the results are discussed in the following section.

Identification of the Purple Complex as the Rapidly Appearing Intermediate

By rapid reaction methods, the spectrum having a broad absorption band in the longer wavelength region was observed upon mixing the enzyme with D-alanine. It was considered that if this spectrum was identical with that of the purple complex, the involvement of the purple complex in the catalytic reaction could be established. However, the identity between the two was not unequivocally accepted, even though the spectra were similar to each other.

Massey and his co-workers (*4*) suggested the formation of two kinds of purple intermediate based on their observations that the increase in absorbance at 550 mμ upon anaerobic mixing of the enzyme with D-alanine was biphasic and that the absorption spectrum observed during the rapid phase (150 msec after mixing) was different from that observed when the slow phase was completed (1.8 sec after mixing). They assigned the rapidly appearing spectrum to a complex of the flavin radical and substrate radical and the slowly appearing spectrum to a charge-transfer complex of the fully reduced enzyme and the imino acid. They considered that the rapidly appearing species was the catalytically important intermediate while the slowly appearing one was not catalytically significant, and the crystallized complex corresponded to the latter species.

In contrast, we reported on the basis of our experimental data that the spectrum observed during the rapid phase of the reaction was not due to a single entity, but was composed of the spectrum of the purple complex and that of the oxidized enzyme. It was considered that the purple intermediate was spectrophotometrically unitary (*5*). However, the presence of a trace amount of oxygen in the reaction mixture caused some ambiguity, and the results could not exclude alternate interpretations (*6*). To substantiate this, we made extensive spectroscopic examinations of the rapidly appearing intermediate under strictly anaerobic conditions (*7*).

The biphasic increase in 550 mμ absorbance reported by Massey *et al.* was carefully checked. The absorbance change at 550 mμ in the anaerobic reaction of the enzyme with D-alanine was obviously biphasic. The results were essentially in accord with that reported by Massey *et al.* However, an analysis of the biphasic change led us to an interpretation different from that of Massey *et al.* To examine the biphasic increase, the visible absorption spectra were recorded using a rapid scan spectrophotometer at different periods after the anaerobic mixing of the enzyme with D-alanine. Figure 4 shows the spectra obtained at the moments when the rapid and slow phases of the above-mentioned biphasic increase in 550 mμ absorbance were completed. As can be seen from the figure, the spectra obtained for the rapid and slow phases and the spectrum of the oxidized enzyme were isosbestic at 340 mμ and 507 mμ. The spectrum obtained for the slow phase was identical with that of the purple complex. The difference spectrum between the spectra obtained for the rapid and slow phases was essentially identical with the difference spectrum of the oxidized enzyme and the purple complex. These results indicated that the observed slower change could be ascribed to the conversion of the oxidized enzyme to the purple species. Such a mechanism of biphasic change was more clearly demonstrated in the complex formation between the enzyme and *o*-aminobenzoate, in which case the enzyme was not reduced and was expected to provide clear evidence.

The above analysis of the biphasic change led us to the conclusion

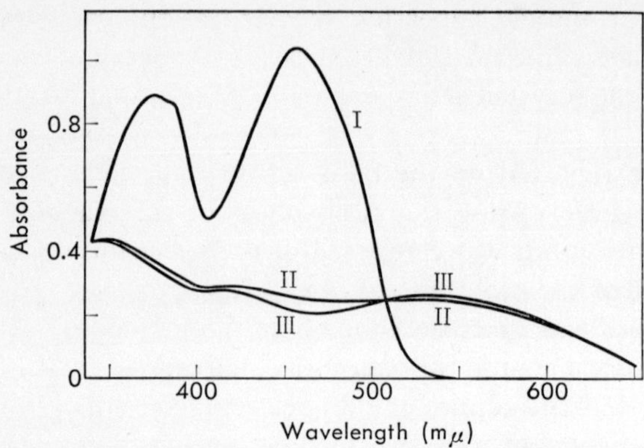

FIG. 4. Spectrophotometrical examination on the identity of the rapidly appearing purple intermediate of D-amino acid oxidase. Changes in absorption spectrum of D-amino acid oxidase (9.2×10^{-5} M with respect to the bound FAD) occurring upon anaerobic addition of D-alanine (final concentration, 1.0×10^{-2} M) at pH 8.3 and 19°C were followed using a rapid scan spectrophotometer (reproduced from the trace recorded on a storage oscilloscope). I: oxidized enzyme; II: 300 msec after mixing; III: 2 sec after mixing.

that the major portion of the molecules of the oxidized enzyme, E_1, was converted rapidly to the purple intermediate, while the rest of the oxidized enzyme molecules, E_2, were converted slowly to the purple intermediate, indicating inhomogeneity of the enzyme preparation.

In interpreting the nature of this inhomogeneity, it should be taken into account that the present enzyme preparation had been highly purified and that it contained no isoenzymes. The only possibility seems to be that the oxidized enzyme was in a monomer-dimer equilibrium (*8*). To verify this point, the results of an ultracentrifugal examination and the kinetic analysis were compared. A gross agreement was found between dimer per monomer and E_1 per E_2, measured by the reaction of the enzyme with *o*-aminobenzoate, when the temperature was lowered.

From these results, it could be concluded that the purple intermediate occurring in the rapid phase of the anaerobic reaction of D-amino acid oxidase with D-alanine was a complex of dimeric form. Accordingly, the purple complex observed in the solution of the equilibrium state of the enzymic reaction is identical with the rapidly appearing purple intermediate observed in stopped-flow experiments.

Electronic Interaction Involved in the Purple Complex

Although the broad absorption band in the vicinity of 550 mμ observed with the purple complex suggested the occurrence of a charge-transfer interaction between the enzyme and substrate, the observation of a small ESR signal with the purple complex raised some doubts regarding the identity of the complex. When Kubo *et al.* (*3*) first observed the purple color on mixing the enzyme with D-alanine in the presence of pyruvate, they observed a small ESR signal and thought the colored entity to be an oxidoreduction intermediate of semiquinoid species. We also found a small ESR signal with our crystallized purple complex and considered the crystal to be that of a complex of the semiquinoid enzyme and the partially modified substrate.

On the other hand, Nakamura *et al.* (*9*) found that the solution of the purple intermediate had no ESR signal, and they distinguished the purple entity from the semiquinoid enzyme that was produced by reducing the enzyme with dithionite. Massey and Gibson (*10*) supported the result and further stated that the purple solution could be converted into the semiquinoid enzyme solution upon irradiation. This was also confirmed by us. We observed by measurements of optical rotatory dispersion (*2*) and circular dichroism (*11*) that an interaction between the isoalloxazine nucleus of the enzyme and the substrate existed. We emphasized our finding that the diamagnetic purple complex was converted into a paramagnetic species upon storage in the dark. This finding could account for the discrepancy regarding the occurrence of a small ESR signal in the purple complex, because the occurrence of a small ESR signal in the aged purple solution or crystals is not surprising.

In considering the electronic interaction between the flavin chromophore of the enzyme and substrate, the observation that the purple complex gradually changed into a paramagnetic species should be especially noted. This was clearly demonstrated by the appearance of the absorption peak at 492 mμ and an ESR signal accompanied by the disappearance of the absorption band in the vicinity of 550 mμ and the negative Cotton effect having a trough at 430 mμ. This indicated the appearance of the semiquinoid enzyme and the disappearance of the interaction between the coenzyme and substrate moieties. This was also demonstrated by anaerobic addition of benzoate. When the interaction between the coenzyme and substrate moieties disappeared, the semiquinoid enzyme formed was converted to the blue entity upon anaerobic addition of excess benzoate (*12*). The semiquinoid enzyme formed from the purple complex was found to be essentially identical with that obtained by dithionite reduction (*9*) or by photoreduction in the presence of EDTA (*13*). From these results, it could be concluded that the purple complex was a molecular complex which dissociates into the semiquinoid enzyme and the substrate radical. This could be expressed as

$$D+A \rightleftarrows (D, A) \rightleftarrows (D^+-A^-) \rightleftarrows D^+ + A^-, \tag{3}$$

where (D, A) means an outer complex and (D^+–A^-) an inner complex according to the definition of Mulliken (*14*). The acceptor anion, A^-, corresponds to the semiquinoid enzyme and the donor cation, D^+, to the substrate radical that is probably decomposed rapidly due to its instability. Stability of the semiquinoid enzyme might be due to the protection of the radical by the protein moiety of the enzyme. The denaturation of the enzyme protein was found to result in the rapid dissociation of the complex shown in Eq. 3. As can be seen from Photo 2 a–d, the addition of trichloroacetic acid to the purple complex crystals produced a carmine red color which was assigned to rhodoflavin produced from the semiquinoid flavin (*15*). This result clearly indicated that in the crystals an interaction represented by (D^+–A^-) in Eq. 3 is occurring.

To verify further the assignment of the purple complex to the (D^+–A^-) species, the complex of the enzyme with *o*-aminobenzoate, which has been known to inhibit the enzyme in competition with the substrate, was examined. In this type of complex, a broad absorption band, which could be regarded as a charge-transfer band, also appeared (*16, 17*) (see Fig. 5). However, the complex neither yielded semiquinoid enzyme upon storage in the dark nor formed rhodoflavin upon treatment with the acid. These results indicated that the complex might be an outer complex as represented by (D, A) in Eq. 3.

It should also be noted that benzoic acid forms a 1:1 complex with this enzyme without the appearance of such a charge-transfer band (see Fig. 3, curve II). Comparing the structure of benzoic acid with that of *o*-aminobenzoic acid, it was obvious that the occurrence of an amino group is essential for *o*-aminobenzoate to form a charge-transfer complex with this enzyme. The possible reasons as to why D-amino acid formed an inner complex whereas the above inhibitor formed an

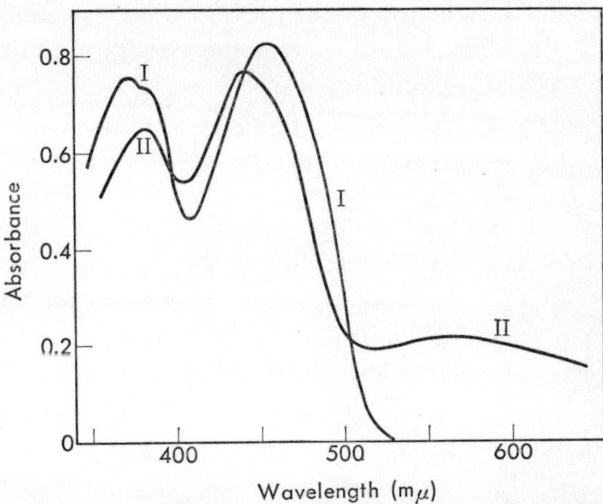

FIG. 5. Change in absorption spectrum of D-amino acid oxidase occurring upon addition of *o*-aminobenzoate. I: oxidized enzyme (7.5×10^{-5} M with respect to the bound FAD); II: oxidized enzyme was mixed with *o*-aminobenzoate (final concentration, 1.0×10^{-3} M). Measurements were made at pH 8.3.

outer complex with this enzyme will be discussed in the next section.

As mentioned before, the purple complex dissociates into the semiquinoid enzyme and the substrate radical, but the dissociation occurred very slowly. Since the dielectric constant of the environment surrounding the sites of charge-transfer interaction influenced the dissociation, it could be postulated that some hydrophobic environment might be present around the sites of charge-transfer interaction. In this connection, it might be mentioned that flavin chromophore was considered to be surrounded by a hydrophobic environment. This contention was based on the characteristics of the absorption spectrum of the enzyme (*17*, *18*), fluorescence properties of the flavin derivatives soluble in organic solvents (*19*), and the results of investigation using a hydrophobic probe, monobenzoylated monoaminostilbene disulfonate (*20*). It should also be mentioned that the R-group of the substrate interacted with a hydrophobic locus of the enzyme protein as evidenced by the data obtained by using a series of straight-chain fatty acids as substrate-substitutes (*17*).

A hydrophobic environment such as supposed to exist arround the sites of charge-transfer interaction could prevent the dissociation of the inner complex and, therefore, afford a favorable condition to facilitate a two-electron type oxidoreduction.

Essential Events Inducing Formation of the Purple Intermediate

The occurrence of the oxidized enzyme-substrate complex prior to the purple intermediate was shown kinetically by Massey and Gibson (*10*) using D-alanine, D-methionine and D-proline as substrates. This was more clearly demonstrated when D-leucine or D-valine was employed (*21*). It is clear that the formation of an equimolar complex between the enzyme and substrate is a primary event inducing the formation of the purple intermediate.

In determining the site of the substrate for the charge-transfer interaction with flavin chromophore, the data on the formation of a charge-transfer complex of this enzyme with *o*-aminobenzoate or Δ^1-piperidine 2-carboxylate are quite suggestive. As described in the previous section, mixing of *o*-aminobenzoate with this enzyme yields a green complex having a broad absorption band in the longer wavelength region, which is regarded as a charge-transfer band. Δ^1-Piperidine 2-carboxylate gave a similar result (*16*). However, mixing of benzoic acid with this enzyme yielded a complex without the appearance of such a charge-transfer band. This indicates that *n*-donor group, *i.e.*, a nitrogen lonc pair, is essential for the formation of the charge-transfer complex. Since the carboxyl group and the nitrogen atom of *o*-aminobenzoate or Δ^1-piperidine 2-carboxylate are located in sterically similar positions to those of the substrate, *e.g.*, D-alanine, which also has a nitrogen lone pair, a similar interaction is expected to occur in the enzyme-substrate interaction.

Accepting the fact that the nitrogen lone pair of the substrate or the above-mentioned inhibitors is the site for the charge-transfer interaction, a question was naturally raised as to why the substrate forms an inner complex while the inhibitor forms an outer complex. This seemed to be an important question, since it might afford a key to an understanding of the essential difference between the substrate and inhibitor in this case. To answer this question, the following studies were made.

It is obvious that the cleavage of the α-C-H bond of the substrate should occur during the oxidation of the substrate catalyzed by this

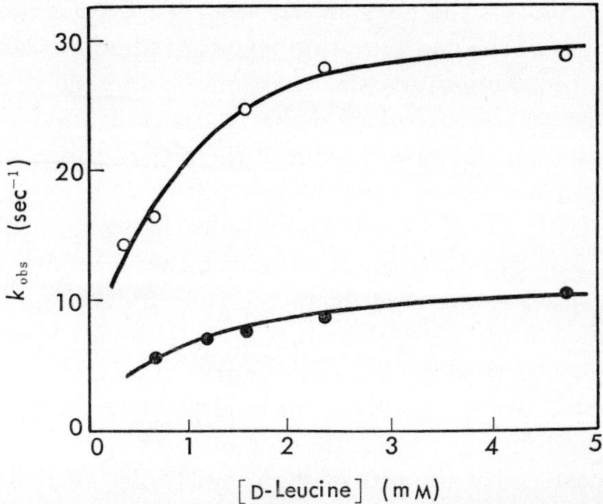

FIG. 6. Kinetic isotope effect on the formation of the purple intermediate of D-amino acid oxidase with D-leucine. With DL-[α-^{1}H] leucine (○) and DL-[α-^{2}H] leucine (●) at various concentrations, the formation of the purple intermediate was followed by measuring the absorbance at 550 mμ with a stopped-flow spectrophotometer at pH 8.3 and 20°C. The enzyme concentration, 1.4×10^{-5} M with respect to the bound FAD.

enzyme. It is also clear that such a cleavage does not occur with the inhibitor. Therefore, it was considered to be worthwhile to examine the relation between the α-C-H bond cleavage and the formation of an inner complex, and an attempt was made to settle the question as to whether the α-C-H bond cleavage occurs before the initiation of the electronic interaction between the substrate nitrogen atom and the isoalloxazine nucleus of the coenzyme or whether it occurs after the initiation.

We examined the kinetic isotope effect using an α-deuterated substrate, since it was known that H-D exchange of the substrate occurs at the α-carbon, but not at the ß-carbon (*22*). The rate of formation of the purple intermediate with DL-[α-^{2}H] leucine was compared with that with DL-[α-^{1}H] leucine. The pseudo-first-order rate constant (k_{obs}) of formation of the purple intermediate was plotted *versus* the concentration of the substrate (*23*) (Fig. 6). As can be seen from the figure, the rate of formation of the purple intermediate levelled off in the range of the substrate concentration investigated, suggesting that the rate is probably controlled by the process $E_{ox} \cdot S \longrightarrow$ purple intermediate. Although L-leucine inhibits the enzymic reaction in competition with the substrate D-leucine, in the concentration investigated, and exact estimation of the rate of reaction, $E_{ox} \cdot S \longrightarrow$ purple intermediate, was not possible, it is reasonable to conclude that the substitution of the α-hydrogen of leucine for deuterium reduced the rate of formation of the purple intermediate to approximately one-third of the original at the substrate concentration range where the rate levelled off.

Such a kinetic isotope effect was not observed in the process of transformation of the purple intermediate to the fully reduced enzyme (*6*). It can be concluded that the cleavage of the α-C-H bond occurs before, or at least at the same time as, the achievement of a strong interaction between the substrate and the isoalloxazine nucleus of the coenzyme.

The data on the effect of pH on the rate of formation of the purple intermediate indicated that the rate of the formation is smaller in acidic than in slightly alkaline ranges (*24*). Since the affinity of the enzyme with substrate is considered to be essentially independent of pH in the range investigated, as judged from the results of the affinity of the enzyme with the inhibitor, the above result can be interpreted to imply that the amino group of the enzyme-bound substrate should be deprotonated in forming the purple intermediate. This inference is in accord with the necessity of a nitrogen lone pair in forming a charge-transfer complex as specified before.

From these results, it might be concluded that the strong interaction between the nitrogen lone pair of the enzyme-bound substrate and the coenzyme occurs in concert with the cleavage of the α-C-H bond of the substrate. It is conceivable that the α-C-H bond cleavage results in the increased charge-transfer interaction. Apart from the simple example of donor-acceptor complex, in which only one electron is transferred from a donor to an acceptor, a two-electron transfer from a donor to an acceptor in usual oxidoreduction reactions of organic compounds occurs concomitantly with rearrangement of chemical structure of reactants. Although direct evidence is lacking as to whether the α-C-H bond cleavage occurs *via* proton, hydrogen or hydride ion transfer, the base-catalyzed abstraction of α-proton of the substrate may be predicted as suggested by Neims *et al.* (*25*). If this is the case, base-catalyzed proton abstraction preceding the oxidation facilitates the donor-acceptor interaction, which lowers the free energy of activation for the flow of two electrons.

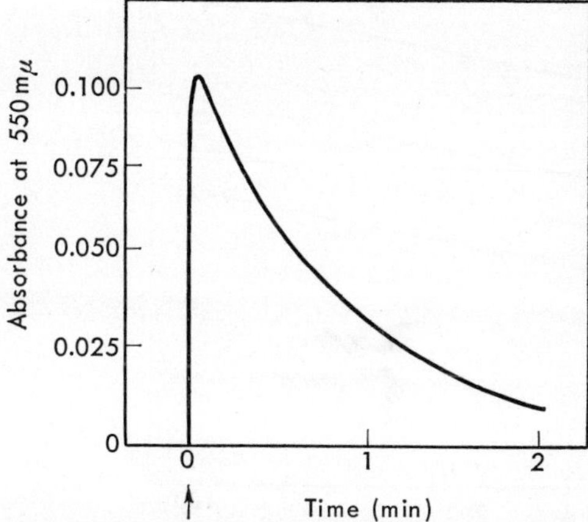

FIG. 7. Conversion of the purple intermediate to the fully reduced enzyme. Transmittance change at 550 mμ due to the reaction of D-amino acid oxidase (1.8×10^{-5} M with respect to the bound FAD) with D-alanine (final concentration, 5.0×10^{-3} M) at pH 8.66 and 25°C was followed by stopped-flow method with 20 mm cell and reproduced in the figure. Arrow shows the time of mixing.

Reaction of the Purple Intermediate with Molecular Oxygen

Under anaerobic conditions, the mixing of the enzyme with D-alanine yields rapidly the purple intermediate which is then converted slowly to the fully reduced enzyme (Fig. 7). To understand the catalytic reaction of this enzyme under aerobic conditions, reactivity of the purple intermediate with molecular oxygen should be elucidated.

As to this problem, Nakamura *et al.* (*9*) pointed out that the purple intermediate reacts with molecular oxygen, since the rate of conversion of the purple intermediate to the fully reduced enzyme is so small that the involvement of the oxidation of the fully reduced enzyme with molecular oxygen cannot explain the large turnover number of this enzyme. The direct reaction of the purple intermediate with molecular oxygen was further confirmed by Massey and Gibson (*10*) by observing a " spike " in the stopped-flow trace at 550 mμ of the reaction of this enzyme with the substrate. The " spike " was interpreted to indicate the rapid formation of the purple intermediate followed by its oxidation with molecular oxygen.

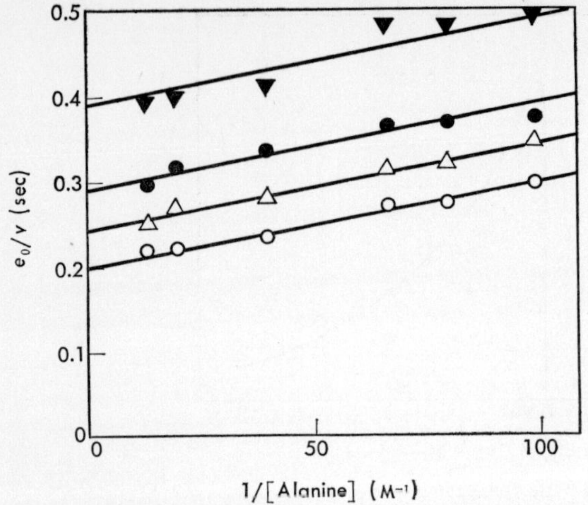

FIG. 8. Lineweaver-Burk plots of the overall rates of oxidation against the concentrations of D-alanine at various oxygen concentrations. The final concentration of the enzyme was 3.4×10^{-7}M with respect to the bound FAD. The final concentrations of oxygen were 520 μM (○), 260 μM (△), 195 μM (●) and 104 μM (▼). The rate of the overall reaction was measured polarographically at pH 8.3 and 25°C. Data reported by Nakamura *et al.* (*26*).

The overall rate of the oxidation of the substrate catalyzed by this enzyme was dependent on the concentrations of both the substrate and oxygen. When the Lineweaver-Burk plots were made of the overall rate against D-alanine concentration at various oxygen concentrations, straight lines parallel to each other were obtained as shown in Fig. 8 (*26*).

Although the detailed mechanism of the reaction of the purple intermediate with oxygen is not known, Nakamura *et al.* (*9*) demonstrated that the reaction was of the second orders and the rate of the reaction was 1.2×10^5 M^{-1} sec^{-1} at pH 8.3 and 20°C. According to them, as well as to Massey and Gibson (*10*), the aerobic reaction of the enzyme with neutral amino acid is to be expressed by the following sequence of reactions.

$$E_{ox}+S \underset{k_{-1}}{\overset{k_{+1}}{\rightleftharpoons}} E_{ox}\cdot S \underset{k_{-2}}{\overset{k_{+2}}{\rightleftharpoons}} E_{purple} \quad (4)$$

$$E_{purple}+O_2 \xrightarrow{k_{+3}} E_{ox}\cdot P+H_2O_2 \quad (5)$$

$$E_{ox}\cdot P \underset{k_{-4}}{\overset{k_{+4}}{\rightleftharpoons}} E_{ox}+P \quad (6)$$

According to these formulae, the initial velocity is

$$\frac{e_0}{v}=\frac{1}{k_{+2}}+\frac{1}{k_{+4}}+\frac{k_{-1}+k_{+2}}{k_{+1}\cdot k_{+2}}\cdot\frac{1}{s}+\frac{k_{+2}+k_{-2}}{k_{+2}\cdot k_{+3}}\cdot\frac{1}{(O_2)}+\frac{k_{-1}\cdot k_{-2}}{k_{+1}\cdot k_{+2}\cdot k_{+3}\cdot s(O_2)}. \quad (7)$$

This formula does not seem to correspond to the data of Fig. 8. However, both formula and data are valid, since the last term of the formula can be practically neglected (*26*, *27*). In addition, the fact that the purple intermediate accumulates at a definite time after the anaerobic mixing of the enzyme with the substrate indicates $k_{+2} \gg k_{-2}$, and Eq. 7 is reduced to

$$\frac{e_0}{v}=\frac{1}{k_{+2}}+\frac{1}{k_{+4}}+\frac{k_{-1}+k_{+2}}{k_{+1}\cdot k_{+2}}\cdot\frac{1}{s}+\frac{1}{k_{+3}(O_2)}. \quad (8)$$

Obviously this equation fits the data shown in Fig. 8.

The spike observed in the trace at 550 mμ upon aerobic mixing of the enzyme with [α-^{2}H] alanine was identical with that observed with [α-^{1}H] alanine. This indicates that the substitution of the α-hydrogen of the substrate with deuterium does not seem to affect the rate of the reaction of the purple intermediate with molecular oxygen (*28*).

It was recently found that in the anaerobic mixing of basic D-amino acids, such as D-arginine, with the enzyme the occurrence of the purple intermediate could not be demonstrated by both spectrophotometric and kinetic methods (*21*). In this case, the enzyme-substrate complex is considered to be labile, and molecular oxygen seems to react with the fully reduced enzyme. It was also noted that the reduction of the oxidized enzyme to the fully reduced form by D-arginine is faster than that by D-alanine (*21*).

Mechanism of Enzyme Action as Followed by the Study of the Purple Intermediate

The fact that the crystallized purple intermediate is composed of equimolar amounts of the enzyme and substrate moieties, strongly supports the view that an enzyme-substrate complex is formed during enzymic reaction and that the mechanism of enzyme action can be brought to light by following out the interaction between the enzyme and substrate. It was noted that the purple intermediate is involved in the equilibrium of the enzyme-substrate reaction under anaerobic conditions. This was verified by the fact that the addition of an excess amount of benzoate to the purple solution yielded the oxidized enzyme-benzoate complex, but not the complex of benzoate with semiquinoid enzyme, which might be produced upon benzoate addition if the purple intermediate could be dissociated to form the semiquinoid enzyme. The interaction between the enzyme and substrate seems to be strong enough to prevent their dissociation from the complex until the enzyme-substrate reaction reaches the stage of formation of $E_{red}\cdot P$ or the stage of reformation of $E_{ox}\cdot S$.

These facts together with the recent result, that $E_{ox}\cdot S$ has two phases depending upon its conformational change (*29*), lead us to infer that the reaction sequence of this enzyme proceeds under anaerobic conditions in the following way.

$$E_{ox}+S \rightleftharpoons (E_{ox}\cdot S)_1 \rightleftharpoons (E_{ox}\cdot S)_2 \rightleftharpoons E_{purple} \rightleftharpoons E_{red}\cdot P \rightleftharpoons E_{red}+P \qquad (9)$$

and under aerobic conditions in the following manner:

$$E_{ox}+S \rightleftharpoons (E_{ox}\cdot S)_1 \rightleftharpoons (E_{ox}\cdot S)_2 \rightleftharpoons E_{purple} \xrightarrow{O_2} E_{ox}\cdot P+H_2O_2 \rightleftharpoons E_{ox}+P+H_2O_2 . \qquad (10)$$

In the presence of excess oxygen, $E_{ox}\cdot P$ is a complex which controls the overall rate. Upon decreasing the concentration of oxygen, E_{purple} would become the rate-controlling compound, and determine the relationship between its concentration and the overall catalytic activity. In this case, the purple intermediate could be called the " Michaelis complex " according to the definition of Chance (*30*, *31*).

Another important event to be discussed in dealing with the present series of data is the process of abstraction of the α-proton of the substrate. This seems to induce a strong charge-transfer interaction between the nitrogen lone pair of the substrate amino group and the isoalloxazine nucleus of the coenzyme. Conceivably, these phenomena occur in concert with each other. It is accepted that the steric fit of these interacting sites can be established through binding of the coenzyme and substrate with the enzyme protein, which subsequently attains the proper fit with some conformational change of the enzyme protein as indicated by $(E_{ox}\cdot S)_1 \longrightarrow (E_{ox}\cdot S)_2$ in Eqs. 9 and 10. Appreciating such a state of affairs, in which the protein moiety of the enzyme plays a cardinal role prior to the formation of the purple intermediate, as well as the subsequent role played by flavin, it seems really relevant that the coenzyme of this enzyme bears the prefix " co."

Although flavin plays essential roles in various oxidoreduction reactions, it is not a simple mediater; its behavior is largely influenced by the properties of protein moiety and/or the structure of the substrate. The present studies have shown that the flavin-substrate interaction is affected to a large extent by at least two factors, namely, the hydrophobic nature and electric state of its circumambience. The hydrophobic circumstance seems to prevent the dissociation of the inner complex into semiquinoid enzyme and substrate radical and, consequently, affords a favorable condition for a two-electron transfer mechanism. The electric field which is established by the occurrence of a positively charged group at the ω-position of the substrate seems to accelerate the electron transfer from the donor to the acceptor, if one takes into account the theoretical view of Yomosa (*32*), according to whom the local electric field determines, in a critical manner, the degree of charge transfer.

The fact that the purple intermediate is not observed with D-arginine may be reasonably explained by the above-mentioned view. Considering that the alkyl group of neutral amino acid interacts with some hydrophobic locus of the enzyme, the presence of such a charged group at the ω-position may expel the substrate from the enzyme at its binding site, resulting in instability of the enzyme-substrate complex. On the other hand, the reduction of the enzyme occurs rapidly because of the influence of the electric field. The situation is exactly the reverse in the case of neutral amino acids.

Probably, the above discussion could be applied, in basic principles, to many other enzymes, especially those catalyzing oxidoreduction processes.

Summary

1. The purple complex, which was isolated by crystallization from the anaerobic solution of D-amino acid oxidase reduced with D-alanine, was composed of equimolar amounts of the enzyme and substrate moieties. The complex was found to be involved in the equilibrium, $E_{ox}+S \rightleftarrows E_{purple} \rightleftarrows E_{red}+P$.

2. By the rapid reaction method, the purple intermediate was studied

spectrophotometrically under strictly anaerobic conditions. Analyzing the spectra obtained for the rapid and slow phases of the reaction observed in the stopped-flow experiments, it was concluded that the purple complex mentioned above is identical with the rapidly appearing purple intermediate.

3. Upon storage in the dark, the purple complex slowly changed into the semiquinoid enzyme. The addition of trichloroacetic acid to the purple complex yielded rhodoflavin. From these results together with other evidences, the purple complex was regarded as being an inner complex, *viz.* a strong charge-transfer complex between the enzyme and substrate moieties.

4. A hydrophobic environment around the flavin chromophore seemed to prevent the dissociation of the inner complex and therefore afford a favorable condition to facilitate a two-electron type oxidoreduction in the reaction catalyzed by this enzyme.

5. As an essential event inducing formation of the purple intermediate, the complex formation between the enzyme and substrate, which subsequently enables the interaction between the nitrogen lone pair of the amino group of the substrate and the isoalloxazine nucleus, was noted. In connection with this interaction, cleavage of α-hydrogen of the substrate was found to be essential for the formation of the purple intermediate as evidenced by the results of kinetic isotope effect using α-deuterated amino acids.

6. Differing from the mode of catalytic oxidation of neutral amino acids such as D-alanine, in which oxygen was known to react with the purple intermediate, catalytic oxidation of basic amino acids such as D-arginine was found to involve oxidation of the fully reduced enzyme with oxygen.

7. Difference in the modes of action between the neutral amino acid and basic amino acid was discussed from the viewpoint of the reaction mechanism of this enzyme.

References

1 K. Yagi and T. Ozawa, *Biochim. Biophys. Acta*, **81**, 29 (1964).
2 K. Yagi, K. Okamura, M. Naoi, N. Sugiura and A. Kotaki, *Biochim. Biophys. Acta*, **146**, 77 (1967).
3 H. Kubo, H. Watari and T. Shiga, *Bull. Soc. Chim. Biol.*, **41**, 981 (1959).

4 V. Massey, G. Palmer, C. H. Williams, Jr., B. E. P. Swoboda and R. H. Sands, *in* " Flavins and Flavoproteins," ed. by E. C. Slater, Elsevier, Amsterdam, p. 133 (1966).
5 K. Yagi, M. Nishikimi, N. Ohishi and K. Hiromi, *J. Biochem.*, **65**, 663 (1969).
6 K. Yagi, M. Nishikimi, N. Ohishi and A. Takai, *in* " Flavins and Flavoproteins," ed. by H. Kamin, University Park Press, Baltimore, Vol. III, p. 239 (1971).
7 K. Yagi, M. Nishikimi and N. Ohishi, *J. Biochem.*, in press.
8 K. Yagi and N. Ohishi, *J. Biochem.*, **71**, 993 (1972).
9 T. Nakamura, S. Nakamura and Y. Ogura, *J. Biochem.*, **54**, 512 (1963).
10 V. Massey and Q. H. Gibson, *Federation Proc.*, **23**, 18 (1964).
11 A. Kotaki, N. Sugiura and K. Yagi, *Biochim. Biophys. Acta*, **151**, 689 (1968).
12 K. Yagi, N. Sugiura, K. Okamura and A. Kotaki, *Biochim. Biophys. Acta*, **151**, 343 (1968).
13 V. Massey and G. Palmer, *Biochemistry*, **5**, 3181 (1966).
14 R. S. Mulliken, *J. Phys. Chem.*, **36**, 801 (1952).
15 K. Yagi, J. Okuda and K. Okamura, *J. Biochem.*, **58**, 300 (1965).
16 V. Massey and H. Ganther, *Biochemistry*, **4**, 1161 (1965).
17 K. Yagi, M. Naoi, M. Nishikimi and A. Kotaki, *J. Biochem.*, **68**, 293 (1970).
18 A. Kotaki, M. Naoi and K. Yagi, *J. Biochem.*, **59**, 625 (1966).
19 K. Yagi, N. Ohishi, M. Naoi and A. Kotaki, *Arch. Biochem. Biophys.*, **134**, 500 (1969).
20 M. Naoi, A. Kotaki and K. Yagi, unpublished results.
21 K. Yagi, M. Nishikimi, N. Ohishi and A. Takai, *J. Biochem.*, **67**, 153 (1970).
22 K. Yagi, N. Ohishi and M. Nishikimi, *Biochim. Biophys. Acta*, **206**, 181 (1970).
23 K. Yagi, M. Nishikimi, N. Ohishi and A. Takai, *FEBS Letters*, **6**, 22 (1970).
24 M. Nishikimi and K. Yagi, unpublished results.
25 A. H. Neims, D. C. DeLuca and L. Hellerman, *Biochemistry*, **5**, 203 (1966).
26 S. Nakamura, Z. Takatsu and Y. Ogura, *Symp. Enzyme Chem.* (*Kanazawa*), **19**, 234 (1968).
27 G. Palmer and V. Massey, *in* " Biological Oxidations," ed. by T.P. Singer, Interscience, New York, p. 263 (1968).
28 K. Yagi, M. Nishikimi, N. Ohishi and A. Takai, unpublished results.
29 M. Nishikimi, M. Osamura and K. Yagi, *J. Biochem.*, **70**, 457 (1971).

30 B. Chance, *in* "Modern Trends in Physiology and Biochemistry," ed. by E. S. Barron, Academic Press, New York, p. 25 (1952).
31 B. Chance, *in* "Flavins and Flavoproteins," ed. by E. C. Slater, Elsevier, Amsterdam, p. 223 (1966).
32 S. Yomosa, *Progr. Theor. Phys. Suppl.*, **4**, 249 (1967).

Received for publication December 27, 1971.

DYNAMIC ASPECTS OF THE ENZYMES IN D-AMINO ACID OXIDASE AND CYTOCHROME P-450 REACTIONS

Yoshihiro Miyake and Toshio Yamano
Department of Biochemistry, School of Medicine, Osaka University, Osaka

Enzymes and proteins which contain heme iron, non-heme iron, copper or flavin participate in biological oxidations as oxidases, oxygenases, oxygen carriers, and dehydrogenases or reductases. In recent years, many of these redox enzymes and proteins have been purified or crystallized, and their reaction mechanisms have been investigated in many ways. When compared in detail, their modes of reaction with oxygen and other electron acceptors are found to be not necessarily independent of each other (*1*). Oxidase and oxygenase reactions partially involve reductase reactions and activation of oxygen. Ferroperoxidase may form Compound III with oxygen whose absorption spectrum is similar to those of oxyhemoglobin and oxymyoglobin (*2*), and the peroxidase-oxidase reaction involves the reduction of oxygen by peroxidatively generated free radicals (*3*). Furthermore, some oxygenases show oxidase reactions (*4–7*). One of the characteristics of these redox enzymes and proteins is that those containing identical prosthetic groups show different biological functions. Some of them are listed in Table I.

No doubt such differences in function arise from characteristic struc-

TABLE I. Main Classes of Redox Proteins

Functions	Prosthetic groups			
	Heme iron	Non-heme iron	Copper	Flavin
Oxygen carrier	Hemoglobin Myoglobin	—	Hemocyanin	—
Oxidase	Cytochrome oxidase	—	Ascorbate oxidase Laccase	L-Amino acid oxidase D-Amino acid oxidase
Oxygenase	Cytochrome P–450 Tryptophan pyrrolase	Metapyrocatechase Pyrocatechase	Tyrosinase Dopamine-β-hydroxylase	Lactate oxygenase Salicylate oxygenase
Peroxidase	Horse-radish peroxidase Paraperoxidase	—	—	NADH peroxidase (*S. faecalis*)
Reductase or electron carrier	Cytochrome *b* Cytochrome *c*	Adrenodoxin Ferredoxin	Plastocyanin Azurin	Succinate dehydrogenase NADH-Cytochrome b_5 reductase

TABLE II. Comparison of Some Properties of D-Amino Acid Oxidase before and after Tryptic Digestion

	Before digestion	After digestion[a]
Residual activity (%)	100	67
Residual protein (%)[b]	100	90
$s_{20,w}$	7.0	6.9
Michaelis constant for D-alanine (mM)	3.3	3.3
Maximum velocity, $\times 10^{10}$ (mole·sec^{-1})	11.4	6.1
Reaction rate, k_3[c] (sec^{-1})	1.82	1.44

[a] The experimental conditions were the same as described for Fig. 2. [b] The amount calculated from the area of sedimentation peaks. [c] The reaction rate was calculated using the equations $E+S \rightleftharpoons ES \xrightarrow{k_3} E+P$ and $k_3 = V_{max}/e_{act}$, where V_{max} and e_{act} represent, respectively, the maximum velocity and moles of the active enzyme in the reaction mixture.

tures of their apoprotein moieties. Therefore, the elucidation of the functional differences among those enzymes and proteins will not be achieved without studying the role of the apoprotein moiety. The interaction of the apoprotein moiety with its prosthetic group may also be important for their functions. The mode of such an interaction may be changed by binding of the substrate and by oxidation-reduction of the prosthetic group. Such changes in the interaction may induce conformational changes of the apoprotein moiety, especially in their active regions.

This review deals with dynamic changes in the enzymes during their functions with special reference to D-amino acid oxidase and cytochrome P-450.

D-Amino Acid Oxidase

Rapid progress in understanding the reaction mechanisms of D-amino acid oxidase was promoted by the work of Kubo *et al.* (*8–10*), who crystallized the enzyme and found an intermediate spectral species characterized by a broad absorption band centering around 550 mμ during the anaerobic reaction with D-alanine. Since then, this intermediate has been generally called the purple intermediate because of its color development. The chemical nature of the bound FAD and substrate in the purple intermediate has been repeatedly discussed (*9, 11–14*). However, the problem has not been fully resolved. Intermediate states of the enzyme other than the purple intermediate have not been detected during the enzyme reaction, although the occurrence of semiquinone forms of the bound FAD has been shown by the reduction with dithionite or by irradiation (*11, 14–17*). In spite of those extensive studies on oxidation-reduction of the bound FAD, information on the apoprotein moiety during catalysis is still not sufficient. Following is some evidence indicating changes in the apoprotein moiety accompanied by binding FAD and substrate, and by oxidation-reduction of the bound FAD.

1. Effects of hydrolytic enzymes on redox states of the enzyme

On addition of Nagarse (a proteinase from *B. subtilis*) to the apoenzyme solution, the enzymatic activity decreases and the 280 mμ species of

Abbreviations: the bound FAD, flavin adenine dinucleotide bound to D-amino acid oxidase apoenzyme; PAL, pyridoxal-5′-phosphate; ESR, electron spin resonance.

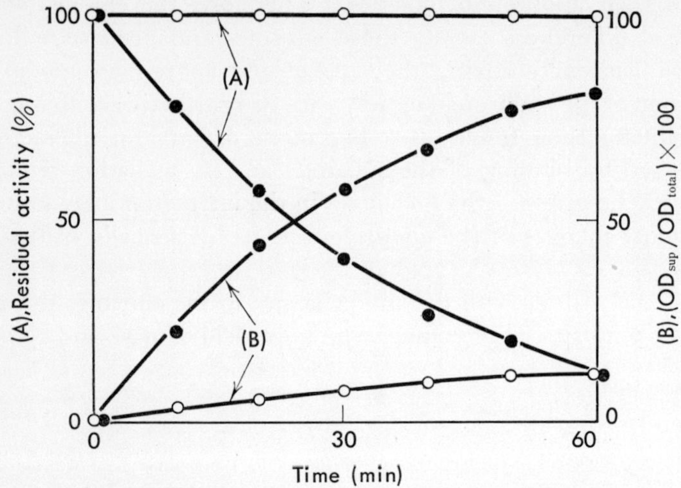

FIG. 1. Digestion of D-amino acid oxidase and its apoenzyme by Nagarse. The apoenzyme (8 mg) was subjected to the action of 0.1 mg Nagarse in 0.1 M pyrophosphate buffer, pH 8.3, (●). In the case of the holoenzyme, 0.2 ml of 0.1 mM FAD solution was added to the above reaction mixture (○). The total volume of the reaction mixture was 3.0 ml, and the reaction was carried out at 30°C. (A) and (B) represent the residual activity and degree of digestion, respectively. OD_{sup} and OD_{total} represent absorbancy of the supernatant after trichloroacetic acid treatment at 280 mμ and that of the solution after complete denaturation.

digested small fragments appear in the supernatant after trichloroacetic acid treatment. In this case, the degree of digestion and inactivation of the apoenzyme were parallel (*25*). When FAD was added to the reaction mixture, the enzymatic activity changed little and very few 280 mμ species appeared, indicating that the holoenzyme was resistant to Nagarse digestion (Fig. 1). On addition of trypsin (obtained from Mochida Pharmaceutical Co.) to the apoenzyme solution, the enzymatic activity rapidly decreased. However, FAD did not show as strong a protective effect as in Nagarse digestion (Fig. 2). As seen in Fig. 2, trypsin digestion resulted in the loss of the enzymatic activity, but the degree of digestion was much less than that of inactivation. Benzoate, which is an inhibitor competing with the substrate, protected the holoenzyme from the trypsin digestion (Fig. 2). The reduced enzyme was also digested by trypsin, whereas benzoate did not show any protective effect for it.

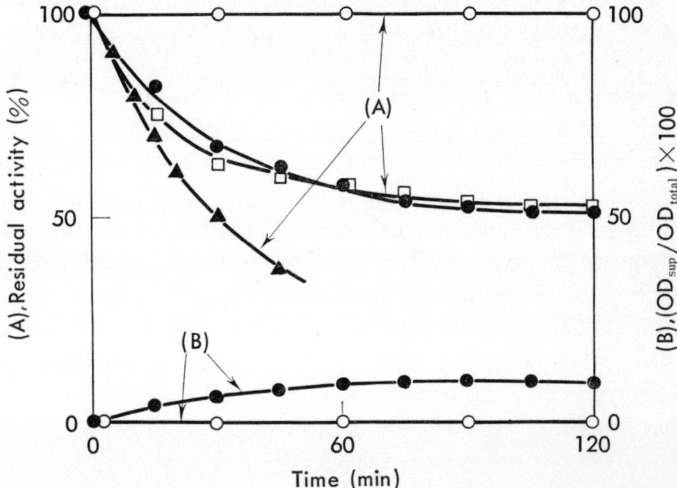

FIG. 2. Digestion of D-amino acid oxidase by trypsin. The oxidized enzyme (●), the reduced enzyme (□), benzoate complex of the oxidized enzyme (○), and the apoenzyme (▲) were subjected to the action of 0.4 mg trypsin. The ratio of the amounts of the enzyme and trypsin was 25. In the case of the reduced enzyme, 0.2 ml of 0.05 M DL-alanine or a small amount of dithionite was added to the reaction mixture, and the reaction was carried out under anaerobic conditions. Other experimental conditions were the same as those given for Fig. 1. (A) and (B) represent the residual activity and degree of digestion, respectively.

With respect to the tryptic digestion, the following results were obtained. The maximum velocity of the enzyme decreased on digestion, while the sedimentation constant, the Michaelis constant, and the molar activity of the enzyme were almost unchanged before and after the digestion (Table II). These results suggest that the enzyme was digested by trypsin in an all-or-none fashion with respect to the enzymatic activity, and that trypsin attacked some amino acid residues, probably lysine and arginine residues which were closely related to the enzymatic activity. Benzoate may mask these residues and protect the oxidized enzyme from the tryptic attack. The fact that benzoate did not show any protective effect on the enzyme reduced by substrate or dithionite against trypsin digestion and that FAD protected the apoenzyme from Nagarse digestion, also suggests that there existed some differences in the conformation of the protein moiety among the oxidized and reduced forms of the enzyme and the apoenzyme.

2. *Lysine residues of the enzyme at the active region*

Concerning lysine residues of the enzyme at the active site, Coffey *et al.* (*18*) and Hellerman and Coffey (*19*) have demonstrated that the substrate in an intermediate state of the enzyme reaction was fixed to a lysine residue to form ε-N-(1-carboxyethyl)-L-lysine by the reduction of the intermediate state of the enzyme with $NaBH_4$. It has also been shown that pyridoxal-5′-phosphate reacted with the apoenzyme and 2 moles of PAL per mole of enzyme were fixed to lysine residues to form a reduced Schiff's base by reduction with $NaBH_4$ (*20*). It has also been reported that the enzymatic activity was lost by modification of the enzyme with succinic anhydride (*21*) and trinitrobenzene sulfonate (*22*). However, it seems unlikely that lysine residues are involved directly in the binding of FAD or the substrate. The enzyme modified by the method of Coffey *et al.* (*18*) retained full enzymatic activity (*23*), and the PAL-apoenzyme still showed 52% enzymatic activity of the native enzyme (*20*). As the Michaelis constant of FAD for the PAL-apoenzyme was ten times larger than that for the native apoenzyme, a lysine residue would be located close to the active site, but would not be involved in it directly. Participation of arginine residue for the enzyme reaction has also been reported from the modification of the enzyme with glyoxal (*24*).

3. *Effect of urea on redox states of the enzyme*

The difference in the conformation between the oxidized and reduced forms of the enzyme has been observed more distinctly in the effect of urea on the two forms of the enzyme (*25*). As shown in Fig. 3, both the apoenzyme and the oxidized enzyme were denatured by the reaction with 5 M urea for 60 min and at 30°C, while the reduced enzyme was quite stable under the same conditions. The reduced enzyme was also stable against Nagarse digestion in the presence of 1 M urea, while the oxidized enzyme was digested even at a concentration of 0.5 M urea (*26*). Though the details concerning the difference in the effect of urea on the two forms of the enzyme are unknown, the oxidized enzyme may have changed its tertiary structure as well as secondary structure on reduction, which resulted in the formation of a conformation stable enough to be resistant to urea denaturation.

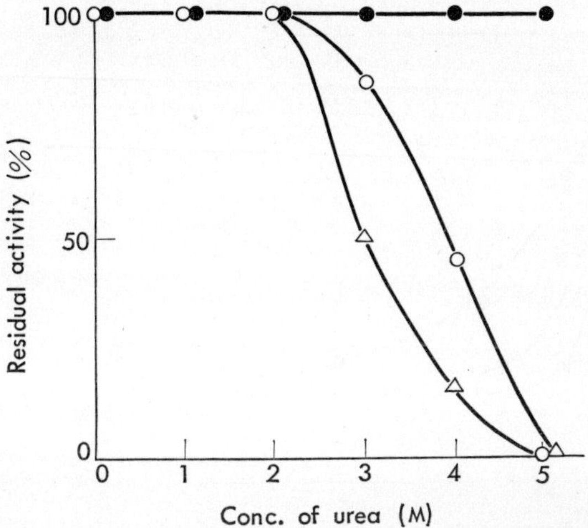

FIG. 3. Effect of urea on oxidized and reduced D-amino acid oxidase and its apoenzyme. The enzyme (8 mg) was incubated at 30°C for 60 min in the presence of several concentrations of urea. The total volume of the reaction mixture was 3.0 ml. In the case of the reduced enzyme, 0.2 ml of 0.05 M DL-alanine or a small amount of dithionite was added to the reaction mixture, and the reaction was carried out under anaerobic conditions. After the reaction, the enzyme solution was diluted to a concentration suitable for measurement of the enzymatic activity. On dilution, the concentration of urea became between 2 mM and 10 mM, and the enzymatic activity was not affected by such a concentration range of urea. ○ oxidized enzyme; ● reduced enzyme; △ apoenzyme. Miyake *et al.* (*25*).

4. *Subunit interaction of the enzyme*

It had been reported that the molecular weight of the enzyme is about 100,000, and 2 moles of FAD are contained per mole of the enzyme (*25*, *27*, *28*). Recent studies have demonstrated that it consists of two subunits whose molecular weights are about 35,000–50,000 (*29–32*). In addition, there exists an equilibrium between the monomer and the dimer (*31*), and benzoate has a strong effect on dimerization of the enzyme (*30*). In fact, the chromatography of the enzyme over Sephadex G-100 column using the small zone method showed that the apoenzyme was eluted as a monomer in the enzyme concentrations shown in Fig. 4 (*33*). While the holoenzyme and the benzoate complex of the enzyme were eluted as a dimer at high enzyme concentrations, they tended to become a monomer

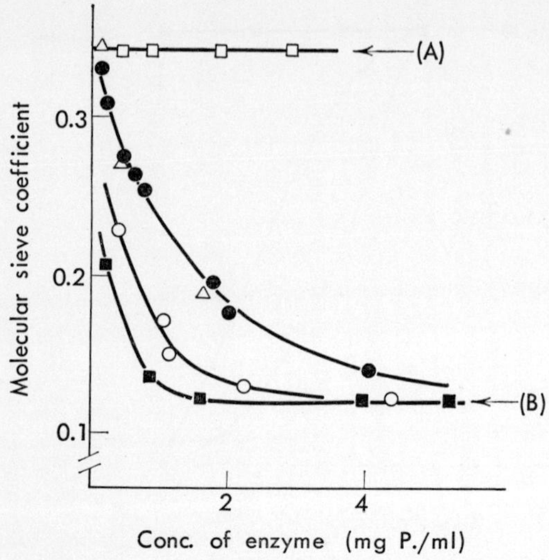

FIG. 4. Effects of D-alanine and benzoate on the molecular sieve coefficient of D-amino acid oxidase. Abscissa represents the concentrations of the enzyme applied to a Sephadex G-100 column (1×40 cm). Each curve was obtained from the chromatography of the enzyme on the column equilibrated with 0.1 M sodium pyrophosphate buffer, pH 8.3, containing 10 μM FAD (●), 10 μM FAD and 0.1 M DL-alanine (○), 10 μM FAD and 0.05 M L-alanine (△), and 10 μM FAD and 2 mM benzoate (■), respectively. The apoenzyme was chromatographed with 0.1 M pyrophosphate buffer, pH 8.3, (□). The molecular weights calculated by using the values of molecular sieve coefficients indicated by the arrows, (A) and (B), were 45,000 and 90,000, respectively. Miyake *et al.* (*33*).

on decreasing the concentration of the enzyme. The benzoate complex of the enzyme was eluted as a dimer up to a much lower concentration of the enzyme than in the case of the holoenzyme. Using a Sephadex G-100 column, it was also found that the substrate, D-alanine, acted as an effector on dimerization of the enzyme (*33*). In the presence of D-alanine, the elution volume of the enzyme began to increase at lower concentrations than that in the absence of D-alanine (Fig. 4). When the enzyme was chromatographed over the column in the presence of D-alanine, the state of the enzyme in the column became somewhat complicated. It was also unclear whether or not reversible conversion between the monomer and dimer occurred during catalysis. At extremely low concentrations of the enzyme for measurements of the enzymatic activity, subunit interaction of the enzyme seemed unlikely. Therefore, it is indicated from these molecular sieve studies that the monomeric form of the enzyme changed the structure during catalysis and that such conformations may have been favorable for the formation of the dimeric form of the enzyme at high concentrations.

The properties of D-amino acid oxidase described above show that the conformation of apoprotein moiety changed on binding with FAD and the substrate, or when accompanied by oxidation-reduction of FAD. D-Amino acid oxidase reaction involves several reaction steps, such as binding and activation of substrate, electron transfer from substrate to FAD, and binding and reduction of oxygen. Accordingly the apoprotein moiety may have changed its form during those reaction processes. However, the evidence is still insufficient, and details on the conformational changes of the enzyme during catalysis remain unknown. Further study on this problem will offer information not only to elucidate the reaction mechanism of D-amino acid oxidase reaction, but to investigate the functional differentiation of oxidase, oxygenase, peroxidase and reductase, among flavin enzymes.

Cytochrome P-450 in Liver Microsomes

According to Mason's classification of oxidases (*34*), the liver microsomal b type cytochrome, P-450, is a hydroxylase of the external mixed function oxidase class having an NADPH dependent electron transfer system as electron donor. Because of its unique character in subcellular particles, extensive studies have been carried out on this hemoprotein. Membrane-bound P-450 is affected by surface active agents, chelating agents, sulfhydryl reagents, and lipophilic substances, being converted into an inactive form, P-420 (*35–38*), and reduced P-450 in microsomes gives an abnormal CO difference spectrum having a peak at 450 mμ (*35, 36*). In addition, the content of this hemoprotein in microsomes is changed by treating animals with certain drugs (*39, 40*). Some of them induce formation of a spectrally different type of P-450. Owing to its instability toward surface active agents, the solubilization and purification of liver cytochrome P-450 has not yet been successful. Recently, Miyake *et al.* (*41*) isolated detergent soluble submicrosomal particles which contain P-450 but no cytochrome b_5. The absolute spectra of P-450 in the submicrosomal particles provide a reasonable explanation for the CO differrence spectrum of reduced P-450 and the difference spectrum of oxidized P-450 minus reduced P-450 in microsomes. Furthermore, the spectral changes of P-450 caused by oxidation-reduction and by the reactions with ligands are quite different from those observed with other well known protoheme proteins. These physicochemical and physiological characteristics of P-450 are probably due to the nature of the apoprotein moiety, especially the structure around the heme binding site. We are concerned here with those specific characteristics of P-450 in connection with its biological function and compared with the corresponding properties of other protoheme proteins.

1. *On the spin states of P-450*

Hashimoto *et al.* (*42*) have found that there exists a component in microsomes which shows an anisotropic ESR signal at the $g=2$ region (Fig. 5). They designated this component microsomal Fe_x. Recently, the microsomal Fe_x was proved by Miyake *et al.* (*41*) to be a manifestation of P-450. The ESR spectrum of microsomal Fe_x is that of a low spin ferric hemo-

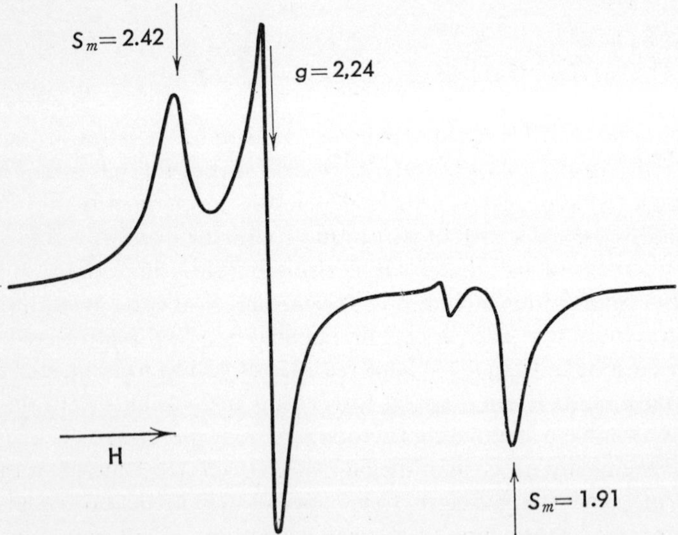

FIG. 5. ESR spectrum of P-450 in rabbit liver microsomes. ESR spectrum was taken with a Varian V-4500 spectrometer equipped with 100 kHz field modulation and a Varian variable temperature accessory at an incident power of 25 mW at a field modulation of 15 gauss; microwave frequency, 9.15 GHz; temperature, near liquid nitrogen.

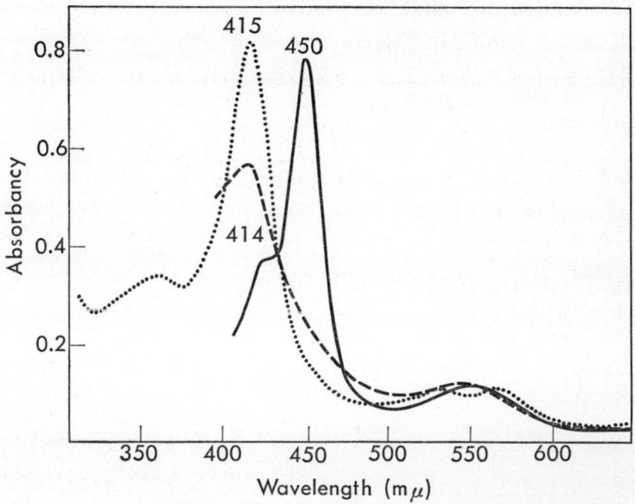

FIG. 6. Absorption spectra of P-450 in submicrosomal particles in oxidized (.....), dithionite-reduced (– – –), and CO-reduced, CO complex of dithionite-reduced, (—) forms. Submicrosomal particles (2.4 mg p./ml) dissolved in 0.1 M phosphate buffer, pH 7.5, containing 50% glycerol, were used. Miyake *et al.* (*41*).

protein and the absorption spectrum of oxidized P-450 has α- and β-bands but almost no charge transfer band in the visible region (Fig. 6). With respect to the reduced P-450, Williams (*43*) suggested that it is in high-spin state judging from its reactivity with CO and oxygen. The absolute spectrum of reduced P-450 is also similar to those of other high-spin ferrous hemoproteins (Fig. 6). Therefore, P-450 is a low-spin protoheme protein in its oxidized state and probably a high-spin ferrohemoprotein in its reduced state. These spin states of P-450 in oxidized and reduced states seem to be unusual for protoheme proteins. Protoheme proteins which have been well investigated may be classified into two groups. One group comprises species which react with oxygen or hydrogen peroxide, and the other includes species which do not react with oxygen but transfer electrons to electron acceptors. The physicochemical characteristics of these two groups are summarized in Table III. Those reacting with oxygen or hydrogen peroxide are in a high-spin state in both oxidized and reduced forms and react with cyanide, fluoride, azide, NO, and CO, while those not reacting with oxygen are in alow-spin

TABLE III. Some Physicochemical Properties of Low and High Spin Hemoproteins

	Cytochrome b_5 (*57, 58*)	
Positions and intensities of absorption bands in visible region (mμ)		
Ferric	532	560
Ferrous	526 (13.4)	556 (25.6)
ESR signals	$g=2.23$, $S_m=3.03$, 1.43 (anisotropic)	
Paramagnetic susceptibility ($\chi_M^{20°C}$, $\times 10^6$)		
Ferric	2,430	
Ferrous	500	
Reactivity for anions (F^-, N_3^-, CN^-)	Negligibly reactive	
Reactivity for CO and O_2	Negligibly reactive	
Biological function	Electron carrier	

The values in parentheses represent millimolar extinction coefficients.

state in both oxidized and reduced forms and react little with those anions and CO. If this classification is applied to P-450, it should be in high-spin states in both oxidized and reduced forms, because P-450 reacts with oxygen and functions as a hydroxylase. However, the oxidized P-450 is apparently in low-spin state as shown in Figs. 5 and 6. At present, the physicochemical and physiological significance of the spin state conversion of P-450 caused by reduction is unknown. It is interesting to note that oxidized cytochrome oxidase in a resting state is also in a low-spin state and the high-spin species appear in the reduction process (*44*). The fact that both P-450 and cytochrome oxidase are membrane-bound and accept electrons from electron transfer systems might be correlated with such a spin state conversion.

2. *Bifunctional feature of P-450*

Besides the hydroxylase reactions in microsomes, it is well known that microsomes consume oxygen upon addition of NADPH even in the absence of substrates which are hydroxylated by P-450. Under anaerobic conditions, P-450 in microsomes is also reduced by NADPH in the absence of the substrates. Staudinger *et al.* (*45*) showed that in microsomes there are two different terminal oxidases having different afin Comparison with Those of Cytochrome P–450

Horse-radish peroxidase (*59–62*)	Cytochrome P–450 (*41*)
497 (10.03) 641.5 (2.84) 558 (11.2)	532 (19) 567 (18.5) 543 (21.5)
$S_m = 6.1$, $g = 2.05$	$g = 2.24$, $S_m = 2.42$, 1.91 (anisotropic)
11,560	—
11,410	—
Reactive	Negligibly reactive except CN^-
Reactive	Reactive
Peroxidase reaction	Hydroxylase

finities for oxygen, and they suggested that one of them is P-450. The following facts strongly support the suggestion by Staudinger *et al.* (*45*), namely, (a) oxygen consumption of microsomes in the presence of NADPH is inhibited by cyanide, and the degree of the inhibition is parallel with that of the modification of the ESR spectrum of P-450 by cyanide (*7*), (b) oxygen consumption by lipid peroxidation of rabbit liver microsomes is very slight even in the presence of NADPH (*46*), (c) hydroxylase reactions in microsomes *via* a cyanide sensitive factor are inhibited by a few millimoles of cyanide (*47*, *48*), but NADPH-dependent oxygen consumption of microsomes in the absence of a substrate is not inhibited by such a low concentration of cyanide, and (d) oxygen consumption of microsomes reduced by NADH is only one tenth that in the presence of NADPH (*7*). Therefore, NADPH-dependent oxygen consumption of microsomes is due to the oxidase action of P-450, and P-450 in microsomes may have a bifunctional feature, hydroxylase and oxidase activities.

Regarding the oxidase action of P-450, a possible reaction mechanism has been proposed (*7*). It is shown in Fig. 7. In this reaction mechanism, low-spin oxidized P-450 is reduced to high-spin reduced P-450 *via* an intermediate, low-spin reduced form. The process of the formation of reduced P-450 by the reduction of oxidized P-450 with excess di-

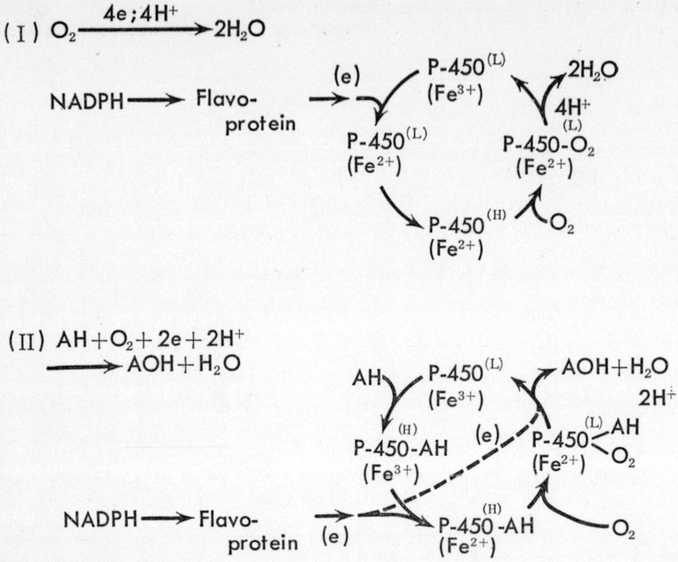

(I) $O_2 \xrightarrow{4e;\,4H^+} 2H_2O$

(II) $AH + O_2 + 2e + 2H^+ \longrightarrow AOH + H_2O$

FIG. 7. A comparison of possible reaction mechanisms of oxidase (I) and hydroxylase (II) actions of P-450 in liver microsomes. (L) and (H) indicate that P-450 is in low and high spin states, respectively.

thionite gives a hyperbolic curve on semilog plots, indicating that the reduction is not a single step reaction, but involves two successive reactions. However, the spectrum of such an intermediate suggesting the low-spin reduced form has not yet been detected. As catalase does not affect the oxygen consumption of microsomes reduced by NADPH, the oxidase reaction of P-450 may involve the transfer of four reducing equivalents to oxygen. The proposed mechanism of the oxidase reaction of P-450 differs from that of hydroxylase reactions of P-450, especially in the spin states in the reduction step. It has been proposed that in hydroxylase reactions in adrenal mitochondria and in bacteria, low-spin oxidized P-450 becomes a high-spin form upon complex formation with a substrate, and is then reduced to high-spin reduced P-450 (*49*, *50*). A comparison of the possible reaction mechanisms for oxidase and hydroxylase reactions is shown in Fig. 7. The significance of the physiological role of the oxidase action of P-450 and the difference in the reaction mechanism between oxidase and hydroxylase reactions of P-450 in the reduction step, however, await further investigation.

3. *Reactions of P-450 with some ligands*

In addition to the unique character of P-450 as described above, the interaction of P-450 with ligands also differs from those of other protoheme proteins. The establishment of P-450 as a protoheme protein (*35*, *36*) originates from findings by Klingenberg (*51*) and Garfinkel (*52*). They found a component showing a peak at 450 mμ in the CO difference spectrum of reduced microsomes. This peak was proved to be due to the Soret band of the CO complex of reduced P-450 from the measurement of its absolute spectrum (*41*, *54*) (Fig. 6). In the high-spin protoheme proteins so far discussed, the position of the Soret band of the CO complex of reduced P-450 is unusual. Not only CO but other ligands interact in a peculiar fashion with reduced P-450. As already shown in Fig. 6, the absolute spectrum of oxidized P-450 is not as unusual as that of a low-spin protoheme protein. Oxidized P-450 complexed with NO, phenylisocyanide, and cyanide also give usual absorption spectra of low spin protoheme proteins. The absorption spectrum of the cyanide complex of reduced P-450 has absorption bands at 432, 532, and 560 mμ. Their positions and intensities are very similar to those of the cyanide complexes of other ferrous hemoproteins, and the abnormality seen in the CO complex is not observed. The phenylisocyanide complex of reduced P-450, on the other hand, has two absorption bands in the Soret region (428 mμ and 455 mμ) and only one band in the visible region (550 mμ) at pH 7.5. At pH 5.4, the band at 428 mμ is greatly increased, and that at 455 mμ is decreased. Such spectral patterns in the Soret region have also been observed in the ethylisocyanide difference spectra of P-450 in microsomes (*53*) and in the absolute spectra of isolated preparations (*41*, *54*). In addition to those observations, the single band at 550 mμ is split at pH 5.4. These results suggest that the absorption spectrum of the phenylisocyanide complex of reduced P-450 at pH 7.5 is a composite of the spectra of two spectral species, one having the Soret band at 428 mμ and two bands, α and β, in the visible wavelength region, and the other having the Soret band at 455 mμ and a single band at 550 mμ. When these spectral properties of the phenylisocyanide complex of reduced P-450 are compared with those of the cyanide and CO complexes of reduced P-450, the spectral pattern of the species having the Soret band at 428 mμ is found to be similar to that of the cyanide complex, while the spectral pattern of the species having the Soret band at 455 mμ is similar to that of the CO complex. These are shown in

TABLE IV. Comparison of Positions and the Molar Extinction Coefficients of Absorption Peaks Shown by Some Complexes of Oxidized and Reduced P-450 (*55*)

	Ligands	Wavelength (mμ)		
Oxidized P-450	—	415 (138)	532 (19)	567 (18.5)
	Nitric oxide	432 (104)	543 (23)	575 (18))
	Phenylisocyanide	430 (133)	545 (23)	—
	Cyanide	436 (135)	553–4 (24)	—
Reduced P-450	—	414 (96)	543 (21.5)	
	Nitric oxide	410	550	
	Carbon monoxide	450 (131)	552 (21)	
	Phenylisocyanide			
	pH 7.5	428, 455†	550	
	pH 5.4	428, 455†	535	553–4
	Cyanide	432	532	560

The values in parentheses represent millimolar extinction coefficients. The underlined values are brief approximations, since in those cases the absorption bands were very broad. † The ratios of absorbancy at 428 mμ and 455 mμ are 1.06 at pH 7.5 and 2.94 at pH 5.4, respectively.

Table IV. Reduced P-450 shows two types of absorption spectra depending upon ligands. Though the factors producing two such spectral species are not known, this type of reaction of reduced P-450 with ligands is not observed with other protoheme proteins, and the factors may be related to the spin state conversion and bifunctional feature of P-450.

We have described the dynamic aspects of D-amino acid oxidase and cytochrome P-450 reactions focusing the problem on the protein moiety. It is now evident that the conformational changes in the apoprotein moiety of D-amino acid oxidase are brought about by substrate-binding and by oxidation-reduction of the FAD moiety. It seems that the structure of the reduced enzyme is more compact than that of the oxidized enzyme because of its resistance to urea denaturation. The compactness of the reduced enzyme might be favorable for the binding of oxygen or electron transfer from FAD to oxygen. The changes in the apoprotein moiety of the oxidized enzyme caused by the substrate and benzoate also might facilitate the oxidation of the substrate. Concerning these relationships between the conformation of the apoprotein moiety and the reaction mechanism, the hypothesis presented by Coffey *et al.* (*18*), Hellerman and Coffey (*19*), and Neims *et al.* (*13*) is very interesting. They suggested the presence of an enzyme nucleophile at the active site, which participates in the activation and the cleavage of the α-CH bond of the substrate. If the structural changes at the active site of the apoprotein moiety would be brought about by the binding of the substrate, it might produce a conformation favorable for the reaction of the substrate with the enzyme nucleophile. However, there is no evidence supporting such a possibility.

As cytochrome P-450 in liver microsomes has not been purified, we could not directly deal with its apoprotein moiety. The unique characteristics of the heme moiety, however, are derived from the specific nature of its apoprotein moiety, in particular, the structure around the heme binding site. In hydroxylase reactions, it was shown that low-spin oxidized P-450 becomes high spin on binding with the substrate (*49*, *50*). Then the question arises as to why low-spin oxidized P-450 becomes high spin before it is reduced by the NADPH-dependent electron transfer system. In this case also, there is no evidence allowing interpretation of the significance of the spin state conversion. It is certain that the spin state conversion plays a role in the P-450 reaction, and the solution of this problem will reveal interesting and important properties of this hemoprotein.

In spite of attempts to elucidate the role of the apoprotein moiety in the reactions of the two redox enzymes, we have been able to deal with the problem only superficially. A microscopic approach, such as the orbital steering theory proposed by Storm and Koshland (*56*), may be necessary for full elucidation.

Summary

Susceptibility of D-amino acid oxidase to proteolytic enzymes and the extent of denaturation with urea are dependent on the states of the enzyme, such as the apoenzyme and the oxidized and reduced holoenzyme. Benzoate and FAD protect the enzyme from proteolytic attack. D-Alanine as well as FAD and benzoate acts as an effector for dimerization of the monomeric form of the enzyme.

A bifunctional feature of cytochrome P-450, oxidase and hydroxylase activities, is also described in connection with its spin state conversion during the reaction. In addition, unusual types of reactions with ligands are discussed.

The properties of the two enzymes are indicative of the dynamic nature of their active sites during the enzyme reactions, and the conformations may be closely correlated with each reaction step involved.

References

1 Y. Miyake, *Protein, Nucleic Acid and Enzyme*, **16**, 255 (1971) (in Japanese).
2 K. Yokota and I. Yamazaki, *Biochim. Biophys. Acta*, **105**, 301 (1965).
3 I. Yamazaki, K. Yokota and R. Nakajima, *in* "Oxidases and Related Redox Systems," ed. by T. E. King, H. S. Mason and M. Morrison, John Wiley and Sons, Inc., New York, p. 485 (1965).
4 O. Hayaishi, Y. Ishimura, T. Nakazawa and M. Nozaki, *in* "Biochemie des Sauerstoffs," ed. by B. B. Hess and Hj. Staudinger, Springer-Verlag, Berlin, p. 196 (1968).
5 S. Takemori, *Seikagaku*, **42**, 1 (1970) (in Japanese).
6 R. H. White-Stevens and H. Kamin, *Biochem. Biophys. Res. Commun.*, **38**, 882 (1970).
7 Y. Miyake, K. Mori and T. Yamano, *Biochem. Biophys. Res. Commun.*, **44**, 564 (1971).
8 H. Kubo, T. Yamano, M. Iwatsubo, H. Watari, T. Soyama, J. Shiraishi, S. Sawada, N. Kawashima, S. Mitani and K. Ito, *Bull. Soc. Chim. Biol.*, **40**, 431 (1958).
9 H. Kubo, H. Watari and T. Shiga, *Bull. Soc. Chim. Biol.*, **41**, 981 (1959).
10 H. Kubo, T. Yamano, M. Iwatsubo, H. Watari. T. Shiga and A. Isomoto, *Bull. Soc. Chim. Biol.*, **42**, 569 (1960).
11 V. Massey and Q. H. Gibson, *Federation Proc.*, **23**, 18 (1964).
12 V. Massey, G. Palmer, C. H. Williams, Jr., B. E. P. Swoboda and R. H. Sands, *in* "Flavins and Flavoproteins," ed. by E. C. Slater, Elsevier, Amsterdam, Vol. VIII, p. 133 (1965).
13 A. H. Neims, D. C. Deluca and L. Hellerman, *Biochemistry*, **5**, 203 (1966).
14 K. Yagi, T. Ozawa, M. Naoi and A. Kotaki, *in* "Flavins and Flavoproteins," ed. by K. Yagi, University of Tokyo Press, Tokyo, p. 237 (1968).
15 T. Nakamura, S. Nakamura and Y. Ogura, *J. Biochem.*, **54**, 512 (1963).
16 T. Yamano, *Arch. Biochem. Biophys.*, **106**, 360 (1964).
17 V. Massey and G. Palmer, *Biochemistry*, **10**, 318 (1966).
18 D. S. Coffey, A. H. Neims and L. Hellerman, *J. Biol. Chem.*, **240**, 4058 (1965).

19 L. Hellerman and D. S. Coffey, *J. Biol. Chem.*, **242**, 582 (1967).

20 Y. Miyake and T. Yamano, *Biochim. Biophys. Acta*, **198**, 438 (1970).

21 K. Yamaji, E. Uehara, K. Aki and T. Yamano, *Koso Kagaku Symposium*, **19**, 228 (1968) (in Japanese).

22 K. Yagi, M. Harada and A. Kotaki, *Biochim. Biophys. Acta*, **122**, 182 (1966).

23 V. Massey, B. Curti, F. Muller and S. G. Mayhew, *J. Biol. Chem.*, **243**, 1329 (1968).

24 A. Kotaki, M. Harada and K. Yagi, *J. Biochem.*, **60**, 592 (1966).

25 Y. Miyake, K. Aki, S. Hashimoto and T. Yamano, *Biochim. Biophys. Acta*, **105**, 86 (1965).

26 K. Aki, Y. Miyake and T. Yamano, *Tokushima J. Exp. Med.*, **11**, 153 (1964).

27 V. Massey, G. Palmer and R. Bennett, *Biochim. Biophys. Acta*, **48**, 1 (1961).

28 P. A. Charwood, G. Palmer and R. Bennett, *Biochim. Biophys. Acta*, **50**, 17 (1961).

29 K. Yagi, M. Naoi, M. Harada, K. Okamura, T. Hidaka, T. Ozawa and A. Kotaki, *J. Biochem.*, **61**, 580 (1967).

30 M. L. Fonda and B. M. Anderson, *J. Biol. Chem.*, **243**, 5635 (1968).

31 S. W. Henn and G. K. Ackers, *Biochemistry*, **8**, 3829 (1969).

32 S. W. Henn and G. K. Ackers, *J. Biol. Chem.*, **244**, 465 (1969).

33 Y. Miyake, T. Abe and T. Yamano, *J. Biochem.*, **70**, 719 (1971).

34 H. S. Mason, *Ann. Rev. Biochem.*, **34**, 595 (1965).

35 T. Omura and R. Sato, *J. Biol. Chem.*, **239**, 2370 (1964).

36 T. Omura and R. Sato, *J. Biol. Chem.*, **239**, 2379 (1964).

37 H. S. Mason, J. C. North and M. Vanneste, *Federation Proc.*, **24**, 1172 (1965).

38 Y. Ichikawa and T. Yamano, *Biochim. Biophys. Acta*, **147**, 518 (1967).

39 S. Orrenius, J. Ericsson and L. Ernster, *J. Cell Biol.*, **25**, 627 (1965).

40 N. E. Sladek and G. J. Mannering, *Mol. Pharmacol.*, **5**, 186 (1969).

41 Y. Miyake, J. L. Gaylor and H, S. Mason, *J. Biol. Chem.*, **243**, 5788 (1968).

42 Y. Hashimoto, T. Yamano and H. S. Mason, *J. Biol. Chem.*, **237**, PC 3843 (1962).

43 R. J. P. Williams, *in* "Structure and Function of Cytochromes," ed. by K. Okunuki, M. D. Kamen and I. Sekuzu, University of Tokyo Press, Tokyo, p. 645 (1968).

44 B. F. van Gelder and H. Beinert, *Biochim. Biophys. Acta*, **189**, 1 (1969).

45 Hj. Staudinger, B. Kerekjarto, V. Ullrich and Z. Zubrycki, *in* "Oxidases and Related Redox Systems," ed. by T. E. King, H. S. Mason and M. Morrison, John Wiley and Sons, Inc., New York, p. 815 (1965).

46 T. E. Gram and J. R. Fouts, *Arch. Biochem. Biophys.*, **114**, 331 (1966).
47 N. Oshino, Y. Imai and R. Sato, *7th Int. Congr. Biochem.*, 725 (1967).
48 J. L. Gaylor and H. S. Mason, *J. Biol. Chem.*, **243**, 4966 (1968).
49 R. Tsai, C. A. Yu, I. C. Gunsalus, J. Peisach, W. Blumberg, W. H. Orme-Johnson and H. Beinert, *Proc. Natl. Acad. Sci. U.S.*, **66**, 1157 (1970).
50 J. A. Whysner, J. Ramseyer and B. W. Harding, *J. Biol. Chem.*, **245**, 5441 (1970).
51 M. Klingenberg, *Arch. Biochem. Biophys.*, **75**, 376 (1958).
52 D. Garfinkel, *Arch. Biochem. Biophys.*, **77**, 493 (1958).
53 Y. Imai and R. Sato, *Biochem. Biophys. Res. Commun.*, **22**, 620 (1966).
54 H. Nishibayashi and R. Sato, *J. Biochem.*, **63**, 766 (1968).
55 Y. Miyake, K. Mori and T. Yamano, *Arch. Biochem. Biophys.*, **133**, 318 (1969).
56 D. R. Storm and D. E. Koshland, Jr., *Proc. Natl. Acad. Sci. U.S.*, **66**, 445 (1970).
57 P. Strittmatter and S. F. Velick, *J. Biol. Chem.*, **221**, 253 (1956).
58 R. Bois-Poltoratsky and A. Ehrenberg, *Eur. J. Biochem.*, **2**, 361 (1967).
59 D. Keilin and E. F. Hartree, *Biochem. J.*, **49**, 88 (1951).
60 D. Keilin and E. F. Hartree, *Biochem. J.*, **61**, 153 (1955).
61 Y. Morita and H. S. Mason, *J. Biol. Chem.*, **240**, 2654 (1965).
62 H. Theorell, *Ark. Kemi Min. Geol.*, **16A**, No. 3 (1942).

Received for publication November 17, 1971.

MECHANISM OF O_2 REDUCTION BY ENZYMIC SYSTEMS

Isao Yamazaki
Biophysics Division, Research Institute of Applied Electricity, Hokkaido University, Sapporo

In his extensive review, Mason (*1*) has claimed that molecular oxygen is metabolized by three broad classes of enzymes, which he named oxygen transferase, mixed-function oxidases, and electron transfer oxidases. In the presence of electron-transfer oxidases and appropriate electron donors the final product of oxygen is hydrogen peroxide or water. In this article we shall describe some aspects of the reduction of oxygen, mostly in the reaction of electron-transfer oxidases. In the early stage of our study we were faced by two moot questions. One was the problem regarding the formation and participation in the oxidase reaction of superoxide radicals that are a one-electron reduced form of molecular oxygen. The other problem was that concerning the formation of a complex between reduced enzyme and oxygen as an active intermediate in the oxidase reaction. These problems may also be more or less of importance in dealing with oxygen metabolizing enzymes in general.

Abbreviations: HRP, horse-radish peroxidase; DHF, dihydroxyfumarate; DKS, diketosuccinate; ESR, electron spin resonance.

When the study of oxygen-consuming oxidation catalyzed by peroxidase (peroxidase-oxidase reaction) was begun some 15 years ago, we were interested particularly in identifying the chemical species that reacts with molecular oxygen. It was generally accepted, without firm experimental evidence, that the reduced forms of oxidase such as ferrous or cuprous enzymes might react with molecular oxygen to form oxygenated forms. The reaction appeared to be similar to that of hemoglobin or hemocyanin with molecular oxygen.

This simple model, however, was found to be inadequate to explain the mechanism of peroxidase-oxidase reaction. After much meandering, these questions have been solved not only in the mechanism of peroxidase-oxidase reaction but also in that of some related oxidase reactions.

Peroxidase-oxidase Reaction

Swedin and Theorell found in 1940 (*2*) that DHF oxidation was catalyzed by a purified preparation of HRP. Since that time, several molecules have been found to be oxidized in the presence of HRP under aerobic conditions. The molecules were indoleacetate (*3*), triose reductone (*4*), reduced pyridine nucleotide (*5*) and 2-methyl 1,4 naphthohydroquinone (*6*). The properties of the peroxidase-oxidase reaction are not always the same for various hydrogen donors nor for various peroxidases from different sources, but common features of the reaction seem to be the necessity of hydrogen peroxide as an initiator, activation by manganous ion and monophenols, and partial inhibition by CO. These results were reported in earlier stages of the study chiefly by Swedin and Theorell (*2*). Controversial results have been reported on the effect of CO, and the mechanism proposed for the peroxidase-oxidase reaction has fluctuated with the interpretation of the results of CO experiments and of the role of an oxyperoxidase-like intermediate which accumulates during DHF oxidation.

On the basis of the results obtained by Swedin and Theorell (*2*), participation of ferrous enzymes in the reaction was suggested by Lemberg and Legge (*7*) and Mason (*8*). The mechanism seemed plausible to us until we found that the reduction of methylene blue was catalyzed by peroxidase, and the reaction was similar to the peroxidase-oxidase reaction (*9*). It was found that methylene blue was reduced by a peroxidase system consuming a stoichiometric rather than catalytic amount of hydrogen peroxide. The reactions could be formulated as follows: for the oxidase reaction,

$$AH_2+O_2 \xrightarrow[(H_2O_2)]{\text{peroxidase}} A+H_2O_2 \qquad (1)$$

and for reduction of methylene blue (MB),

$$2\,AH_2+MB+H_2O_2 \xrightarrow{\text{peroxidase}} 2\,A+MBH_2+2\,H_2O\,. \qquad (2)$$

The results suggested that molecular oxygen and methylene blue were reduced by the same chemical species formed during the reaction. It was supposed that, as peroxide would be a reduction product of oxygen, a catalytic role of hydrogen peroxide in Reaction (1) could be substituted for the stoichiometric uptake in the reduction of methylene blue. The idea that the chemical species might be free radicals was then introduced. The mechanism was consistent with that of the peroxidase reaction proposed by George (*11*) and Chance (*12*).

Using cytochrome *c* as an electron acceptor it was possible to classify hydrogen donors of the peroxidase reaction into redogenic and oxidogenic (*13*). During the peroxidase reaction, the redogenic donors generated radicals with reducing ability and the oxidogenic donors, radicals with oxidizing ability. The difference in the redox properties could be easily explained on the basis of the chemical structure of donor molecules (Fig. 1). The crucial evidence for the formation of such free radicals was

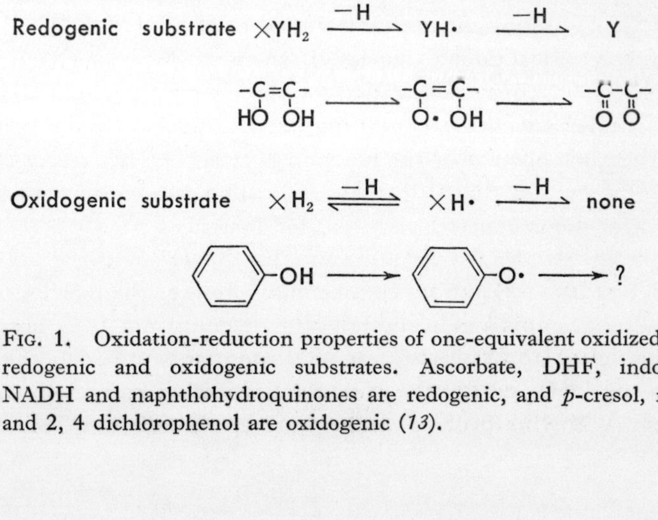

Redogenic substrate $\times YH_2 \xrightarrow{-H} YH\cdot \xrightarrow{-H} Y$

$$\underset{HO\;\;OH}{-C=C-} \longrightarrow \underset{O\cdot\;\;OH}{-C=C-} \longrightarrow \underset{O\;\;\;O}{-\overset{}{C}-\overset{}{C}-}$$

Oxidogenic substrate $\times H_2 \underset{}{\overset{-H}{\rightleftharpoons}} \times H\cdot \xrightarrow{-H}$ none

$$C_6H_5\text{-}OH \longrightarrow C_6H_5\text{-}O\cdot \longrightarrow\ ?$$

FIG. 1. Oxidation-reduction properties of one-equivalent oxidized forms of redogenic and oxidogenic substrates. Ascorbate, DHF, indoleacetate, NADH and naphthohydroquinones are redogenic, and *p*-cresol, resorcinol and 2, 4 dichlorophenol are oxidogenic (*13*).

furnished using an ESR spectrometer equipped with a flow apparatus (*14*).

$$2\,AH_2 + H_2O_2 \xrightarrow{\text{peroxidase}} 2\,AH\cdot + 2\,H_2O \tag{3}$$

The free radicals decayed rapidly through disproportionation or dimerization,

$$2\,AH\cdot \longrightarrow A + AH_2 \text{ (or AH–AH)} \tag{4}$$

and the free radical observed during continuous flow at a moderate rate was found to reach a steady state concentration (*15*).

Physiologically important molecules such as NADH and indoleacetate are also substrates for the peroxidase reaction, but observing ESR signals during the peroxidatic oxidation of these molecules has been unsuccessful. This may be mostly due to the rapid dismutation of these free radicals and partly due to the slow reaction of the donor molecules with peroxidase. Even in these cases the stoichiometric formation of such intermediate free radicals was strongly supported by the results obtained using ferric *o*-phenanthroline complex as an electron acceptor, that is, a scavenger of free radicals of indoleacetate (*16*) and NADH (*17*).

$$2\,AH_2 + H_2O_2 + 2\,Fe^{3+} \xrightarrow{\text{peroxidase}} 2\,A + 2\,Fe^{2+} + 2H_2O \tag{5}$$

Under aerobic conditions the free radicals of DHF, NADH and indoleacetate would seem to react with oxygen at a fairly high rate. The most direct evidence for the reaction is shown in Fig. 2. Although very rapid oxidation of DHF took place in the initial stage of the reaction when oxygen was present, an ESR signal of the DHF free radical was not observed. It suddenly appeared when most of the oxygen in the solution had been consumed, and at a time when slow oxidation that was observed under anaerobic conditions set in again. Rough calculation showed the rate constant of the reaction between DHF free radicals and oxygen to be greater than $10^8\ \text{M}^{-1}\ \text{sec}^{-1}$. The rate constants have been estimated for particular free radicals, for instances $5\times10^6\ \text{M}^{-1}\ \text{sec}^{-1}$ for menadione (*17*) and $2.0\times10^9\ \text{M}^{-1}\ \text{sec}^{-1}$ for NAD (*19*).

The free radical mechanism could explain most of the puzzling features of the peroxidase-oxidase reaction. But it could not fully account for the formation of an oxyperoxidase-like intermediate during the aerobic oxidation of DHF or for the irregular inhibition caused by CO. In connection with this problem there were two conflicting views on the

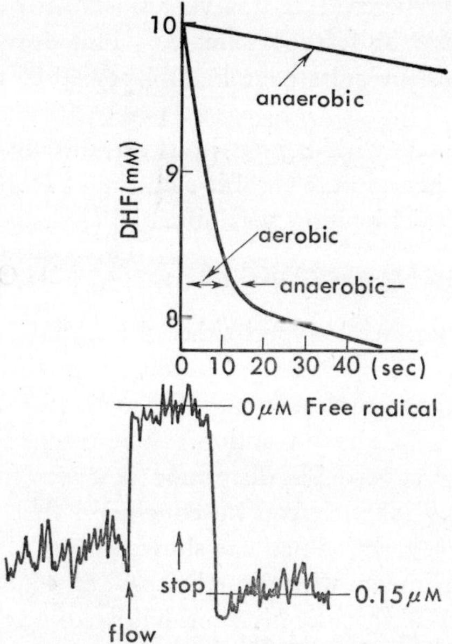

FIG. 2. Free radical of DHF during the peroxidatic reaction (10 mM hydrogen peroxide, pH 4.8). The upper diagram shows the oxidation of DHF measured spectrophotometrically at 342 nm. The aerobic reaction started from an oxygen-saturated solution, and the anaerobic from a nitrogen-saturated solution. The aerobic solution changed to anaerobic after rapid oxidation of about 2 mM of DHF. The lower diagram shows the fluctuation of the ESR signal due to the free radical of DHF. The time of stopping of the flow was almost simultaneous with that of the start of the reaction in the upper diagram (*18*).

mechanism of the reaction of ferroperoxidase with oxygen. It was suggested by many workers that ferroperoxidase combines with oxygen to form oxyperoxidase, or peroxidase Compound III has an oxyhemoglobin type of structure (*1*, *20*). This idea, however, was not supported by experimental evidence. The experiments indicated that ferroperoxidase was oxidized by oxygen to ferriperoxidase without the formation of an oxy-ferrous complex (*21–23*). The problem has been solved satisfactorily by a series of experiments of Wittenberg *et al.* (*26*) and ourselves (*24*, *25*). Ferroperoxidase actually combined oxygen to form oxyperoxidase that had been termed Complex III (*27*) or Compound III (*28*). Oxyperoxidase was reactive with various hydrogen donors and was reduced to ferriperoxidase. Oxyperoxidase was very unstable, especially in the presence of hydrosulfite or ferroperoxidase. This appeared to be the reason that contradictory results have been reported in the past (*25*).

The reaction of ferroperoxidase with oxygen, however, did not account for the formation of oxyperoxidase in the aerobic solution of DHF. For that reaction, the following stoichiometry was obtained (*18*).

$$2\,\mathrm{DHF} + \mathrm{H_2O_2} + 2\,\mathrm{Fe_p}^{3+} + 2\,\mathrm{O_2} \longrightarrow 2\,\mathrm{DKS} + 2\,\mathrm{Fe_p}^{3+}\mathrm{O_2}^- + 2\mathrm{H_3O^+} \quad (6)$$

The results suggested the formation of oxyperoxidase through the reaction between ferriperoxidase and superoxide anion, because the reduction of peroxidase by the DHF radical was not fast enough to account for the rapid formation of oxyperoxidase in the above reaction. The mechanism was recently confirmed (*61*) using superoxide dismutase that was found by McCord and Fridovich (*29*). Now, it can be concluded that oxyperoxidase is formed through three reaction paths as shown in Fig. 3. It was confirmed by titrimetric experiments with dimethyl *p*-phenylenediamine that oxyperoxidase was at the three-equivalent oxidized level above ferriperoxidase (*30*, *31*).

In general, the possibility cannot be neglected that oxyperoxidase is formed through the reaction of ferroperoxidase with oxygen in the peroxidase-oxidase reaction. Unlike the free radical of DHF, that of indoleacetate was an efficient reductant of peroxidase (*18*). It should be also noted here that indoleacetate reduced oxyperoxidase at a considerably high rate (*31*, *32*). From these points of view it can be safely said that the ferro-oxyferro cycle must be involved to some extent in the peroxidase-oxidase reaction when indoleacetate is used as an electron donor.

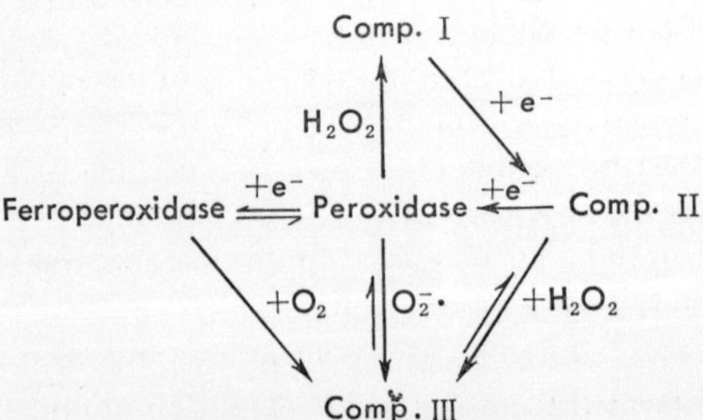

FIG. 3. Five redox forms of HRP (*18*).

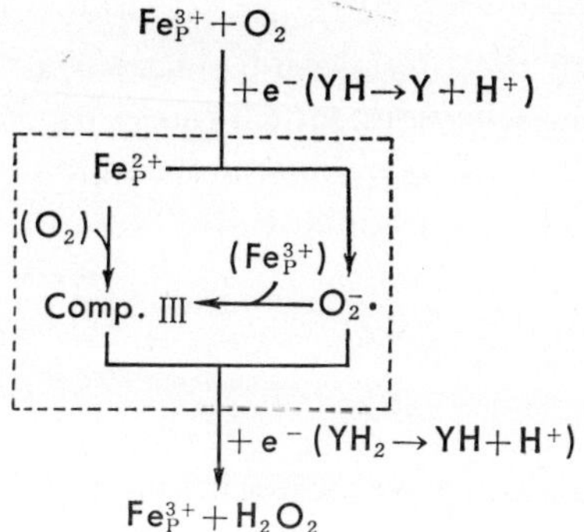

FIG. 4. A mixed mechanism of the peroxidase-oxidase reactions (*36*).

The typical ferro-oxyferro cycle is a case of tryptophan pyrrolase reaction (*33*). The similarity between the reactions of HRP-indoleacetate and tryptophan pyrrolase-tryptophan has been discussed elsewhere (*34*, *35*). It might be concluded that the peroxidase-oxidase reaction has multiple faces and the mechanism differs from substrate to substrate. Figure 4 shows a mixed mechanism which was reported in the first ISOX meeting (*36*). It is clearly shown in this figure that a strongly reducing equivalent formed during the peroxidase reaction is transferred to molecular oxygen through a mixed mechanism of two different paths, the ratio being dependent upon the properties of free radical species and experimental conditions. The mechanism can also account for the fact that CO inhibition depends upon experimental conditions.

It would appear pertinent to add here two more features characteristic of peroxidase-oxidase reactions. HRP was transformed into a green hemoprotein releasing one mole of CO per mole of the enzyme during its catalytic oxidation of indoleacetate (*37*). The green hemoprotein had an absorption spectrum similar to choleglobin which was found by Lemberg and his coworkers (*38*) to be an intermediate in the transformation of the heme of hemoglobin into biliverdin. Thus, it is very likely that the protoheme of HRP decomposed oxidatively at the α-methenyl carbon. The reaction occurred under physiological conditions and was specific both for HRP isozymes and for electron donors (*37*).

The other striking feature of peroxidase-oxidase reactions is oscillatory consumption of oxygen when the reaction is carried out in an open system with constant supply of oxygen (*39*, *40*). Stable and sustained oscillations were observed in the oxidation of NADPH, the oxidation product of which was reduced back again in the presence of glucose-6-phosphate and its dehydrogenase (*41*). The mechanism which causes oscillations is complicated, but it can be said that the formation and decomposition of oxyperoxidase are closely related to the regulatory function in the peroxidase-oxidase reaction.

Milk Xanthine Oxidase

Milk xanthine oxidase is a complicated flavoprotein which contains 8 atoms of iron, 2 atoms of molybdenum, and 2 FAD per molecule. The mechanism of the intramolecular electron transfer has been studied extensively by Palmer *et al.* (*42*, *43*). It was suggested by them that the reducing equivalents travel from substrate to molybdenum, to flavin, to iron, and finally to the acceptor. On the other hand, it has been reported (*44*, *45*) that the enzyme can catalyze the reduction of cytochrome *c* by xanthine only when oxygen is present. This might imply that cytochrome *c* does not directly react with the enzyme. Considerable efforts have been given by Handler and his colleagues (*46*, *47*) to the elucidation of the mechanism of cytochrome *c* reduction by the xanthine oxidase systems. They claimed that the superoxide anion radical is an intermediate that reduces cytochrome *c*.

The formation of the superoxide anion in the xanthine oxidase reaction has been recently confirmed in two different ways. From xanthine oxidase reaction solutions Knowles *et al.* (*48*) observed an ESR signal which was attributable to superoxide. Bray *et al.* (*49*) have also obtained crucial evidence employing $^{17}O_2$ that the species observed by ESR is indeed the superoxide anion. On the other hand, McCord and Fridovich (*29*) found that erythrocuprein catalyzes the dismutation of superoxide anions and, consequently, inhibits the reduction of cytochrome *c* caused by xanthine oxidase. This finding has made it possible to analyze the mechanism of oxygen reduction by various oxidases (*50*).

It seems beyond doubt that the superoxide anion is formed during xanthine oxidase reactions. Our interest has been in confirming whether the formation of such half-reduced species of the acceptor be a main or side reaction. Using *p*-benzoquinone as an electron acceptor instead of oxygen, a quantitative measurement of the mechanism of electron transfer from xanthine oxidase to acceptor became possible (*51*). Figure 5

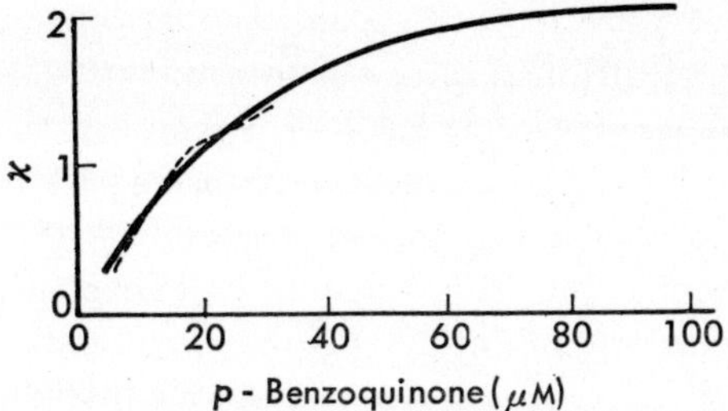

FIG. 5. Dependence of kappa upon the concentration of *p*-benzoquinone in the xanthine oxidase reactions. Kappa is defined by Eq. 7. Dotted line shows the values calculated according to Eq. 7, and solid line those obtained by measuring steady state concentrations of *p*-benzosemiquinone with ESR (*51*).

shows that the mechanism depended greatly upon the concentration of *p*-benzoquinone. Two-electron reduction of *p*-benzoquinone predominated in the low level of the acceptor, while typical one-electron reduction was observed in the high level of the acceptor. When oxygen was used as an electron acceptor, the formation of superoxide anions was measured using a sufficient amount of lactoperoxidase as a scavenger of the radical. The lactoperoxidase reacted with a superoxide anion to form Compound III (Fig. 3). The experimental conditions suitable for this purpose were restricted, but it was concluded that in xanthine oxidase reactions, the mechanism of oxygen reduction is identical with that of *p*-benzoquinone reduction.

Handler *et al.* (*52*) suggested that iron atoms of xanthine oxidase might be the electron-transfering agent closest to the external acceptor such as oxygen. Komai *et al.* (*53*) however, have shown that the reduction of oxygen requires direct electron transfer from a reduced form of flavin to oxygen in the xanthine oxidase reaction.

Reduction of Oxygen by Enzymes

Electron transfer oxidases catalyze reduction of oxygen to hydrogen peroxide or to water. Mason (*1*) has classified these oxidases into two groups, two-electron and four-electron transfer oxidases. Since it was found that xanthine oxidase catalyzes the formation of superoxide anions, a group of one-electron transfer oxidases must be added to his classification. The important problem might then be to confirm whether the formation of superoxide anions is a main or side reaction. For this purpose, quantitative analysis of the electron transfer reaction was needed. The details were discussed elsewhere (*54*). It has been concluded that when molecules are oxidized or reduced by enzymes, electron transfer reactions are classified into three mechanistic categories: a one-electron transfer, a two-electron transfer and a mixed mechanism. When oxygen is reduced by one-electron transfer oxidases the primary product of oxygen is a superoxide radical which is freed from the oxidase. The radical HO_2 is a weak acid and exists in a dissociated state at physiological pH. The superoxide anions dismutate to oxygen and hydrogen peroxide at a considerable speed (*55*, *56*), and the steady state concentration of the radical is not very high at neutral pH. Consequently, the apparent reduction product of oxygen is hydrogen peroxide even in a one-electron transfer reaction.

It is evident that measurement of ESR is the most direct physical technique for detecting free radicals in solution, but it is still difficult to use it for kinetic study, especially when the formation of superoxide anions is involved. More accurate quantitative analysis can be made by using a superoxide radical scavenger, such as cytochrome *c* or peroxidase (*54*). Yamazaki and Piette (*15*) have introduced a parameter, kappa, that characterizes the type of electron transfer reactions. When a sufficient amount of cytochrome *c* is used as a scavenger, kappa can be defined as

$$\kappa = \frac{\text{rate of cytochrome } c \text{ reduction}}{\text{rate of overall oxidase reaction}}\,. \qquad (7)$$

The value of kappa is 0 for a typical two-electron transfer mechanism (or a four-electron one). When the rate of overall reaction is calculated in terms of mole concentration of a two-equivalent redox substrate, the value of kappa will then be 2 for a typical one-electron transfer mechanism. In order to discriminate between reductions of cytochrome *c* by superoxide anions and through other reaction pathways, (including direct reaction with the enzyme), the use of superoxide dismutase proves highly efficient as demonstrated by McCord and Fridovich (*57*), Fridovich (*58*) and Massey *et al.* (*50*).

Based on the results so far obtained by many workers it may be concluded that flavoprotein oxidases are classified into two groups having two-electron transfer and mixed mechanisms. A mixed mechanism implies that kappa deviates from the extreme case, 0 or 2 beyond experimental errors, and the reaction proceeds in both one-electron and two-electron transfer mechanisms at the same time. No flavoprotein that catalyzes a compulsory one-electron reduction of oxygen has been reported. In the case of xanthine oxidase, one-electron transfer appears to occur under limited experimental conditions (*51*, *58*). It should be noted here that when *p*-benzoquinone is used as an electron acceptor, a compulsory one-electron transfer is observed in the reaction of flavoprotein dehydrogenases isolated from electron transport systems (*59*).

When oxygen is reduced in biological systems, a four-electron transfer mechanism becomes very important. In this case, the primary product of oxygen is water. Identification of the mechanism, in general, may not be very difficult. The formation of hydrogen peroxide as a reaction product can be excluded by confirmation that hydrogen peroxide is neither accumulated nor consumed by the reaction. The formation of superoxide anions inevitably results in the formation of hydrogen peroxide. Consequently, when the formation of hydrogen peroxide is not detected, it may be concluded that the reaction is neither a one-electron nor two-electron transfer type, but is very likely to be a four-electron type. A three-electron transfer mechanism may imply that the primary reduction products of an oxygen molecule are water and hydroxyl radicals (OH·), the latter being extremely reactive. The formation of free hydroxyl radicals will result in the destruction of enzyme proteins or in the initiation of complicated chain reactions that induce the formation of hydrogen peroxide. It is unlikely that such a three-electron transfer mechanism is involved in ordinary oxidase reactions. The types of oxy-

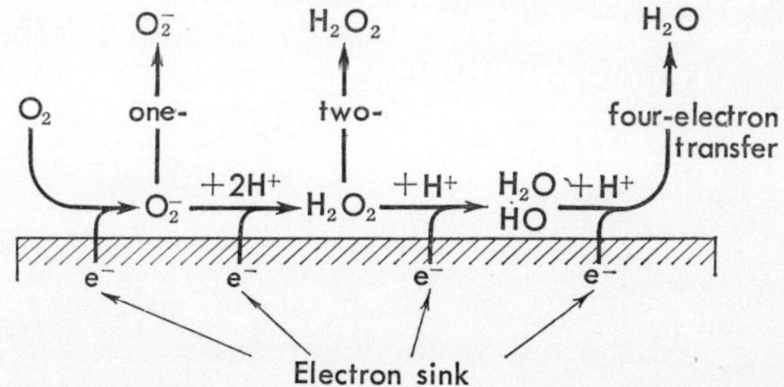

FIG. 6. Schematic representation of the types of oxygen reduction on the surface of oxidases. Manner and position of protonation in this figure have no definite meaning. Horizontal movement of intermediates on the surface of oxidases should be assumed to be the course of reaction time.

TABLE I. Mode of Electron Transfer from Oxidases to Oxygen and Their Specificity for the Electron Acceptor[a]

1. One-electron transfer mechanism
   Typical case has not been reported.
2. Mixed mechanism; oxygen-facultative[b]
   Xanthine oxidase, aldehyde oxidase, old yellow enzyme and dihydroorotate dehydrogenase.
3. Two-electron transfer mechanism; oxygen-obligative
   D-amino acid oxidase, L-amino acid oxidase, glucose oxidase and glycolate oxidase.
4. Four-electron transfer mechanism; oxygen-obligative
   Cytochrome oxidase, ascorbate oxidase and laccase.

[a] This classification is based mostly on the papers of Mason (*64*) and Massey *et al.* (*50, 65*) and further information is needed in order to confirm this idea. [b] In most cases ferricyanide is a good electron acceptor.

gen reduction on the surface of oxidases may be schematized as in Fig. 6.

The mechanism by which electron-transfer oxidases reduce oxygen will thus be classified into four categories as shown in Table I. This table presents an interesting comparison between the electron transfer mechanism and the acceptor specificity of oxidases. Mason has stated in his review (*1*) that the oxidases are *oxygen-obligative* if they reduce oxygen only, but if other electron acceptors can serve as substrate, they are *oxygen-facultative*. Oxygen-obligative oxidases reduce oxygen according to a two- or four-electron transfer mechanism, while oxygen-facultative oxidases do so according to a one-electron transfer or mixed mechanism. This seems reasonable because a specific interaction between the enzyme and acceptor will make it possible to hold the intermediate complex until a two- or four-electron transfer is completed.

General Mechanism of Oxygen Activation

It might be said that the sluggish activity of oxygen toward many physiological molecules (H_2A) arises from the low redox potential of the first one-electron reduction of oxygen

$$O_2 + e^- \rightleftharpoons O_2^-$$

and from the high redox potential of one-electron oxidation of H_2A.

General consideration of the one-electron redox potential in a two-electron redox system was made by Michaelis (*60*) for organic molecules and by George (*62*) for oxygen. According to Michaelis, normal potentials of the first (E_1) and second (E_2) oxidations are

$$E_1 = E_m - RT/2F \ln K_s$$
$$E_2 = E_m + RT/2F \ln K_s \, ,$$

where E_m is the normal potential of the overall two-electron oxidation process and K_s is the semiquinone formation constant,

$$K_s = \frac{(HA)^2}{(H_2A)(A)} \qquad \text{for oxidation of } H_2A \text{ and}$$

$$K_s = \frac{(HO_2)^2}{(H_2O_2)(O_2)} \qquad \text{for reduction of } O_2 \, .$$

The activation of substrates by oxidases might be considered as movement of the potential of the first one-electron reduction of oxygen to the positive side, and of the first one-electron oxidation of H_2A to the negative side, as shown in Fig. 7. This can be achieved by stabilizing the

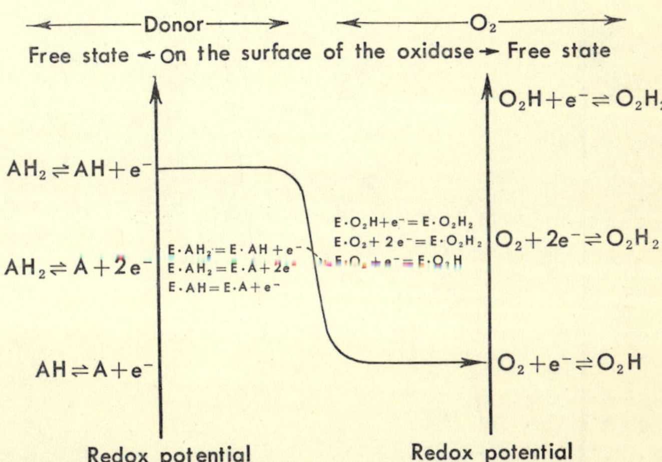

FIG. 7. Tentative mechanism for activation of electron donors and oxygen by oxidases. The ordinate position of each reaction indicates redox potential. Electron transfer from donor to oxygen occurs through the solid line in the free state, and through the dotted line on the surface of oxidases. Proton participation is not included in this figure (*30*).

semiquinone intermediate at the surface of the oxidases, namely, increasing the semiquinone formation constant, K_s. Figure 7 shows a simple case in which the E_m of both substrates in the bound state on the oxidase is the same as in the free state. This means that in this case the affinities of the oxidase for the reduced and oxidized forms of a substrate molecule are equal. Electron transfer from the donor to oxygen becomes much easier on the surface of oxidases than that in the free state because of the decrease in the potential barrier, as well as the possible proximity effect of both molecules.

During the formal stepwise addition of four electrons to molecular oxygen, two very reactive radical intermediates, superoxide and hydroxyl, are produced. When oxygen is reduced to water in the free state, the formation of such unstable radicals appears to be rate-limiting. It seems likely that peroxidase stabilizes the two intermediates and removes the barrier of their formation. There are five redox forms from +2 to +6 in HRP in terms of the effective charge on the heme iron. This might be a good example showing how oxygen is activated by oxidases. It was possible to prepare these redox forms in fairly stable states and to

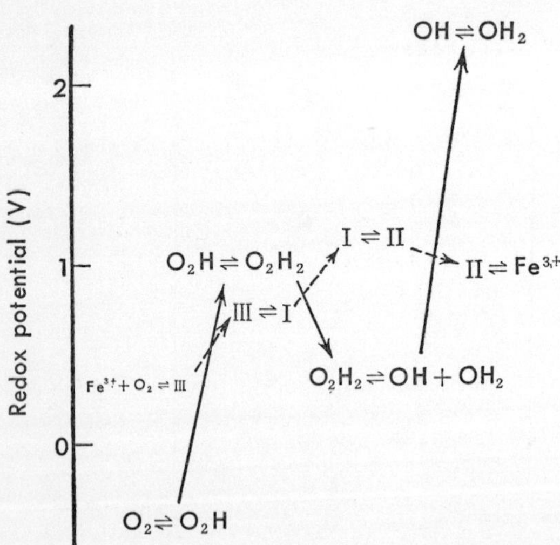

FIG. 8. Assumed redox potentials of four single-equivalent steps from oxygen to water in the free state (solid line) and in the bound state to HRP (dotted line). Here, 10^{-10} M is used as the dissociation constant of oxyperoxidase (*30*).

crystallize them from ammonium sulfate solution (*63*). But it is very difficult and still not possible to measure the normal potential of each single-electron transfer step. Figure 8 shows very approximate redox potentials of four one-electron steps from oxygen to water. It can be roughly said that three one-electron processes from oxyperoxidase to ferriperoxidase take place at relatively high redox potentials. This accords with the fact that oxyperoxidase can be readily reduced to ferriperoxidase by various electron donors.

HRP might behave like an oxidase if oxyperoxidase were reduced to ferrous form instead of ferric, or HRP were reduced to a ferrous enzyme by a given electron donor. During the oxidation of NADH or DHF, oxyperoxidase is formed even in the presence of a trace of hydrogen peroxide, which is insufficient to explain the formation of oxyperoxidase from the reaction between Compound II and hydrogen peroxide. Thus, as shown in Fig. 3, the formation of oxyperoxidase needs the reduction of HRP or oxygen that occurs only at a fairly low redox potential. The oxidation-reduction potential for the ferric-ferrous HRP was found to be -0.271 at pH 7 and 30°C (*22*) and that for the O_2/O_2^- was supposed to be -0.45 V at neutral pH (*62*). The reduction is indeed caused by the free radical of donor molecules that is produced by one-electron oxidation catalyzed by HRP.

Summary

It is concluded that the superoxide anion is the primary product of oxygen reduction in xanthine oxidase reactions and is involved as an active intermediate in the oxygen-consuming oxidation catalyzed by peroxidase. In the presence of suitable electron acceptors such as cytochrome *c*, the superoxide anion acts as an electron donor. It has been also suggested that it oxidizes some organic molecules to produce one-electron oxidized radicals or reacts with ferriperoxidase to form oxyperoxidase.

When oxygen is reduced in the presence of electron-transfer oxidases, the type of oxygen reduction may be classified into four categories: four-electron, two-electron, one-electron and mixed mechanisms. It will be possible to identify the mechanism for most oxidases, and it is beyond doubt that the mechanism depends on the environmental factors of the catalytic site at which oxygen is reduced. Of these, the ability of stabilizing intermediates which occur in the process of oxygen reduction appears to have great influence on the mode of electron transfer from the enzyme to oxygen. The mechanism of oxygen activation by oxidases might be explained in terms of such a stabilizing ability for oxygen radicals on the surface of the enzyme.

References

1 H. S. Mason, *Adv. Enzymol.*, **19**, 79 (1957).
2 B. Swedin and H. Theorell, *Nature*, **145**, 71 (1940).
3 R. H. Kenten, *Biochem. J.*, **59**, 110 (1955).
4 I. Yamazaki, K. Fujinaga, I. Takehara and H. Takahashi, *J. Biochem.*, **43**, 377 (1956).
5 T. Akazawa and E. E. Conn, *J. Biol. Chem.*, **232**, 403 (1958).
6 M. H. Klapper and D. P. Hackett, *J. Biol. Chem.*, **238**, 3743 (1963).
7 R. Lemberg and J. W. Legge, "Hematin Compounds and Bile Pigments," Interscience Publications, New York, pp. 419 (1949).
8 H. S. Mason, "Proc. Int. Symposium on Enzyme Chemistry," Maruzen Co. Ltd., Tokyo, pp. 220 (1958).
9 I. Yamazaki, K. Fujinaga and I. Takehara, *Arch. Biochem. Biophys.*, **72**, 42 (1957).
10 I. Yamazaki, *J. Biochem.*, **44**, 425 (1957).
11 P. George, *Nature*, **169**, 612 (1952).
12 B. Chance, *Arch. Biochem. Biophys.*, **41**, 416 (1952).
13 I. Yamazaki, "Proc. Int. Symposium on Enzyme Chemistry," Maruzen Co. Ltd., Tokyo, pp. 224 (1958).
14 I. Yamazaki, H. S. Mason and L. H. Piette, *Biochem. Biophys. Res. Commun.*, **1**, 336 (1959); *J. Biol. Chem.*, **235**, 2444 (1960).
15 I. Yamazaki and L. H. Piette, *Biochim. Biophys. Acta*, **50**, 62 (1961).
16 I. Yamazaki and H. Souzu, *Arch. Biochem. Biophys.*, **86**, 294 (1960).
17 T. Ohnishi, H. Yamazaki, T. Iyanagi, T. Nakamura and I. Yamazaki, *Biochim. Biophys. Acta*, **172**, 357 (1969).
18 I. Yamazaki and L. H. Piette, *Biochim. Biophys. Acta*, **77**, 47 (1963).
19 E. J. Land and A. J. Swallow, *Biochim. Biophys. Acta*, **234**, 34 (1971).
20 P. George, *Adv. Catalysis*, **4**, 367 (1952).
21 H. Theorell, *Adv. Enzymol.*, **7**, 265 (1947).
22 H. A. Harbury, *J. Biol. Chem.*, **225**, 1009 (1957).
23 B. Chance, *in* "Oxidases and Related Redox Systems," ed. by T. E. King, H. S. Mason and M. Morrison, Academic Press, New York, p. 504 (1965).
24 I. Yamazaki and K. Yokota, *Biochem. Biophys. Res. Commun.*, **19**, 249 (1965).

25 I. Yamazaki, K. Yokota and M. Tamura, *in* " The Chemistry of Hemes and Hemoproteins," ed. by B. Chance, R. W. Estabrook and T. Yonetani, Academic Press, New York, p. 319 (1966).
26 J. B. Wittenberg, R. W. Noble, B. A. Wittenberg, A. Antonini, M. Brunori and J. Wyman, *J. Biol. Chem.*, **242**, 626 (1967).
27 D. Keilin and E. F. Hartree, *Biochem. J.*, **49**, 88 (1951).
28 P. George, *J. Biol. Chem.*, **201**, 427 (1953).
29 J. M. McCord and I. Fridovich, *J. Biol. Chem.*, **244**, 6049 (1969).
30 I. Yamazaki, H. Yamazaki, M. Tamura, T. Ohnishi, S. Nakamura and T. Iyanagi, *Adv. Chem. Series*, **77**, 290 (1968).
31 M. Tamura and I. Yamazaki, *J. Biochem.*, **71**, 311 (1972).
32 K. Yokota and I. Yamazaki, *Biochem. Biophys. Res. Commun.*, **18**, 48 (1965).
33 Y. Ishimura, M. Nozaki, O. Hayaishi, T. Nakamura, M. Tamura and I. Yamazaki, *J. Biol. Chem.*, **245**, 3593 (1970).
34 I. Yamazaki, *in* " Biological and Chemical Aspects of Oxygenases," ed. by K. Block and O. Hayaishi, Maruzen Co. Ltd., Tokyo, p. 433 (1966).
35 I. Yamazaki, R. Nakajima, K. Miyoshi, R. Makino and M. Tamura, *in* " 2nd Int. Symp. Oxidases and Related Oxidation-Reduction Systems," Memphis, (1971) preprint.
36 I. Yamazaki, K. Yokota and R. Nakajima, *in* " Oxidases and Related Redox Systems," ed. by T. E. King, H. S. Mason and M. Morrison, Academic Press, New York, p. 485 (1965).
37 H. Yamazaki, S. Ohishi and I. Yamazaki, *Arch. Biochem. Biophys.*, **136**, 41 (1970).
38 R. Lemberg, *Rev. Pure Appl. Chem.*, **6**, 1 (1956).
39 I. Yamazaki, K. Yokota and R. Nakajima, *Biochem. Biophys. Res. Commun.*, **21**, 582 (1965).
40 I. Yamazaki and K. Yokota, *Biochim. Biophys. Acta*, **132**, 310 (1967).
41 S. Nakamura, K. Yokota and I. Yamazaki, *Nature*, **222**, 794 (1969).
42 G. Palmer, R. C. Bray and H. Beinert, *J. Biol. Chem.*, **239**, 2657 (1964).
43 R. C. Bray, G. Palmer and H. Beinert, *J. Biol. Chem.*, **239**, 2667 (1964).
44 B. L. Horecker and L. A. Heppel, *J. Biol. Chem.*, **178**, 683 (1949).
45 M. M. Weber, H. M. Lenhoff and N. O. Kaplan, *J. Biol. Chem.*, **220**, 93 (1956).
46 I. Fridovich and P. Handler, *J. Biol. Chem.*, **237**, 916 (1962).
47 J. M. McCord and I. Fridovich, *J. Biol. Chem.*, **243**, 5753 (1968).
48 P. F. Knowles, J. F. Gibson, F. M. Pick and R. C. Bray, *Biochem. J.*, **111**, 53 (1969).
49 R. C. Bray, F. M. Pick and D. Samuel, *Eur. J. Biochem.*, **15**, 352 (1970).
50 V. Massey, S. Strickland, S. G. Mayhew, L. G. Howell, P. C. Engel,

R. G. Matthews, M. Schuman and P. A. Sullivan, *Biochem. Biophys. Res. Commun.*, **36**, 891 (1969).
51 S. Nakamura and I. Yamazaki, *Biochim. Biophys. Acta*, **189**, 29 (1969).
52 P. Handler, K. V. Rajagopalan and V. Aleman, *Federation Proc.*, **23**, 30 (1964).
53 H. Komai, V. Massey and G. Palmer, *J. Biol. Chem.*, **244**, 1692 (1969).
54 I. Yamazaki, *in* "Advances in Biophysics," ed. by M. Kotani, University of Tokyo Press, Tokyo, Vol. II, p. 33 (1971).
55 J. Rabani and S. O. Nielsen, *J. Phys. Chem.*, **73**, 3736 (1969).
56 D. Behar, G. Czapski, J. Rabani, L. M. Dorfman and H. A. Schwarz, *J. Phys. Chem.*, **74**, 3209 (1970).
57 J. M. McCord and I. Fridovich, *J. Biol. Chem.*, **245**, 1374 (1970).
58 I. Fridovich, *J. Biol. Chem.*, **245**, 4053 (1970).
59 T. Iyanagi and I. Yamazaki, *Biochim. Biophys. Acta*, **216**, 282 (1970).
60 L. Michaelis, *in* " The Enzymes," ed. by J. B. Summer and K. Mirback, Academic Press, New York, No. 2, Part 1, p. 1 (1951).
61 T. Odajima and I. Yamazaki, to be published.
62 P. George, *in* " Oxidases and Related Redox Systems." ed. by T. E. King, H. S. Mason and M. Morrison, Wiley, New York, Vol. I, p. 3 (1965).
63 R. Nakajima and I. Yamazaki, to be published.
64 H. S. Mason, *Science*, **125**, 1185 (1957).
65 R. W. Miller and V. Massey, *J. Biol. Chem.*, **240**, 1466 (1965).

Received for publication November 2, 1971.

ACTION MECHANISMS OF CYTOCHROME OXIDASE INHIBITORS

Yutaka Orii and Shinya Yoshikawa
Department of Biology, Faculty of Science, Osaka University, Osaka

Information on the chemical nature of the active sites of an enzyme molecule can be obtained by studying the properties of chemical reagents acting as potent and reversible inhibitors of the enzyme, and the properties of bonding between the inhibitors and enzyme. In connection with this, cytochrome oxidase was identified as a haematin derivative long before it was isolated as a pure enzyme. This was first suggested by the fact that the CO-inhibited respiration of yeast cells was restored by illumination with bright light, and was confirmed by the similarity of the photochemical action spectrum of the CO-inhibited respiration to the light absorption spectrum of carbonyl hemoglobin (*1*).

Cytochrome oxidase contains copper in addition to heme *a* as the prosthetic group. A possible involvement of copper in cytochrome oxidase was first pointed out by Keilin and Hartree as early as in 1938 (*2*), and the equimolar relationship between them was later established by Takemori in 1960 (*3*). The light insensitive nature of CO-copper protein complexes seems to exclude the possibility of the binding site of carbon monoxide, or of molecular oxygen, on cytochrome oxidase being copper. However, the valency change of the prosthetic copper, depending on the redox state of the heme iron, has been shown either by EPR studies (*4*, *5*) or by spectrophotometric studies in which the absorption band at 830 nm was tentatively ascribed to the binding of cupric copper to the protein moiety (*6*). As the inhibitors of cytochrome oxidase Takemori used potassium ferrocyanide, salicylaldoxime, and ethylxanthate, which supposedly react with the copper forming inactive complexes (*3*).

Heme *a* is an electronically conjugated system which consists of the porphyrin nucleus with some side groups attached to its periphery and the central iron. The formyl side group at position 8 deserves notice for its versatile chemical reactivity. Takemori *et al.* examined the inhibitory effects of carbonyl reagents, such as hydrazine, phenylhydrazine, sodium bisulfite, and hydroxylamine, on cytochrome oxidase, and found that the first two reagents competed with cytochrome *c* against the oxidase (*7*). Therefore, the participation of heme *a*, or the heme iron and the formyl group if viewed microscopically, and the copper in the enzymic activity seems to have been established, although the details of the role of each functional group and a mode of interaction among them, if any, still remain to be investigated.

Because of multiplicity in the functional groups of cytochrome oxidase, the inhibition data cannot be unequivocally interpreted. First, one difficulty is the versatile reactivities of the inhibitors so far tested. For example, it is well known that cyanide and hydroxylamine combine with either copper, heme iron or carbonyl groups, and that azide combines with heme iron as well as with copper. Therefore, even if one of these inhibitors is used in an attempt to attack the heme iron, the sites other than the original target might be blocked, and the binding site would be identified only after the inhibition data are accounted for under consideration of the change of physicochemical properties which are specifically ascribed to respective functional groups. Secondly, if there are some interactions among the active sites, the inhibition data would have rather complicated features, and this possibility is highly probable. The formation of cyanhydrin and Schiff's base between the formyl group of heme *a* in cytochrome oxidase and cyanide or an ε-NH_2 group of lysyl residues in the protein moiety were shown to occur more readily when the heme iron was in the ferrous rather than the ferric state (*8*, *9*), and contrary to this, cyanhydrin formed with free heme *a* in an aqueous medium was shown to decompose readily as the heme iron was converted from the ferric to ferrous state (*10*). These results clearly indicate that the chemical reactivity of the formyl group is dependent on the redox state of the heme iron. There are several cases in which the experimental results are best interpreted by assuming a certain kind of interaction among the functional groups.

Taking these problematical points into consideration, the theoretical treatments of the inhibition data of cytochrome oxidase and their significance and limitation will be briefly reviewed.

Classification of Inhibitors Based on Their Modes of Action

Even though the chemical nature of the active sites of an enzyme is unknown, their number can be assessed by inhibition studies if appropriate precautions are taken in interpreting the data. In the cytochrome oxidase reaction, ferrocytochrome *c* is a substrate of cytochrome oxidase, and its oxidation in air is usually followed by observing the absorbance decrease at 550 nm with time. This reaction apparently obeys the first order kinetics, and its rate constant is proportional to the concentration of the free enzyme as far as the total concentration of cytochrome *c* is kept constant. In the presence of an inhibitor, however, the oxidation process deviates from the first order kinetics, and usually the apparent rate constant decreases as the reaction proceeds, approaching a limiting value which corresponds to the activity of the inhibited system. The degree of inhibition of the reaction, H, is defined as

$$H = 1 - V_i/V_0, \tag{1}$$

where V_i and V_0 refer to the rate constants in the presence and absence of the inhibitor, respectively. If only an inactive 1 : 1 cytochrome oxidase-inhibitor complex is formed, a $\log [H/(1-H)] - \log (I)$ plot is expected to give a straight line with a slope of 1. However, this would be the simplest case because cytochrome oxidase contains at least three kinds of functional groups which are supposed to react with inhibitors of versatile reactivities, forming inactive complexes of different compositions and properties. In fact, the double logarithmic plots obtained with the inhibitors so far examined showed different characteristics suggesting different modes of inhibition as illustrated in Fig. 1.

Azide and cyanide gave straight lines with a slope of 1, and these re-

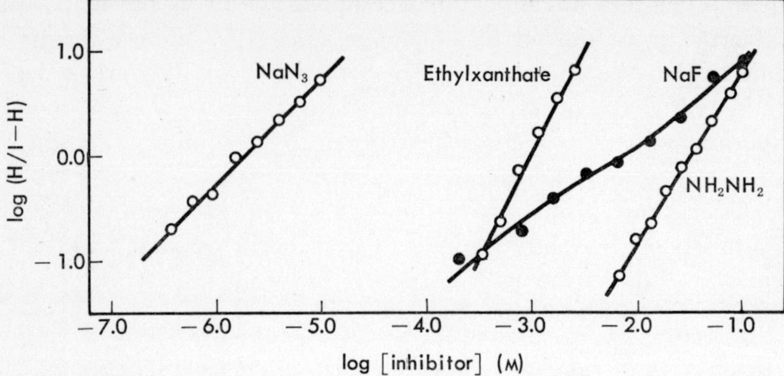

FIG. 1. Action of the inhibitors of cytochrome oxidase. The slopes of the plots are 1.0, 2.0 and 1.6 for inhibition by azide, ethylxanthate and hydrazine, respectively. The solid line for the fluoride inhibition is a theoretical curve which was obtained by calculation using $K_1=1.43\times10^2\ \text{M}^{-1}$, $K_2=4.71\times10^4\ \text{M}^{-2}$, $A_1=0.59$, and the equation: $H=[(1-A_1)K_1(I)+K_2(I)^2]/[1+K_1(I)+K_2(I)^2]$.

sults are interpreted to indicate that each molecule of cytochrome oxidase combines with one molecule of the inhibitor forming a completely inactive complex.

Hydrazine, semicarbazide, EDTA and salicylaldoxime gave apparently straight lines with slopes between 1 and 2, and these results are expected if it is assumed that cytochrome oxidase has more than one binding site for the inhibitor and that any enzyme-inhibitor complex is inactive irrespective of its stoichiometry. As a general consideration, let us assume that cytochrome oxidase has q binding sites for an inhibitor. The degree of inhibition is expressed by

$$H=\sum_{i=1}^{q}K_i(I)^i/[1+\sum_{i=1}^{q}K_i(I)^i], \qquad (2)$$

where K_i is an apparent equilibrium (association) constant for the reaction, $E+iI\rightleftharpoons EI^i$.* E signifies an inhibitor-free oxidase irrespective of whether it has been bound with a substrate or not. The slope of the $\log[H/(1-H)]-\log(I)$ plot in this case is given by

* The existence of quasi-equilibrium is implicitly assumed in the present treatment. Detailed discussions on aspect were made elsewhere (*12*).

$$n = \frac{d \ln [H/(1-H)]}{d \ln (I)} = \sum_{i=1}^{q} iK_i(I)^{i-1} \Big/ \sum_{i=1}^{q} K_i(I)^{i-1}. \tag{3}$$

This equation shows that, when the inhibitor concentration is diminished, the slope approaches 1 and, when it is increased to infinity, the slope becomes close to q. Practically, the plot will be either concave upwards or straight with an apparent slope between 1 and q. The upper limit of the slopes found with the inhibitors of this class was smaller than 2, indicating that cytochrome oxidase would have at least two binding sites. Accordingly, the mode of reaction of cytochrome oxidase with the inhibitor may be schematically represented as follows:

$$\begin{array}{ccccc} & E & \overset{K_1^*}{\longleftrightarrow} & E^1I & \\ K_2^* & \updownarrow & aK_1^* & \updownarrow & aK_2^*. \\ & E^2I & \longleftrightarrow & EI_2 & \end{array}$$

In this scheme E^1I and E^2I denote 1 : 1 enzyme-inhibitor complexes which are microscopically distinguishable, and K_1^* and K_2^* are apparent equilibrium (association) constants for the reactions which involve different binding sites. An interaction constant, a, larger (smaller) than 1 signifies that a blockage of either one of the binding sites with an inhibitor increases (decreases) the reactivity of the other toward the second molecule of the inhibitor. In the case of $a>1$, the K_i values in Eq. 2 for $q=2$ can be determined from the inhibition data, and they are correlated to the microscopic kinetic parameters in the above scheme according to Eqs. 4 and 5.

$$K_1 = K_1^* + K_2^* \tag{4}$$

$$K_2 = aK_1^*K_2^* \tag{5}$$

From these equations, Eq. 6 is derived.

$$a \geq 4K_2/K_1^2 \tag{6}$$

Therefore, when $4K_2/K_1^2>1$ or $K_1^2-4K_2<0$, a is always larger than unity. The K_1 and K_2 values, which were experimentally determined with the inhibitors of present concern, are summarized in Table I. The $K_1^2-4K_2$ values were calculated for each of them. All of the inhibitors examined, except those of the $n=1$ type and hydroxylamine, were found to give negative values of $K_1^2-4K_2$. Thus, an attractive interaction between two binding sites is suggested. It has not been decided, however, which two out of the three functional groups in cytochrome oxidase are involved in the complex formation.

Ethylxanthate gave a slope of 2. This would represent the extreme case of the above one; that is, the interaction between the two binding sites is so strong that no intermediate 1: 1 complex but a 1: 2 cytochrome oxidase-inhibitor complex is formed.

The inhibition of cytochrome oxidase by hydroxylamine and fluoride is unique in that the double logarithmic plots gave curves which were concave downward. Such results are best accounted for by assuming that the enzyme molecule combined with not less than $r+1$ inhibitor molecules is completely inactive, whereas the one combined with up to r inhibitor molecules still retains residual activity. We have already shown that in such a case the $\log[H/(1-H)]-\log(I)$ plot is not always concave upwards, and has two asymptotes with slopes of 1 and $q-r$ in the lower and higher concentration ranges of the inhibitor, respectively (*11*, *12*). Therefore, if the plot is concave downward in any restricted range of $\log(I)$, it is regarded as a sure sign of a residual activity of an enzyme-inhibitor complex even though an upwardly concave portion is found outside this range. With hydroxylamine and fluoride, the slope $q-r$ was 1. Thus, $q=2$ and $r=1$ were suggested for these cases. The degree of inhibition H for a general case of this category is expressed as follows:

$$H=[\sum_{i=1}^{q}(1-A_i)K_i(I)^i]/[1+\sum_{i=1}^{q}K_i(I)^i] \tag{7}$$

$$A_i \neq 0 \qquad (i=1,2,3,\ldots,r)$$

$$A_i = 0 \qquad (i=r+1,r+2,r+3,\ldots,q)$$

$$A_i \neq 1,$$

where the residual activity ratio A_i is defined as the ratio of the reaction rate due to all the enzyme species combined with i inhibitor molecules to that of the free enzyme at the same concentration. Table I summarizes the modes of inhibition so far observed and their thermodynamic parameters. This table clearly shows that most of the inhibitors, except for the classical ones like azide and cyanide, combine with cytochrome oxidase at more than one binding site, and that there is an attractive interaction between these binding sites.

The inhibitors of cytochrome oxidase are tentatively classified according to their modes of action as follows:

TABLE I. Classification of Cytochrome Oxidase Inhibitors

Type	Inhibitor	K_1 (M^{-1})	K_2 (M^{-2})	RAR†	Interaction
$n=1$	Cyanide	1.0×10^6			
$n=1$	Azide	5.5×10^5			
$n=1$	Bisulfite	2.4×10^2			
$n=1$ & 1	Hydroxylamine	1.9×10^6	8.3×10^9	0.88	
$n=1$ & 1	Fluoride	1.4×10^2	4.7×10^4	0.59	Attractive
$n=1$ & 2	Hydrazine	8.6	6.6×10^2	0	Attractive
$n=1$ & 2	Semicarbazide	8.1	2.3×10^2	0	Attractive
$n=1$ & 2	EDTA	9.7	8.8×10^2	0	Attractive
$n=1$ & 2	Salicylaldoxime	4.3×10^2	2.1×10^5	0	Attractive
$n=2$	Ethylxanthate		1.2×10^6	0	Attractive

† Residual activity ratio.

Group I ($n=1$): Inhibitors of this group form completely inactive 1:1 enzyme-inhibitor complexes.

Group II ($n=1$ & 1): The 1:1 complexes still retain some residual activity and only 1:2 complexes are completely inactive.

Group III ($n=1$ & 2): Both 1:1 and 1:2 complexes are formed and they are all inactive.

Molecular State of Cytochrome Oxidase and Modes of Inhibition

Cytochrome oxidase in solution is dimeric and dissociates into active monomeric species when it is treated with either sodium dodecyl sulfate, guanidine hydrochloride or alkali with a concomitant nearly twofold increase of activity (*13–15*). Based on this, it is suggested that the catalytic activity of one of the monomeric halves of the dimer is depressed and that the full activity appears when the dimer is dissociated. Therefore, if each monomeric half has one binding site of an identical chemical nature and the two in the dimer are functioning together, it may be unnecessary to postulate two binding sites of different chemical nature for the interpretation of the inhibition data. Then, the inhibitory effects of cyanide ($n=1$), hydroxylamine ($n=1$ & 1), and hydrazine and EDTA ($n=1$ & 2) were examined under the conditions where the monomeric species were predominent, and the same results as those obtained with dimeric oxidase were gained. Therefore, the model of the multiple binding sites of different chemical nature may be regarded as being valid.

Combined Effect of Inhibitors on Cytochrome Oxidase

If the inhibitors combine with cytochrome oxidase at the same binding site, competition is expected to occur between two of them acting upon the oxidase simultaneously. Let us suppose that I and I' are inhibitors of the $n=1$ type, and that the apparent equilibrium constants, K_1' and $*K_1'$, for the reactions between E and I' are obtained in the absence and presence of I, respectively. The relationship that a plot of $K_1'/*K_1'$ against the inhibitor concentration (I) gives a straight line with a slope which corresponds to an apparent equilibrium constant for the reaction between E and I in the absence of I' has been derived (*16*). When azide was used as I', and cyanide or bisulfite as I, linear relationships were obtained as shown in Fig. 2. The slopes correspond to $3.7\times10^6\ \text{M}^{-1}$ for cyanide and $4.2\times10^2\ \text{M}^{-1}$ for bisulfite, and they are comparable with the apparent equilibrium constants of $1.0\times10^6\ \text{M}^{-1}$ and $2.4\times10^2\ \text{M}^{-1}$ which were obtained in the absence of azide, respectively. Therefore, it is suggested that the binding site on cytochrome oxidase for azide, cyanide and bisulfite is the same.

With hydroxylamine as I, an $n=1$ and 1-type inhibitor, a linear plot was also obtained to suggest that either one of the hydroxylamine-binding sites is that for azide.

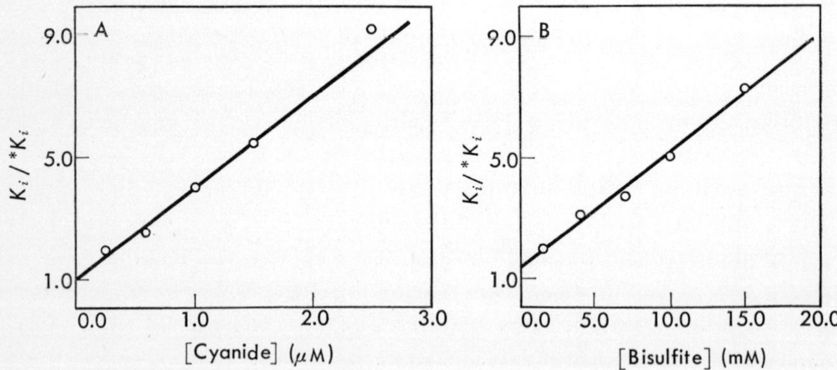

FIG. 2. Effects of cyanide and bisulfite on the apparent equilibrium constant for the reaction between azide and cytochrome oxidase. The slopes correspond to $3.7\times10^6\ \text{M}^{-1}$ for cyanide (A) and $4.2\times10^2\ \text{M}^{-1}$ for bisulfite (B).

Nature of Progressive Inhibition

As mentioned previously, the oxidation of ferrocytochrome *c* deviates from the first order kinetics in the presence of inhibitors of cytochrome oxidase. This anomalous behavior disappeared when the oxidative reaction was started by adding the solution containing cytochrome oxidase—that had been preincubated with the inhibitor in the presence of a substrate amount of ferrocytochrome *c*—to a mixture which contained ferrocytochrome *c* and the inhibitor at the same concentration as those in the preincubation mixture. When, on the contrary, the pretreated oxidase is added to a mixture containing only ferrocytochrome *c*, a semilogarithmic plot showed a progressive increase in the slope with time approaching an asymptote. The slopes of the asymptotes and of the straight semilogarithmic plot of the above three cases were practically the same as long as the inhibitor concentration in the reaction mixtures was kept constant (*11*). Such a progressive inhibition during the cytochrome oxidase reaction and its disappearance by the pretreatment of oxidase with the inhibitor in the presence of the substrate were first noticed by Wainio and Greenlees with respect to the cyanide inhibition (*17*), and further studied by us with hydroxylamine (*11*). These results suggest that the kinetically inactive complexes are formed between cytochrome oxidase in the functioning state and the inhibitors, and that the complex formation is reversible.

Based on the assumption that the progressive inhibition reflects a gradual formation of an enzyme-inhibitor complex, we obtained the rate constant for the complex formation by analyzing a curved semilogarithmic plot as follows. The curved plot is usually expressed by an empirical formula,

$$-d(\text{ferrocyt. } c)/dt = (A+Be^{-kt})\ (\text{ferrocyt. } c)\,, \tag{8}$$

where A, B and k are positive constants which are independent of the reaction time. The former two, however, are dependent on the inhibitor concentration which is virtually unchanged during the reaction, and k corresponds to an apparent rate constant of the complex formation. The integrated form of Eq. 8 is

$$\ln C_t = \ln C_o - \frac{B}{k}(1-e^{-kt}) - At\,, \tag{9}$$

where C_t and C_o represent the concentrations of ferrocytochrome c at

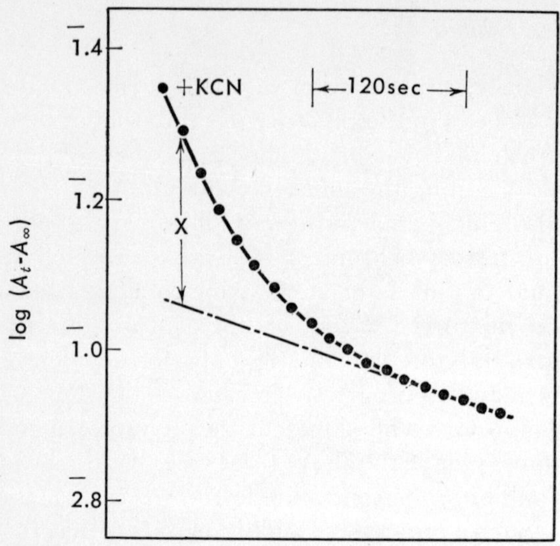

FIG. 3. A semilogarithmic plot of the ferrocytochrome *c* concentration against the time of reactions. Oxidation was followed in the presence of 5.85×10^{-6} M cyanide. The broken line is an extrapolation of the straight line portion of the plot.

reaction times of t and zero, respectively. Thus, the asymptote is expressed by

$$\ln C_\infty = \ln C_o - B/k - At\,, \tag{10}$$

so that the difference between $\ln C_t$ and $\ln C_\infty$, X becomes

$$X = (B/k) \times e^{-kt}\,. \tag{11}$$

The logarithmic form of the above equation is represented by

$$\ln X = \text{constant} - kt\,. \tag{12}$$

Therefore, it is apparent that if the values of X at different time intervals, which are estimated graphically as depicted in Fig. 3, are plotted against t, the rate constant is obtained from the slope of the linear plot.

If we assume a reaction formula,

$$E + I \underset{k_2}{\overset{k_1}{\rightleftharpoons}} EI$$

for the complex formation in this case, the rate constants, k_1 and k_2 are correlated to k by

$$k = k_1(I) + k_2 \,. \tag{13}$$

Thus, a plot of k against (I) would give a straight line with a slope showing the value of k_1 and an intercept representing the value of k_2. Figure 4 shows the k *versus* (I) plots obtained with the inhibitors so far examined. Unexpectedly, with most of the inhibitors except semicarbazide, hydrazine and ethylxanthate, a saturation phenomenon of k in higher concentration ranges of the inhibitors was observed. Even with these three inhibitors, however, it is possible that, beyond the concentration ranges examined, k would reach a plateau. This phenomenon can simply be explained by the following scheme.

$$E \xrightarrow{k_0} E^*, \quad E^* + I \underset{k_2}{\overset{k_1}{\rightleftharpoons}} E^*I \,,$$

where E^* denotes an isomer of E having a higher affinity for the inhibitors; that is, at higher concentrations of the inhibitors the reaction of the isomeric oxidase with the inhibitor would be limited by its forma-

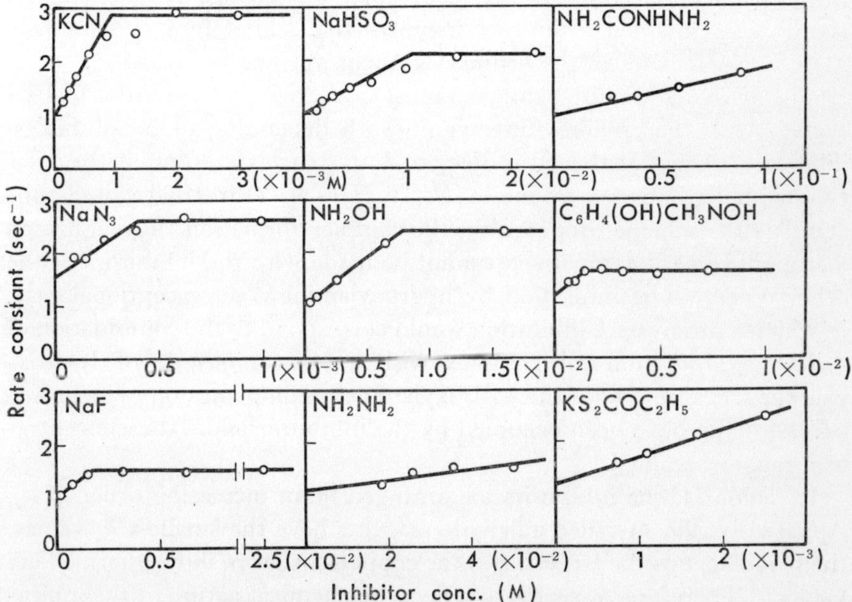

FIG. 4. Relations between overall rate constants for the cytochrome oxidase-inhibitor complex formation and inhibitor concentrations.

TABLE II. Kinetic Parameters for Inhibitor Binding

Inhibitor	Reactive with	k_1 (M^{-1} sec^{-1})	k_2 (sec^{-1})	k_0 (sec^{-1})	k_1/k_2 (M^{-1})
Semicarbazide	X†	6.9×10^{-2}	1.2×10^{-2}		6.0
Hydrazine	X	9.6×10^{-2}	1.1×10^{-2}		9.0
Bisulfite	X	1.0	1.0×10^{-2}		1.0×10^{2}
Salicylaldoxime	Y	2.8	1.2×10^{-2}	1.6×10^{-2}	2.3×10^{2}
Ethylxanthate	Y	2.9	1.4×10^{-2}		2.1×10^{2}
Fluoride	Z	3.1	9.8×10^{-3}	1.5×10^{-2}	3.2×10^{2}
Hydroxylamine	X, Y, Z	1.8×10	9.6×10^{-3}	2.4×10^{-2}	1.9×10^{3}
Cyanide	X, Y, Z	2.2×10^{3}	9.0×10^{-3}	2.9×10^{-2}	2.4×10^{5}
Azide	Y, Z	2.5×10^{3}	1.6×10^{-2}	2.6×10^{-2}	1.6×10^{5}

† X, formyl group; Y, copper; Z, heme iron.

tion from the original enzyme. The apparent rate constants for the conversion of E to E^* are almost the same, as shown in Table II, irrespective of the inhibitors examined, being of the order of 10^{-2} sec^{-1}. These are much smaller than those for the conformational change of physiological significance obtained with other enzymes (*18*). Therefore, it is unlikely that E^* participates in the ordinary cytochrome oxidase reaction.

The k_2 values are also similar in most cases, being of the order of 10^{-2} sec^{-1}. Only the k_1 values differ significantly depending on the inhibitors. The coincidence of the ratio of k_1 to k_2 for each inhibitor of the $n=1$ type with K_1 obtained by the $\ln[H/(1-H)]-\ln(I)$ method would substantiate the scheme proposed for the complex formation, although such a straightforward comparison cannot be made with the inhibitors of the other types. The inhibition by hydroxylamine is an exceptional case, in which a progressive inhibition would correspond to the combination of a 1:1 oxidase-inhibitor complex with another molecule of hydroxylamine, because one of the hydroxylamine binding sites on cytochrome oxidase must have been occupied by the inhibitor below the concentration range examined.

In Table II, the inhibitors are arranged in an increasing order of k_1. Apparently, the so-called aldehyde reagents have the smallest k_1 values, the heme ligands the largest, and the copper-chelators show intermediate values. Therefore, k_1 seems to reflect the chemical nature of the inhibitors, but it is not certain whether or not the reactions occurred exactly in the manner discussed above. The possibility that all of these inhibitors are bound to the heme iron cannot be excluded. Infrared spectrophotometry, NMR and magnetochemical techniques would be helpful for precise identification of the chemical nature of the inhibitor-binding sites.

Nature of Kinetically Inactive Complex of Cytochrome Oxidase with Cyanide

Cyanide is a powerful inhibitor of cytochrome oxidase, and its extent of inhibition of the cell respiration was found to be independent of the oxygen pressure. Based on this, Warburg proposed that the trivalent iron was the species with which cyanide combined to form an inactive complex, because if cyanide had combined with the divalent iron a competition between cyanide and molecular oxygen must have occurred, contrary to the results (*19*).

The dissociation constant of the ferricytochrome oxidase-cyanide complex, which was determined from analyses of spectral changes, was 9.2×10^{-5} M at pH 7.4 and 20°C (*8*). As this value differs significantly from the inhibition constant of 10^{-6} M (see Table I), this complex cannot be the entity of a kinetically inactive complex. Nor can this be the ferrocytochrome oxidase-cyanide complex, because the latter's dissociation constant is 7.4×10^{-4} M at pH 7.5 and room temperature (*20*).

Among several possibilities to account for such discrepancies, the one that the inactive complex is not formed from the static species of cytochrome oxidase seems to be most probable. Cytochrome oxidase is

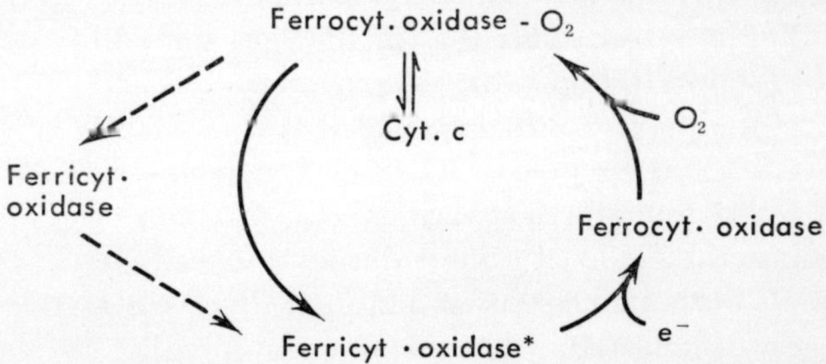

FIG. 5. Cyclic change of cytochrome oxidase during its function.

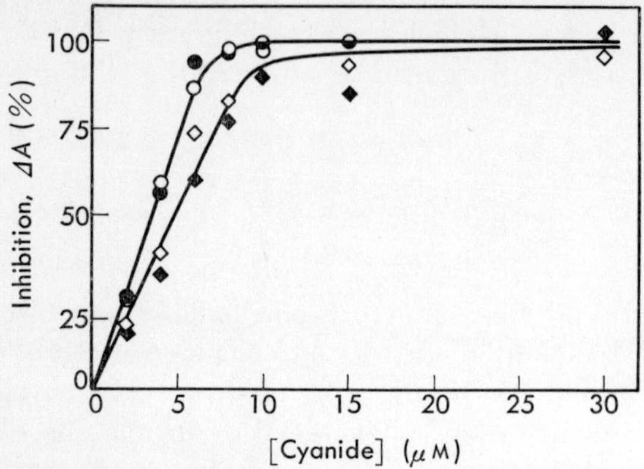

FIG. 6. Effect of cyanide on the activity of cytochrome oxidase and accompanying spectral changes. The oxygen uptake was measured by both manometric (◆) and spectrophotometric methods (◇). The period of steady state was measured by the latter method. The absorbance difference was measured either at 444–414 nm (○) or at 603–630 nm (●).

supposed to cycle through another type of ferricytochrome oxidase, ferrocytochrome oxidase and the oxygenated form during its function (*23*), as illustrated in Fig. 5, but no attention has been paid so far as to whether these species, except ferrocytochrome oxidase, participate in the inactive complex formation.

Accordingly, we tried to find the spectral change, if any, caused by the complex formation by comparing the activity and spectral properties of a reaction system which contained cytochrome oxidase, cytochrome *c*, ascorbate and cyanide. Usually the reaction was started by addition of ascorbate. As shown in Fig. 6, the oxygen uptake was half inhibited at a cyanide concentration of 5×10^{-6} M, which is comparable with the inhibition constant so far discussed. On the other hand, difference spectra of the above reaction mixtures at various concentrations of cyanide were recorded against the oxidized sample during the steady state. Absorbance differences both at the peak and trough increased as the cyanide concentration was increased. The maximum absorbance change occurred at 10^{-5} M cyanide, and the percent changes determined in the Soret and α-band regions were plotted against the cyanide concentration as depicted in Fig. 6. Half the maximum absorbance change occurred at 3.5×10^{-6} M cyanide. This is in good agreement with the value obtained from the activity measurements. Thus, the reaction of cytochrome oxidase with cyanide that caused such a spectral change is suggested to be associated with the formation of the kinetically inactive complex.

Immediately after the addition of ascorbate, ferrocytochrome oxidase appeared at first, and then it was converted to the oxygenated form. This initial reduction was not inhibited by cyanide. The difference spectrum of the steady state against this immediate state showed a trough at 447 nm and a peak at 428 nm, which is ascribed to the oxygenated form. When cyanide was present this peak appeared at 431–433 nm, and its magnitude increased with the cyanide concentration. If cyanide was added to the oxygenated compound which had been prepared by aeration of dithionite-reduced oxidase, the Soret peak shifted from 426 to 428 nm. Therefore, it is likely that the peak migration from 428 to 431–433 nm observed with difference spectra corresponded to the complex formation between the oxygenated compound and cyanide. Recently, it was shown that the oxygenated compound prepared by the aeration method differs from that which participates actually in the cytochrome oxidase reaction in its activity and spectral properties (*22*). Therefore, further studies on the complex formation between the functioning species, not the static ones, of cytochrome oxidase and the ligands would be helpful in confirming the nature of the kinetically inactive complexes.

Summary

The inhibitors of cytochrome oxidase were classified according to their modes of action as follows:

Group I ($n=1$), those forming completely inactive 1:1 oxidase-inhibitor complexes, *e.g.*, cyanide, azide and bisulfite. Group II ($n=1$ & 1), those forming 1:1 and 1:2 complexes, of which 1:1 complexes still retain some residual activity whereas 1:2 complexes are completely inactive; examples are hydroxylamine and fluoride. Group III ($n=1$ & 2), those forming 1:1 and 1:2 complexes which are both inactive; this group includes hydrazine, semicarbazide, EDTA and salicylaldoxime. Ethylxanthate is also included here as an extreme case ($n=2$).

The inhibitors of Group I were shown to combine at the same binding site on cytochrome oxidase. This site is also one of the two hydroxylamine-binding sites.

Most of the inhibitors react with isomeric oxidase which has a higher affinity for them to form inactive complexes.

The discrepancies between the dissociation constants of the cyanide complexes of ferri- and ferrocytochrome oxidase and the inhibition constant indicated that neither of them is kinetically inactive. The complex between oxygenated oxidase and cyanide was also regarded as being kinetically inactive.

References

1 O. Warburg, "Heavy Metal Prosthetic Groups and Enzyme Action," Clarendon Press, Oxford (1949).
2 D. Keilin and E. F. Hartree, *Nature*, **189**, 354 (1938).
3 S. Takemori, *J. Biochem.*, **47**, 382 (1960).
4 H. Beinert, D. E. Griffiths, D. C. Wharton and R. H. Sands, *J. Biol. Chem.*, **237**, 2337 (1962).
5 H. Beinert and G. Palmer, *J. Biol. Chem.*, **239**, 1221 (1964).
6 D. E. Griffiths and D. C. Wharton, *J. Biol. Chem.*, **236**, 1850 (1961).
7 S. Takemori, I. Sekuzu and K. Okunuki, *J. Biochem.*, **48**, 569 (1960).
8 Y. Orii and K. Okunuki, *J. Biochem.*, **55**, 37 (1964).
9 R. Lemberg, *Proc. Roy. Soc.*, **B159**, 429 (1964).
10 Y. Yanagi, I. Sekuzu, Y. Orii and K. Okunuki, *J. Biochem.*, **71**, 47 (1972).
11 S. Yoshikawa and Y. Orii, *J. Biochem.*, **68**, 145 (1970).
12 S. Yoshikawa, *Ann. Rept. Biol. Works, Fac. Sci., Osaka Univ.*, **18**, 83 (1970).
13 Y. Orii and K. Okunuki, *J. Biochem.*, **58**, 561 (1965).
14 Y. Orii and K. Okunuki, *J. Biochem.*, **61**, 388 (1967).
15 Y. Orii, Y. Matsumura and K. Okunuki, The Second International Symposium on Oxidases and Related Oxidation-Reduction Systems (1971).
16 S. Yoshikawa and Y. Orii, *J. Biochem.*, **71**, 873 (1927).
17 W. W. Wainio and J. Greenlees, *Arch. Biochem. Biophys.*, **90**, 18 (1960).
18 G. G. Hammes and P. R. Schimmel, *in* "The Enzymes," ed. by P. D. Boyer, Academic Press, New York, Vol. II, p. 116 (1970).
19 O. Warburg, *Biochem. Z.*, **189**, 354 (1927).
20 S. Yoshikawa and Y. Orii, submitted to *J. Biochem.*
21 Y. Orii and K. Okunuki, *J. Biochem.*, **53**, 489 (1963).
22 Y. Orii and T. E. King, to *FEBS Letters*, **21**, 199 (1972).

Received for publication December 20, 1971.

NATURE OF THE REACTION INTERMEDIATE COMPOUNDS OF PEROXIDASE

Yuhei Morita
Research Institute for Food Science,
Kyoto University, Kyoto

Since one green and two red compounds of peroxidase (EC 1.11.1.7) were found by Keilin and Mann (*1*) and Theorell (*2*) in the course of the reaction of horse-radish peroxidase (HRP) with hydrogen peroxide, the nature of these compounds were extensively studied by Chance, George and other workers (*3*, *5*). At present, it is well recognized that the peroxidatic reaction proceeds according to the following equations.

$$\text{Peroxidase} + \text{ROOH} \longrightarrow \text{Compound I (green)} \qquad (1)$$

$$\text{Compound I} + \text{AH} \longrightarrow \text{Compound II (red)} + \text{A} \qquad (2)$$

$$\text{Compound II} + \text{AH} \longrightarrow \text{Peroxidase} + \text{A}\,, \qquad (3)$$

where ROOH and AH denote hydroperoxide and a hydrogen donor, respectively. The two intermediate compounds in these reactions exhibit well-defined absorption spectra as shown in Fig. 1, and many investigators have proposed possible structures for them. Evidence for the

The following abbreviations are used in the text: HRP, horse-radish peroxidase; JRP, Japanese-radish peroxidase; EPR, electron paramagnetic resonance.

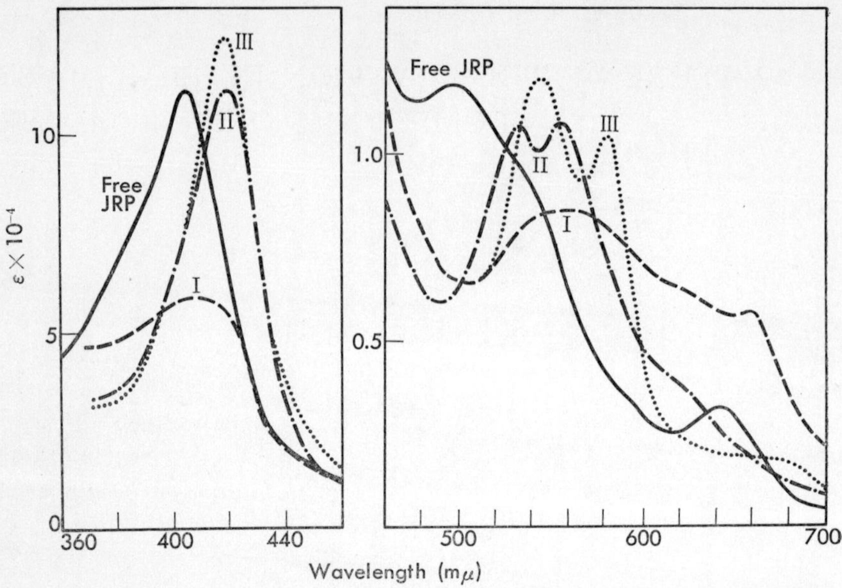

FIG. 1. Absorption spectra of Compounds I, II and III formed by reaction of Japanese-radish peroxidase (isoenzyme No. 3) with hydrogen peroxide.

speculative structures of the compounds was obtained by examining their a) electronic spectra, b) oxidation-reduction equivalents, c) dissociation of proton, and d) magnetic properties. Electronic spectra of hemoproteins in visible and Soret regions are derived principally from electronic transitions in the π-orbitals of porphyrin perturbed by an iron atom and environmental amino acid residues. The similarities of spectra between peroxidase and other hemoproteins of better-known structures, such as myoglobin and hemoglobin, can be often provided to suggest the electronic structures of peroxidase and its derivatives. Moreover, a certain relationship between the spectra and the spin state of the iron has been proposed by George *et al.* (*5*), but it should be noticed that the relationship is valid in the heme iron of the same valence state. The oxidizing equivalent of the heme compounds also provides valuable information about the electronic structure. In the study on this point titration with monovalent oxidants and reductants is usually employed. Simultaneously, the release and accumulation of proton with redox titration will set a limit on the structure of compounds protonated or depro-

TABLE I. Predicted μ_{eff}-values of Heme Compounds with Iron in Various Oxidation States

Oxidation state	Number of unpaired electrons	Magnetic moment μ_{eff}
Fe^{2+}	4	*4.90*–5.64β
Fe^{2+}	0	0
Fe^{3+}	5	*5.92*
Fe^{3+}	1	*1.73*–2.8
Fe^{4+}	4	*4.90*
Fe^{4+}	2	2.5–3.6 (*2.83*)
Fe^{5+}	3	*3.87*

The values in italic type are those predicted for spin-only μ_{eff} (after Ehrenberg, *7*).

tonated (*6*). However, it should be noted that the oxidizing equivalent of the compound molecule does not directly give insight into the electronic structure of the heme iron itself but the oxidation state of the whole molecule. The determination of magnetic properties is a most direct method for elucidating the electronic configuration of the heme iron atom, and for this purpose three methods can be now applied to hemoproteins. The first method is the determination of magnetic susceptibility. The predicted paramagnetic moments for different numbers of unpaired electrons are shown in Table I (*7*). As seen in the table, however, it is sometimes very difficult to estimate the number of the unpaired electrons from the observed magnetic moment, because the magnetic moment is not always given as a critical spin-only value but has a range of value affected by the orbital contribution. Furthermore, it is noticed that the mixture of more than two species of spin states will give only the averaged value. This makes it difficult to know with certainty the true spin state especially in the case of intermediate spin values. We cannot decide whether the iron should be in the true intermediate spin state or in a mixed state of high and low spin, which often occurs in the ferric heme iron. In addition, as a special case of a mixture of spin states, we can suppose a mixture of a critical single spin state of heme iron and a free radical in which the odd electron resides also on the ligand atom or other part of the protein. As the second method for determining magnetic properties, electron paramagnetic resonance (EPR) spectroscopy is useful for isolating single species of mixed spin states (*8*), although it is very difficult to observe the EPR absorption for heme compounds having even numbers of odd electrons on the iron atom. The third method, Mössbauer spectrometry, is very useful as it provides valuable information which complements the other two methods (*9*, *10*). Thus, the Mössbauer spectrum can be observed for all electronic species of the iron atom, and the parameters provide information about the 3d electron density and asymmetric distribution of electrons in the 3d orbitals. The Mössbauer parameters, therefore, will be expected to give the most valuable clue to the elucidation of the electronic structure of heme iron in peroxidase compounds, although this method alone cannot give decisive information about the 3d electron density (*11*).

Previous Studies on the Compounds

Now, it may be worthwhile to review the previous studies on the characteristics of the hydroperoxide compounds of peroxidase, which are necessary for the understanding of their structures. The experiments on the peroxidase compounds have always been performed with reference to the similarities in their properties to the hemoproteins of better-known structures, such as hemoglobin and myoglobin. First, the relationship between the free enzyme and the compounds in their oxidizing equivalents were already shown by Eqs. 1 to 3. Compounds I and II were found to

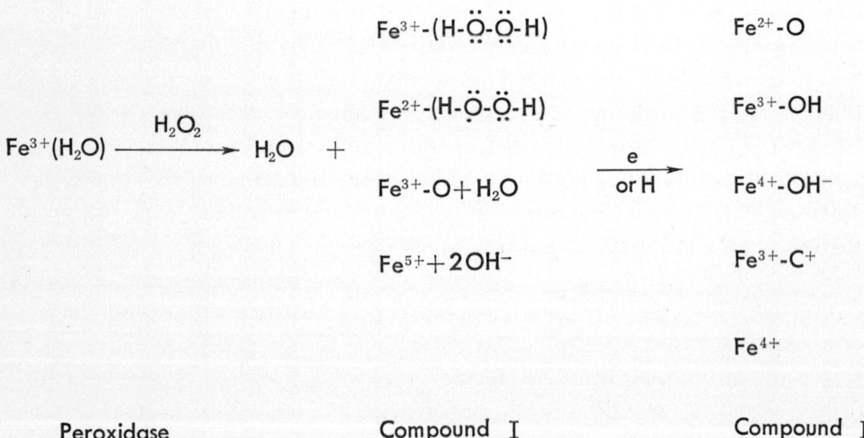

FIG. 2. Possible structures of peroxidase-peroxide Compounds I and II (after Mason, *4*).

retain two and one oxidizing equivalent, respectively, in excess of that of free peroxidase, which has ferric heme iron. The possible structures of Compounds I and II, some of which are shown in Fig. 2, therefore, have been proposed by several workers (*4*). The protonated and deprotonated forms or their resonance structures are possible (*12*), but they are merely speculative; the release or accumulation of protons in Eqs. 1 to 3 has not been studied for peroxidase unlike the case of metmyoglobin (*6*). It should be noted in reviewing the proposed electronic structures that there has been much confusion in the formal representation of the structure of heme iron and the sixth ligand oxygen. The charge sign was often used for the valence state of the iron atom, on the one hand, but in some other cases it was used for the charge distribution on the atoms or net charge of the heme and ligands. Moreover, the effective oxidation number referring to metallic iron was expressed by George and Irvine (*6*) with a plus sign, *i.e.*, 5+, 4+, *etc.* It is, therefore, preferable to assign different symbols to the net charge, the number of electrons or the valence state of iron and the spin state, as proposed by Peisach *et al.* (*13*). At any rate, the possible forms of Compound I are to be classified into two groups. In the first group the structure is such that the peroxide molecule is retained in the whole molecule as postulated by Chance (*3*) who used the term " complex." On the other hand, in the second group, the structure does not contain the original form of peroxide as the oxidizing entity, which has been already transferred to the iron atom or to some other part, at least partly, as proposed by George (*14*). For the latter structure, " complex " is not a pertinent term, instead of which the term " compound " may be suitable. Brill and Williams (*15*) proposed a compromised equilibrium system involving a mixture in which one form had the oxidizing entity in the peroxide molecule attached to the heme iron and the other form had one oxidation equivalent in the porphyrin ring; in both forms ferric iron is assigned. Peisach *et al.* (*13*) also presented postulated forms of Compounds I and II as ferric and ferrous iron states, respectively. However, this speculative proposal was based on the similarity of the visible bands to the well-known heme compounds. It should be emphasized that the electronic spectrum shows only the energy difference between the ground and excited states, providing no information about the absolute levels of the states. Although some interrelationships between spectra and spin states can be found in the hemoproteins, they would not be valid for the

TABLE II. Magnetic Properties of Metmyoglobin and Horse-radish Peroxidase and Their Peroxide Compounds at 20°C (after Ehrenberg, *7*).

	Magnetic suscepti-bility $\chi \cdot 10^6$	Magnetic moment μ_{eff}	Assumed valence of Fe	Assumed number of unpaired electron
	(cgsemu)	(β)		
Metmyoglobin	13,690	5.73	3	5 ($\rightleftharpoons$ 1)
Metmyoglobin–H_2O_2 Compound III	3,400	2.93	4	2
Metmyoglobin–MeO_2H Compound III	3,000	2.76	4	2
Horse-radish peroxidase	11,560	5.27	3	5 ($\rightleftharpoons$ 1)
Horse-radish peroxidase–MeO_2H Compound I	6,500	3.99	5, 3	3 (5 $\rightleftharpoons$ 1)
Horse-radish peroxidase–MeO_2H Compound II	5,040	3.53	4	2

compounds having very different electronic structures. In order to solve the problem, it is necessary to elucidate the localization of an oxidizing entity within the compound molecule. The valence state of the heme iron atom, *i.e.*, the number of 3d electrons, can be determined only by studying the magnetic properties. The magnetic susceptibilities of HRP and its peroxide compounds were first measured by Theorell and Ehrenberg (*16*). The theoretically induced susceptibilities of HRP and its peroxide compounds are shown in Table II together with those of horse heart metmyoglobin and the peroxide compound (*16, 17*). Theorell and Ehrenberg concluded simply that Compound I of HRP had three odd electrons in the molecule because the magnetic moment of the compound fitted the spin-only value predicted for three odd electrons. This suggested the possibility of a quinquevalent heme iron, which has been often confused with the term " the oxidation state of 5+," proposed by George (*14*). Alternatively, Brill and Williams (*15*) proposed an equilibrium system of a mixture of five odd electrons (high spin) and one odd electron (low spin) in ferric heme iron, as described previously, but Ehrenberg (*7*) questioned this proposal based on his later experimental results on catalase (*18*).

On the other hand, the magnetic susceptibility of Compound II seems to be more complicated. Although the compound of metmyoglobin showed a magnetic susceptibility very similar to the spin-only value for two odd electrons (*16*, *17*), Compound II of HRP exhibited a much higher value. For the higher value Theorell and Ehrenberg (*16*) provided an explanation that it was not spin-only value but that it involved orbital contribution. They concluded that Compound II must have two odd electrons and the possible electronic structure should be a ferryl form, in which the iron is in a quadrivalent low-spin state. On the other hand, Peisach *et al.* (*13*) supposed, based upon the optical properties of the compound, ferrous diamagnetic heme iron with the oxidizing equivalent residing on the oxygen atom attached to the iron. Besides the explanations or assumptions reviewed above many proposals have been presented to explain the structures of Compounds I and II (*12*, *19*). They were only speculative, however, especially in respect to the valence state or the number of 3d electrons of iron. As has been pointed out, any one value of magnetic moment allows us to envisage several possibilities for the electronic structures, and some other methods are necessary to set a limit to the kind of structure. For this purpose, EPR spectroscopy is very useful for detecting the odd electrons and isolating single spin states. All ferric heme compounds hitherto investigated exhibited a signal of either a high-spin or low-spin nature. However, it is well known that neither Compound I nor Compound II has exhibited the EPR absorption due to the heme iron. In the measurement of these compounds, Morita and Mason (*20*) found a signal at $g=2.003$ caused by a free radical. Similar observations have been also reported by other workers (*21*, *22*). However, the amount of free radical was too small to relate the radical to either Compound I or Compound II determined spectroscopically, as the maximal concentration of the radical formed during the reaction should not exceed 2% even after the correction for the effect of microwave power saturation. Therefore, it was concluded that the participation of the free radical might be only a few percent in the magnetic moments of the compounds. Moreover, neither high-spin or low-spin ferric iron of both compounds could be expected. This means that both compounds should have essentially an even number of odd electrons on the iron. At least this must be the case for Compound II.

Examination of Structure by Determining the Mössbauer Effect

Now, Mössbauer spectrometry can be used to determine the valence state of the iron, *i.e.*, the number of 3d electrons. As Compound II of

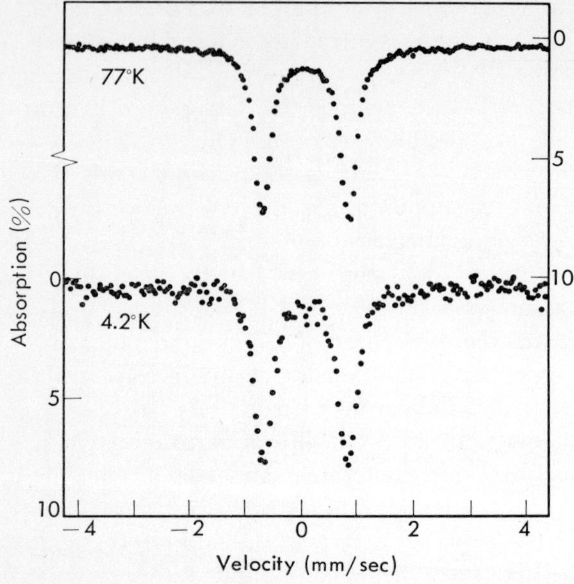

FIG. 3. Mössbauer spectra of metmyoglobin-hydrogen peroxide compound at 77 and 4.2°K.

peroxidase resembles the metmyoglobin-peroxide compound, which is most probably in the quadrivalent state, measurement was made of the Mössbauer spectra of Japanese-radish peroxidase a (JRP-a) and horse heart metmyoglobin as well as their hydrogen peroxide compounds (*23–25*). The spectra of the peroxide compounds are shown in Figs. 3 and 4, and the parameters are summarized in Table III together with those of Compound III of JRP-a, which will be discussed later. The spectrum of the metmyoglobin-peroxide exhibited a pair of sharp symmetric absorption lines over the whole temperature range down to 4.2°K, and the isomer shift and quadrupole splitting were found to be 0.08 and 1.52 mm/sec, respectively. The parameters were independent of temperature. The isomer shift was about 0.2 mm/sec more negative than those of the usual ferric low-spin compounds. This indicates that the charge density at the iron nucleus in the peroxide compound is higher than those in ferric low-spin compounds. As the 3d electrons screen the 3s electrons from the nuclear electrostatic potential and the removal of 3d electron increases the charge density at the iron nucleus, it can be con-

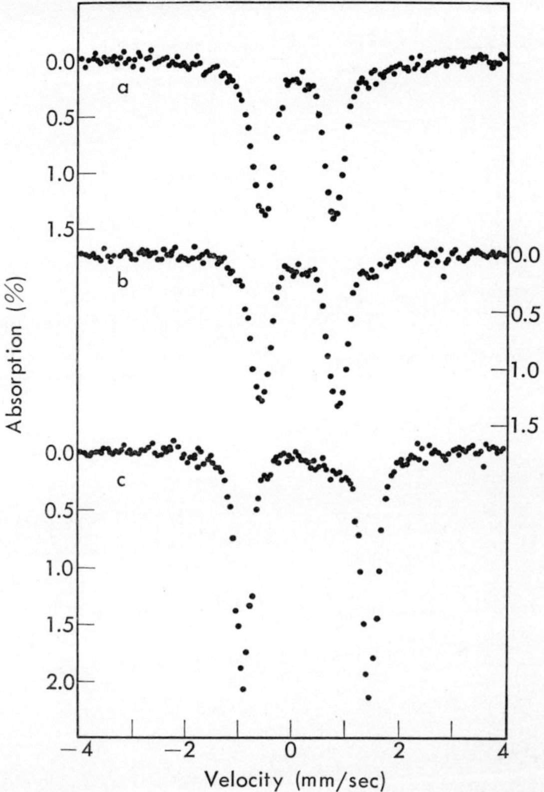

FIG. 4. Mössbauer spectra of Japanese-radish peroxidase-hydrogen peroxide compounds at 77°K. (a) Compound I, (b) Compound II, (c) Compound III.

cluded that the iron ion in the peroxide compound is in a higher valence state than in ferric iron. This fact indicates that it is compatible with the quadrivalent iron. Another possibility might be suggested that the d electrons of the iron ion might be located largely on the vacant orbitals of the ligand, reducing the effective density of the d electrons at the nucleus. Such a reduction of the d electron density was interpreted for a ferrous low-spin compound, oxyhemoglobin, by Lang and Marshall (*9*), and the cyanide compound of ferric peroxidase may be an example of such a case (*10*, *26*, *27*). However, the value of the isomer shift of the peroxide compound is too low to account for the appearance of ferrous iron, and

TABLE III. Mössbauer Parameters of Horse Heart Metmyoglobin, Japanese-radish Peroxidase and Their Hydrogen Peroxide Compounds

	Temperature (°K)	Quadrupole splitting (mm/sec)	Isomer shift[a] (mm/sec)
Horse heart metmyoglobin[b]	243	1.33	0.40
	195	1.33	0.40
	77	Hyperfine structure	
	4.2	Hyperfine structure	
H_2O_2 compound[b]	195	1.52	0.07
	77	1.54	0.09
	4.2	1.50	0.08
Japanese-radish peroxidase[c]	243	1.67	0.31
	195	1.67	0.31
	120	2.12	0.33
	77	2.20	0.37
	4.2	Hyperfine structure	
H_2O_2 Compound I[d]	195	1.38	0.04
	77	1.33	0.10
H_2O_2 Compound II[d]	195	1.44	0.07
	77	1.46	0.11
H_2O_2 Compound III[d]	195	2.33	0.24
	77	2.37	0.29

[a] Isomer shifts include the second-order Doppler shift due to the temperature difference between source and absorber. The temperature of the source was 293°K.
[b] Ref. *25*. [c] Ref. *26*. [d] Ref. *23*.

the Mössbauer spectrum has no magnetic interaction at temperatures down to 4.2°K, unlike the case of ferric iron in which EPR should give some absorption band. The idea that the myoglobin-peroxide compound may be a ferryl compound was first propounded by George, and it was supported by kinetic and stoichiometric experiments with reference to the oxidation number and proton release or accumulation as well as magnetic susceptibility measurements (*6*, *16*, *17*, *28*). The information from the Mössbauer effect is also compatible with this interpretation, *i.e.*, a) a more negative isomer shift, b) a sharp symmetric absorption without magnetic interaction down to 4.2°K, and c) temperature-independent parameters, both isomer shift and quadrupole splitting. These properties cannot be observed in ferric compounds. In conclusion, the heme iron in the metmyoglobin-peroxide compound should be in a quadrivalent low-spin state $(d\varepsilon)^4$ and possibly have a ferryl form, FeO. The quadrivalent low-spin state has a singlet ground state, which gives no observable EPR signal (*29*). The Mössbauer spectrum with no magnetic hyperfine splitting at 4.2°K can be explained by this electronic configuration. Moreover, it is pointed out that in the $(d\varepsilon)^4$ state the magnetic susceptibility should be temperature-dependent, which reduces to zero at 0°K. Recently Nakano *et al.* (*30*) confirmed, with such an experiment, the fact that the metmyoglobin-peroxide has a $(d\varepsilon)^4$ structure with a distorted ligand field.

In close analogy with the parameters, the Mössbauer effect showed that the heme iron in Compound II of peroxidase should exist in the same state as that of metmyoglobin-peroxide compound. Therefore, it is very reasonable to conclude that the heme iron in Compound II of peroxidase is essentially in a quadrivalent low-spin state $(d\varepsilon)^4$, although the microstructure in the ligand or the heme environment might be different from that of metmyoglobin compound, *e.g.*, in the ligand at the fifth coordination position, the protonation of the ferryl oxygen or the resonance structure of the porphyrin ring, which should affect the visible light absorption spectrum or the magnetic susceptibility. The Mössbauer effect of Compound I of JRP-a showed a surprising result, however. As seen in Fig. 4 and Table III, the spectra at all temperatures tested were almost the same as those of Compound II. This means that the electronic state of the heme iron is almost the same for both Compounds I and II, and it can be concluded that the heme iron in Compound I should be also in a quadrivalent low-spin state $(d\varepsilon)^4$. The question is now raised as to why the magnetic moment of the compounds, which is higher than the predicted value for two odd electrons, was explained by introducing the term orbital contribution only for Compound II and not for Compound I. We should not hastily conclude that the measured value of magnetic moment for Compound I simply fitted to the predicted value for three odd electrons, which will reside on the iron in an imaginary ferric intermediate-spin state $(d\gamma)^1(d\varepsilon)^4$. Moreover, the possibility that the intermediate value of the magnetic moment might be derived from a thermal mixture of high- and low-spin states of the quadrivalent iron should not be the case, because it was found that the Mössbauer spectrum and the visible absorption spectrum of Compound I were essentially temperature-independent (*23*, *31*). On the basis of the present discussion, it can be concluded that the iron in Compound I as well as Compound II has only two odd electrons and the higher magnetic moment can be attributed mainly to orbital contribution and, partly, to the presence of a free radical. Similar experimental results have recently been reported for HRP-peroxide compounds (*31*, *32*).

The question is then raised concerning the location of one oxidation equivalent within the Compound I molecule. It must reside on some part other than iron itself. In JRP-a and HRP, the formation of a free radical was observed, and the odd electron could be located on the ferryl oxygen, porphyrin ring or the protein part. As will be discussed later, free radical formation was significant in the case of a metmyoglobin-peroxide compound (*33–35*) and a cytochrome *c* peroxidase-peroxide compound (*36*, *37*). However, the maximum amount of the free radical found in the HRP-peroxide compound by EPR spectroscopy was too small to be attributed to one oxidation equivalent (*20*). It seems, therefore, probable that the oxidation state may not be in a free radical form but in a mono-oxygenated or dehydrated form. As has been pointed out by Brill and Williams (*15*, *38*), it is most probable that the methene carbon or other part of the porphyrin ring may be modified, so that a large distortion of the symmetry of the porphyrin causes a marked change in the absorption spectrum. Alternatively, Dolphin *et al.* (*39*) recently proposed that one oxidizing equivalent in Compound I may be located in the porphyrin ring in the form of a π-cation radical, and they presented a possible explanation to account for the failure of eliciting EPR signals. On the other hand, in the case of the metmyoglobin-peroxide compounds, one oxidation equivalent resides as a free radical on the protein part, which is detectable by EPR. Cytochrome *c* peroxidase is in the same situation. The " ES compound " retaining two equivalent oxidation numbers has one equivalent of free radical and may have the quadrivalent iron.

Discussion on the Reaction Mechanism

The reaction mechanism of peroxidase can be described by the scheme presented in Fig. 5. Thus, peroxidase reacts with hydrogen peroxide to form Compound I, in which the two oxidation numbers are divided into two parts, *i.e.*, the quadrivalent iron and, probably, porphyrin. When the hydrogen donor is added, Compound I is reduced by one equivalent to form Compound II, in which the quadrivalent iron remains.

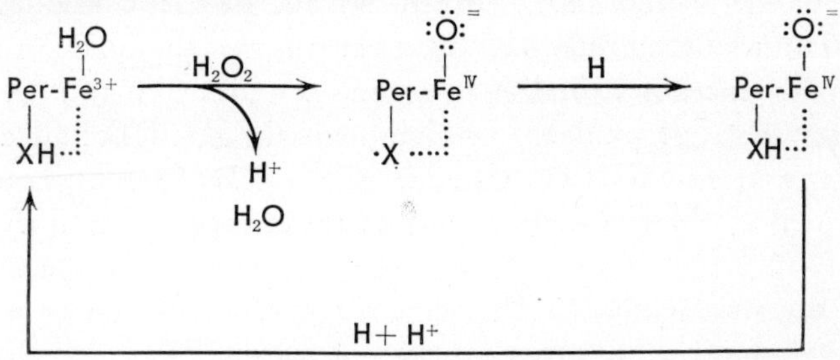

H_2O — Per-Fe^{3+} — XH $\xrightarrow{H_2O_2}$ (H^+, H_2O) :Ö:$^{=}$ — Per-Fe^{IV} — ·X $\xrightarrow{H}$:Ö:$^{=}$ — Per-Fe^{IV} — XH; H + H^+

FIG. 5. Reaction mechanism of peroxidase.

Then one more equivalent reduction causes the transition of the quadrivalent iron to the ferric enzyme. A small amount of free radical is formed most probably on the protein during these reactions, but it seems to participate insignificantly in the overall peroxidatic reaction in the case of JRP and HRP. However, this does not mean the complete exclusion of the role of the free radical in the reaction, as will be discussed later. At least in the case of cytochrome *c* peroxidase, a free radical in the protein part retains one oxidation number and contributes to the oxidation of the hydrogen donor. The peroxidatic reaction of metmyoglobin may proceed in the same manner.

Now the following questions may be raised.

1. Where is the peroxide-reacting site?
2. Where is the hydrogen donor-reacting site?
3. Does the free radical have any role in the reaction?
4. Is there any reaction route other than *via* two intermediate compounds?

Concerning Question 1, it has long been thought that a peroxide molecule may combine with iron at the sixth coordination position to replace the water molecule. The present discussion concludes that the two oxidation numbers of the peroxide are divided into two parts, iron and, probably, porphyrin, in Compound I, but this does not completely exclude the possibility that the complex may be formed in the first step. However, as long as the iron was found to change its valence state, it is possible to assume a site at any part other than the iron. Recent studies on the three-dimensional structure of cytochrome *c* molecules show a typical example in which the oxidation and reduction of heme iron occurs by an electron flow through a beautifully organized three-dimensional structure of the protein (*40*). In this regard, Brill and Weinryb (*41*) suggested that a methionine residue near the heme may participate directly in the reaction with the hydrogen donor. We can assume a possible structure of peroxidase in which the hydrogen donor and hydrogen peroxide react indirectly with the heme iron. In the course of such a reaction sequence, the free radical may possibly contribute to the reaction (*42*).

The last question concerns the alternative route of the peroxidatic reaction, which does not involve the usual formation of Compounds I and II. In this regard, Morita *et al.* (*43*) found that JRP-c, the basic isoenzyme No. 16 of JRP, exhibited a higher activity in the presence of cyanide. JRP-c has a high affinity for cyanide as has been found by Yamazaki *et al.* (*44*), and the cyanide compound turned to Compound I very slowly when hydrogen peroxide was added (*45*). Moreover, no Compound II was observed during the course of the decomposition of Compound I, although the cyanide-free enzyme formed typical Compounds I and II. These facts suggest that Compound I would not be formed in a significant amount during the reaction in the presence of cyanide, although the overall reaction proceeds more rapidly than in the absence of cyanide. In this case it seems reasonable to assume the other form of the intermediate compound involving a free radical or the other oxidation state. It is interesting to note that a similar effect of cyanide on JRP-c was observed in the oxidatic reaction of peroxidase on dihydroxyfumarate (*46*). In this regard, it should be pointed out that a glutathione peroxidase containing no heme or other metallic prosthetic group was recently isolated (*47*). The finding of the oxidatic activity of apoperoxidase on indolyl-3-acetic acid by Siegel and Galston (*48*) is also worth reinvestigating, although their results were recently questioned by Ku *et al.* (*49*). The difference in behavior between the isoenzymes of peroxidase was pointed out by Morita *et al.* (*50*, *51*). The most significant difference is found in the specific activity, *i.e.*, the rate constant for a hydrogen donor, guaiacol, which seems to be correlated with the contents of tyrosine in the isoenzymes. Thus, the rate constants, k_4, for the isoenzymes varied by about tenfold. A similar observation with turnip peroxidase was reported by Hosoya (*52*) and with HRP by Kay *et al.* (*53*).

A Comment on Compound III

Finally, a brief comment should be made on the nature of Compound III of peroxidase. The compound was first found by Keilin and Mann (*1*) upon the addition of a large excess of hydrogen peroxide to HRP, and later it was confirmed to be formed by the reaction of Compound II with hydrogen peroxide (*54*). The formation of Compound III could be observed in the course of the oxidatic reaction of peroxidase on dihydroxyfumarate (*46*, *55*), and it was assumed to be the active intermediate form of the enzyme in the oxidatic reaction. The effective oxidation number of Compound III was found to be 6+, and possible structures have been proposed (*4*, *13*, *54*, *56*). The oxygenated form of ferroperoxidase, formally designated as $Fe^{II}O_2$, is most commonly accepted on the basis of its similarity in the visible spectrum to the spectra of oxyhemoglobin and oxymyoglobin, while Peisach *et al.* (*13*) and Yamazaki *et al.* (*56*) proposed a ferric form, $Fe^{III}O_2^-$. However, the close similarity in the visible spectra and Mössbauer effects between the peroxidase-peroxide Compound III and oxyhemoglobin or oxymyoglobin suggested strongly a similar structure of these three compounds. It is to be noted that the latter two compounds were found to be diamagnetic. Although the magnetic susceptibility of peroxidase Compound III has not been determined, the Mössbauer spectrum of the compound, as shown in Fig. 4, with sharp symmetrical absorption lines at very low temperatures, supports the suggestion that it must be a diamagnetic ferrous compound. The large quadrupole splitting could be explained by the localization of $d\varepsilon$ electrons on the vacant π orbital of the oxygen molecule situated in parallel to the heme plane (*9*, *57*). This phenomenon should cause a large localization of charge on oxygen, but it does not mean the formation of an odd electron on the oxygen molecule. Recent studies on the synthetic oxygen carriers as a model of myoglobin by Ibers (*58*) show that the oxygen molecule should be π-bonded in the form of O_2^- to the carrier metal complex which can combine the oxygen molecule reversibly. Again, this does not mean the radical form of oxygen but the localization of the negative charge. Yamazaki *et al.* (*56*) proposed a free radical form, $Fe^{3+}\cdot O_2^-$, to explain the higher reactivity of peroxidase compound than oxyhemoglobin and oxymyoglobin, but the Mössbauer effect claims the similarity in the electronic structure of iron in these compounds. The higher reactivity should be attributed rather to the other part, probably the protein, which linked to the electron donor and iron atom. At present little is known about the structure of peroxidase and its peroxide compounds. In conclusion, it is essential to scrutinize more in detail the chemical and physical structures of peroxidase for a complete understanding of the reaction mechanism.

Summary

The structures of peroxidase-peroxide compounds are discussed. The magnetic properties of the compounds provide the most valuable information about the electronic configurations of heme iron in the compounds, and Mössbauer effect gives direct evidence of the valence state of the compounds complementing the magnetic susceptibility and electron paramagnetic resonance. The heme iron in Compounds I and II are found to be both quadrivalent, and a reaction mechanism is proposed. A brief comment is made on the structure of Compound III, which is in a ferroxy form.

Acknowledgments
The author should like to express his thanks to Drs. Yutaka Maeda, Takenobu Higashimura and Chiaki Yoshida for their experimental support and discussions.

References

1 D. Keilin and T. Mann, *Proc. Roy. Soc. London*, **122B**, 119 (1937).
2 H. Theorell, *Enzymologia*, **10**, 250 (1942).
3 B. Chance, *in* " The Enzymes," ed. by J. B. Sumner and K. Myrbäck, Academic Press, New York, Vol. II, Part 1, p. 428 (1951).
4 H. S. Mason, *Advan. Enzymol.*, **19**, 79 (1957).
5 P. George, J. Beetlestone and J. S. Griffith, *in* " Haematin Enzymes," ed. by J. E. Falk, R. Lemberg and R. K. Morton, Pergamon Press, Oxford, p. 105 (1961).
6 P. George and D. H. Irvine, *Biochem. J.*, **60**, 596 (1955).
7 A. Ehrenberg, *Svensk Kem. Tidskr.*, **74**, 53 (1962).
8 A. Ehrenberg, *Ark. Kemi*, **19**, 119 (1962).
9 G. Lang and W. Marshall, *J. Mol. Biol.*, **18**, 358 (1966); *Proc. Phys. Soc.*, **87**, 3 (1966).
10 Y. Maeda, *J. Phys. Soc. Japan*, **24**, 151 (1968).

11 L. R. Walker, G. K. Wertheim and V. Jaccarino, *Phys. Rev. Letters*, **6**, 98 (1961).
12 B. C. Saunders, A. G. Holmes-Siedle and B. P. Stark, "Peroxidase," Butterworths, London, pp. 108 (1964).
13 J. Peisach, W. E. Blumberg, B. A. Wittenberg and J. B. Wittenberg, *J. Biol. Chem.*, **243**, 1871 (1968).
14 P. George, *Biochem. J.*, **54**, 267 (1953).
15 A. S. Brill and R. J. P. Williams, *Biochem. J.*, **78**, 253 (1961).
16 H. Theorell and A. Ehrenberg, *Arch. Biochem. Biophys.*, **41**, 442 (1952).
17 A. S. Brill, A. Ehrenberg and H. den Hartog, *Biochim. Biophys. Acta*, **40**, 313 (1960).
18 H. F. Deutsch and A. Ehrenberg, *Acta Chem. Scand.*, **6**, 1522 (1952).
19 A. S. Brill, *in* "Comprehensive Biochemistry," ed. by M. Florkin and E. H. Stotz, Elsevier Publishing Co., Amsterdam, Vol. XIV, p. 447 (1966).
20 Y. Morita and H. S. Mason, *J. Biol. Chem.*, **240**, 2654 (1965).
21 B. Chance and R. R. Fergusson, *in* "A Symposium on the Mechanism of Enzyme Action," ed. by W. D. McElroy and B. Glass, The Johns Hopkins Press, Baltimore, p. 389 (1954).
22 J. F. Gibson, D. J. E. Ingram and P. Nicholls, *Nature*, **181**, 1398 (1958).
23 Y. Maeda and Y. Morita, *Biochem. Biophys. Res. Commun.*, **29**, 680 (1967).
24 Y. Maeda and Y. Morita, *in* "Structure and Function of Cytochromes," ed. by K. Okunuki, M. D. Kamen and I. Sekuzu, University of Tokyo Press, Tokyo, p. 523 (1968).
25 Y. Maeda, Y. Morita and C. Yoshida, *J. Biochem.*, **70**, 509 (1971).
26 Y. Maeda, T. Higashimura and Y. Morita, *Biochem. Biophys. Res. Commun.*, **29**, 362 (1967).
27 Y. Maeda, T. Higashimura and Y. Morita, *Ann. Rept. Res. Reactor Inst., Kyoto Univ.*, **2**, 67 (1969).
28 P. George and D. H. Irvine, *Nature*, **168**, 164 (1951); *Biochem. J.*, **52**, 511 (1952); **55**, 230 (1953); **58**, 188 (1954).
29 H. Kamimura, *J. Phys. Soc. Japan*, **11**, 1171 (1956).
30 N. Nakano, M. Tamura and A. Tasaki, personal communication.
31 T. H. Moss, A. Ehrenberg and A. J. Bearden, *Biochemistry*, **8**, 4159 (1969).
32 A. J. Bearden, A. Ehrenberg and T. H. Moss, *in* "Structure and Function of Cytochromes," ed. by K. Okunuki, M. D. Kamen and I. Sekuzu, University of Tokyo Press, Tokyo, p. 528 (1968).
33 N. K. King and M. E. Winfield, *J. Biol. Chem.*, **238**, 1520 (1963).
34 N. K. King, F. D. Looney and M. E. Winfield, *Biochim. Biophys. Acta*, **88**, 235 (1964).

35 N. K. King, F. D. Looney and M. E. Winfield, *Biochim. Biophys. Acta*, **133**, 65 (1967).

36 T. Yonetani, H. Schleyer and A. Ehrenberg, *J. Biol. Chem.*, **241**, 3240 (1966).

37 T. Yonetani, H. Schleyer and B. Chance, *in*" Hemes and Hemoproteins," ed. by B. Chance, R. W. Estabrook and T. Yonetani, Academic Press, New York, p. 293 (1966).

38 R. J. P. Williams, *in* " Hemes and Hemoproteins," ed. by B. Chance, R. W. Estabrook and T. Yonetani, Academic Press, New York, p. 557 (1966).

39 D. Dolphin, A. Forman, D. C. Borg, J. Fajer and R. H. Felton, *Proc. Natl. Acad. Sci., U.S.*, **68**, 614 (1971).

40 R. E. Dickerson, T. Takano, O. B. Kallai and L. Samson, presented at the Wenner-Gren Symposium, Stockholm, Aug. 1970.

41 A. S. Brill and I. Weinryb, *Biochemistry*, **6**, 3528 (1967).

42 P. Nicholls, *in* " Hemes and Hemoproteins," ed. by B. Chance, R. W Estabrook and T. Yonetani, Academic Press, New York, p. 307 (1966).

43 Y. Morita, C. Yoshida and S. Ida, *Agr. Biol. Chem.*, **33**, 436 (1969).

44 I. Yamazaki, R. Nakajima, H. Honma and M. Tamura, *Biochem. Biophys. Res. Commun.*, **27**, 53 (1967).

45 Y. Morita and K. Kameda, *Mem. Res. Inst. Food Sci., Kyoto Univ.*, **24**, 1 (1962).

46 Y. Morita and K. Kameda, *Mem. Res. Inst. Food Sci., Kyoto Univ.*, **23**, 1 (1961).

47 J. Flohé, E. Schaich, W. Voelter and A. Wendel, *Z. Physiol. Chem.*, **352**, 170 (1971).

48 B. Z. Siegel and A. W. Galston, *Science*, **157**, 1557 (1967).

49 H. S. Ku, S. F. Yang and H. K. Pratt, *Plant Physiol.*, **45**, 358 (1970).

50 Y. Morita, C. Yoshida, I. Kitamura and S. Ida, *Agr. Biol. Chem.*, **34**, 1191 (1970).

51 Y. Morita, C. Yoshida and Y. Maeda, *Agr. Biol. Chem.*, **35**, 1074 (1971).

52 T. Hosoya, *J. Biochem.*, **47**, 369, 794 (1960).

53 E. Kay, L. M. Shannon and J. Y. Lew, *J. Biol. Chem.*, **242**, 2470 (1967).

54 P. George, *J. Biol. Chem.*, **201**, 427 (1953).

55 B. Swedin and H. Theorell, *Nature*, **145**, 71 (1940).

56 I. Yamazaki, K. Yokota and M. Tamura, *in* " Hemes and Hemoproteins," ed. by B. Chance, R. W. Estabrook and T. Yonetani, Academic Press, New York, p. 319 (1966).

57 M. Gouterman and M. Zerner, *in* " Hemes and Hemoproteins," ed. by B. Chance, R. W. Estabrook and T. Yonetani, Academic Press, New York, p. 589 (1966).

58 J. A. Ibers, *in* "Structural Chemistry and Molecular Biology," ed. by A. Rich and B. Davidson, W. H. Freeman and Co., San Francisco, p. 633 (1968).

Received for publication July 9, 1971.

ELECTRONIC STRUCTURES OF IRON IONS IN HEMOPROTEINS

Jinya Ōtsuka
Department of Physics, Faculty of Engineering Science, Osaka University, Osaka

Hemoproteins are a group of proteins which contain hemes as prosthetic groups. A heme is a complex of an iron ion with porphyrin, in which the iron ion is situated nearly at the center of the porphyrin ring (Fig. 1). The 5th and 6th coordination positions of the heme iron, the directions of which are normal to the plane of heme, are occupied by intrinsic or extrinsic ligands. The 5th ligand is identified as a nitrogen atom of histidine residue in myoglobin, hemoglobin and cytochromes.

In respect to their biological functions, hemoproteins are classified into three groups; reversible oxygen carriers (myoglobin and hemoglobin), electron carriers which reversibly change the oxidation state of the heme iron from ferrous to ferric state (cytochromes), and hydroperoxidases (catalases and peroxidases). For all hemoproteins, it is generally believed that the prosthetic group heme is the active site and an especially important role is played by the iron ion whose physical or chemical behavior can be studied in detail theoretically and experimentally through its magnetic and optical properties (Sections 1 and 2). These properties offer much information about the electronic structures of the iron ion,

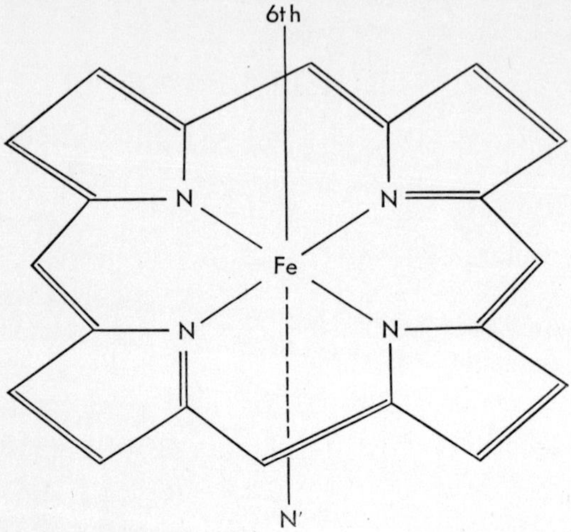

FIG. 1. Structure of heme. N′ shows the nitrogen atom of proximal histidine residue (5th ligand).

through which we can also obtain insights into the surroundings of the iron ion. The electronic structure thus clarified may give clues to the elucidation of the biological functions of hemoproteins. As an example, the problem of oxygen binding to myoglobin or hemoglobin is treated in Section 3 of this paper. Moreover, X-ray diffraction analysis has advanced to be able to determine the atomic coordinates of apoprotein part in myoglobin and hemoglobin also. Therefore, we may expect that in the future, the problems discussed in the present report will be extended to cover the apoprotein part based on the explicit atomic configurations, especially to explain "heme-heme interaction" in hemoglobin. The first step in such an approach is an interpretation of the thermal equilibrium between two electronic states, which is described in Section 4 of this paper.

1. *Ligand Field Theory*

The electronic structures of the transition metal ion in a complex compound such as heme have been extensively studied, and a theoretical method called the "ligand field theory" has been established with successful explanation of optical absorption spectra (d→d transition) and properties. In this section, the ligand field theory is outlined to describe the electronic structures in hemoproteins.

In a free iron ion, five *d*-orbitals are degenerate (equal in orbital energy) and the eigenstates are characterized by the total spin angular momentum $\vec{S}$ and the total orbital angular momentum $\vec{L}$, if the spin-orbit interaction is not considered. The notation for an eigenstate, which has become standard, is to write S, P, D, F, . . . for $L=0$, 1, 2, 3, . . . and to write the numerical value of the multiplicity $2S+1$ as a superscript at the left of the symbol for the L value. Thus for example ^6S (sextet S) indicates a state in which $L=0$ and $S=5/2$. For a definite configuration d^N (the number of *d*-electrons is N), the ground state has the maximum S allowed by the Pauli principle, and the maximum L consistent with this value of S. This is called Hund's rule. Let us consider an example of Hund's rule. The ion Fe^{3+} has five 3*d* electrons; the electronic configuration in which five electrons with parallel spin occupy five different orbitals has a maximum value of $S=5/2$ and the value of L becomes zero (^6S). This is the ground state in Fe^{3+}.

On the other hand, the 3*d* electrons of Fe^{3+} in hemoprotein are subject to the local electric field produced by neighboring ions or atoms. Let us consider the effect of this field on 3*d* orbitals. For the following discussions, the z axis is taken perpendicularly to the heme plane and through the iron ion, and the x, y axes are taken to be parallel to two nitrogen atoms which face each other in the porphyrin ring (see Fig. 1.) If the origin of this coordinate system is chosen to be the iron ion, then, five independent *d*-orbitals may be taken as

$$\left.\begin{array}{ll} u=\dfrac{1}{2\sqrt{3}}R(r)\dfrac{3z^2-r^2}{r^2} & \xi=R(r)\dfrac{yz}{r^2} \\ & \eta=R(r)\dfrac{xz}{r^2} \\ v=\dfrac{1}{2}R(r)\dfrac{x^2-y^2}{r^2} & \zeta=R(r)\dfrac{xy}{r^2} \end{array}\right\} \quad (1\text{–}1)$$

where $R(r)$ is a function of $r=\sqrt{x^2+y^2+z^2}$ only. These orbitals are shown in Fig. 2. Since the nitrogen atoms of porphyrin or histidine residue (5th ligand) are negatively charged, the electron in the orbital which is directed towards these nitrogen atoms may have higher energy. Thus, if we consider only the nearest neighbors of the iron ion, the energies of the electrons in these five orbitals may be different as shown in

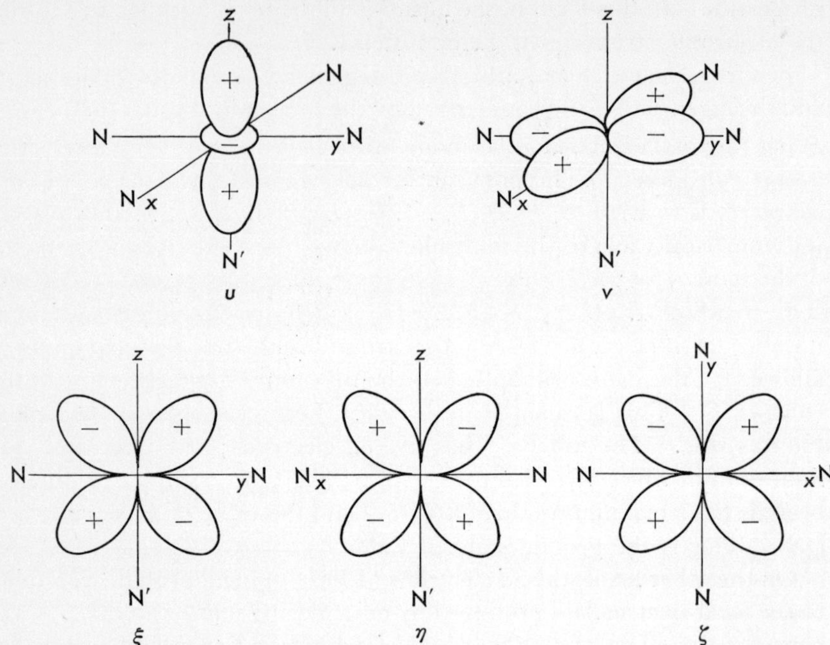

FIG. 2. Five *d*-orbitals. N shows the nitrogen atom of the porphyrin ring and N′ the nitrogen atom of proximal histidine residue.

Fig. 3. In other words, *d*-level is split into four sub-levels differing in energy. In the following, the orbitals for these sub-levels are denoted by the symbols a_1, b_1, e and b_2, that is,

$$\left.\begin{array}{ll} a_1; & u \\ b_1; & v \\ e; & \begin{cases} \xi \\ \eta \end{cases} \\ b_2; & \zeta \end{array}\right\} \qquad (1\text{–}2)$$

These symbols show the transformation properties under the symmetry operations* for the system constructed from the iron ion, five nitrogen

* These symmetry operations are: rotation around the z axis by $2\pi/4$, rotation around the z axis by $2\pi/2$, two kinds of reflection in a plane passing through the z axis and the identity operation. These operations form a group called C_{4v}.

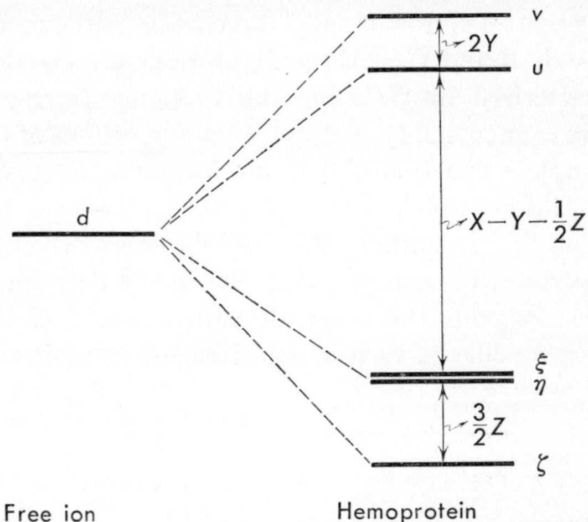

FIG. 3. Splitting of *d*-level of the iron ion in hemoprotein. These energy separations are expressed in terms of three parameters X, Y and Z.

atoms mentioned above, and the 6th ligand. Let the energy differences among these sub-levels be expressed in terms of three parameters X, Y and Z as shown in Fig. 3. These parameter values are determined by experimental data in the ligand field theory.

Now, in quantum theory, one component of orbital angular momentum, usually taken as L_z, and the square of the total orbital angular momentum L^2 are constant in a central field. In a noncentral field, the plane of the orbit will move about; the angular momentum components are no longer constant and the expectation value of $\vec{L}$ becomes zero. The iron ion in hemoprotein is exposed to such a noncentral field. Therefore, the eigenstate of the iron ion in hemoprotein is characterized by the total spin angular momentum S and the symbol Γ which shows the transformation property of the wave function of this state under the symmetry operation. The notation for an eigenstate is to write the numerical value of the multiplicity $2S+1$ as a superscript at the left of the symbol Γ.

If the iron ion under consideration has N electrons in the d shell ($N=5$ for Fe^{3+} and $N=6$ for Fe^{2+}), the electron configuration with which we are concerned has the form $e^l b_2{}^m a_1{}^n b_1{}^{N-(l+m+n)}$ where $0 \leq l \leq 4$, $0 \leq m \leq 2$, $0 \leq n \leq 2$ and $0 \leq N-(l+m+n) \leq 2$. The main task undertaken by the author in the first paper (*1*) is the calculation of interaction energies caused by the Coulomb interaction among electrons in these configurations.

These can be expressed in terms of only three basic parameters called Racah's parameters A, B and C. All configurations are classified into various types characterized by $S\Gamma$, and the Coulomb interaction has nonvanishing matrix elements only among the configurations of the same $S\Gamma$, so that the complete matrix of the Coulomb interaction can be reduced according to different $S\Gamma$.

In the second paper (*2*), the author calculated the energy levels and the eigen-functions of various $S\Gamma$ states for $d^5(Fe^{3+})$ and $d^6(Fe^{2+})$ in order to obtain the basis for analyzing the magnetic and optical data of hemoproteins. The energy values of various states arising from $d^5(Fe^{3+})$ and

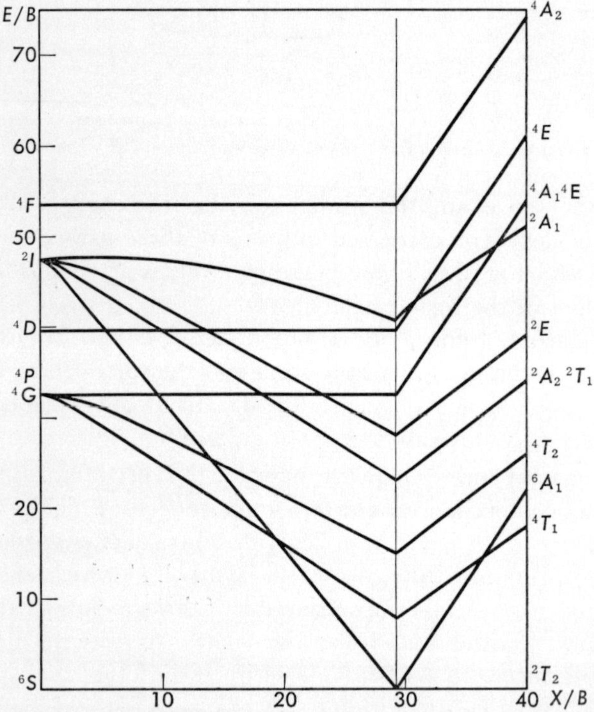

FIG. 4. Relative energy values of various electronic states arising from d^5 configuration in the case of $Y=Z=0$. Racah's parameter A appears only in diagonal elements with a common factor within a definite d^5 system, and these energy differences are independent of A. The value of C/B is kept unaltered from that of free Fe^{3+} ion, *i.e.*, 4.729, and energy values are expressed in B units.

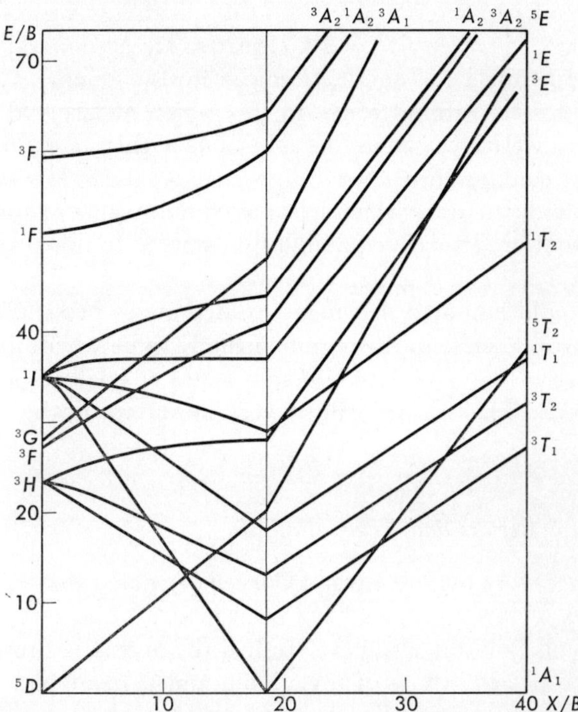

FIG. 5. Relative energy values of various electronic states arising from d^6 configuration in the case of $Y=Z=0$. The value of C/B is kept unaltered from that of free Fe^{2+} ion, *i.e.*, 4.406, and energy values are expressed in B units.

$d^6(Fe^{2+})$ configurations are shown in Figs. 4 and 5, respectively, where the values of parameters Y and Z are taken to be zero for simplicity.* The energy levels at $X=0$ correspond to those of free ions, where Hund's rule holds. It should be noted that even the lowest excited state is higher in energy than the ground state by the order of 2×10^4 cm^{-1} at this value of X. Such energy differences among the states are due to the difference in effectiveness of the Coulomb interaction among the electrons.

When the d-level is split into two sub-levels (u and v belong to one

* If both the 5th and 6th ligands are equivalent to the four nitrogen atoms of porphyrin in the effect on the iron ion, this situation is realized (cubic symmetry). Energy diagrams in this situation had been already calculated (Y. Tanabe and S. Sugano, *J. Phys. Soc. Japan*, **9**, 766 (1954)).

level, and ξ, η and ζ to another level) and the splitting energy X is large enough, the state, in which five or six electrons are accommodated into low-lying three orbitals ξ η and ζ, becomes lowest in energy. Thus, $^2T_2$ and $^1A_1$ state are ground states in the region of large X value in Figs. 4 and 5, respectively. Since the spin value of this state is minimum within a definite configuration d^5 or d^6, this state is called a low spin state. On the other hand, the state which has a maximum value of spin with a definite configuration d^N is called a high spin state. In other words, the ground state in a free ion is a high spin state.

Finally, it should be noted that many excited states become closer in energy to the ground state in the region of the X value where the energy difference between high spin and low spin states is small in both ferric and ferrous ions. This feature will be very important in the following discussions.

2. *Experimental Results and Their Analysis*

Since the pioneer work by Pauling and Coryell (*3*), many studies on magnetic properties of hemoproteins have been made. The author of this paper and his collaborators also have studied the magnetic properties of hemoproteins, especially those of myoglobin and hemoglobin, with the purpose of inquiring into the electronic structures of their iron ions. In this section, those studies are summarized.

Gibson *et al.* (*4*) made EPR measurements on a single crystal of Mb(Fe^{3+}) H_2O and found that the lowest doublet which manifested signals of EPR was widely separated from other sub-levels of the spin sextet normal state $^6A_1$ of Fe^{3+} and the anisotropy in the *xy* plane was negligible. Thus, if the splitting of the sextet state is assumed to be described by a spin Hamiltonian DS_z^2, the value of D is larger than 5 cm^{-1}, but it was unsuccessful to determine an exact value of D from EPR measurements. Since this sextet state shows such a large splitting, the value of D can be determined exactly by the measurement of the temperature dependence of magnetic susceptibility in the low temperature region. Tasaki *et al.* (*5*) measured magnetic susceptibilities on Mb(Fe^{3+})H_2O at temperatures ranging from 77° to 4.2°K, and obtained results showing that the value of D was 10 cm^{-1} in this compound.

This parameter D is related to the energy levels discussed in the preceding section. In the field where the *d*-level is split into four sub-levels as shown in Fig. 3, the level of the quartet state $^4T_1$ in Fig. 4 is split into two levels, one of which is the orbitally non-degenerate state denoted by $^4A_2$, and the other is the orbitally twofold degenerate state denoted by ^4E. The state $^4A_2$ is approximately expressed by one electronic configuration $e^2(^3A_2)b_2{}^2a_1$. The state ^4E is approximately expressed by a linear combination of two electronic configurations $e^3(^2E)b_2a_1(^3B_2)$ and $e^3(^2E)b_2b_1(^3A_2)$. In this case, the spin degeneracy of the ground state $^6A_1$ is removed by the effect of the splitting of $^4T_1$ states through the spin-orbit interaction. The spin Hamiltonian $DS_z{}^2$ shows this splitting of spin states of the ground state $^6A_1$. Thus, with the use of the second order of perturbation calculation, the parameter D is obtained in the following form considering only the lowest states of $^4A_2$ and ^4E,

$$D=\frac{1}{5}a^2\left\{\frac{1}{\varepsilon(^4A_2)-\varepsilon(^6A_1)}-\frac{1}{\varepsilon(^4E)-\varepsilon(^6A_1)}\right\}, \qquad (2\text{–}1)$$

where a is the coefficient of the spin-orbit interaction $a\sum_i \vec{l}_i\cdot\vec{s}_i$ and $\varepsilon(^{2S+1}\Gamma)$ shows the state energy of the $S\Gamma$ state. This equation gives the relation between the D value and the ligand field parameters X, Y and Z.

If the values of the splitting energies Y and Z are large enough, the intermediate spin state ($S=3/2$ for Fe^{3+} and $S=1$ for Fe^{2+}) can be lowest. However, experimental results hitherto obtained show that there is no hemoprotein of the intermediate spin state. This fact indicates that the values of Y and Z are small compared with that of X. On the other hand, there is some evidence that the energy difference between the high spin state and low spin state is very small. For example, George *et al.* (*6*) have shown that the molecules of high spin type and those of low spin type are thermally mixed in many derivatives of ferric myoglobin and hemoglobin in room temperature range from the analysis of the temperature dependence of magnetic susceptibility and absorption spectrum. Therefore, the value of X is first assumed to be $29.70B$ where the energy value of the high spin state is equal to that of the low spin state in the case of $Y=Z=0$. This assumption is justified in later discussions in the present report (see Section 4). As seen in Fig. 4, the energy of the quartet state $^4T_1$ is lowest relative to the ground state at this value of X. This may give one reason for the unusually large value of D, which was first pointed out by Kotani (*7*). Another reason why D is large may be that the level of $^4A_2$ is considerably lower than that of ^4E. The calculated

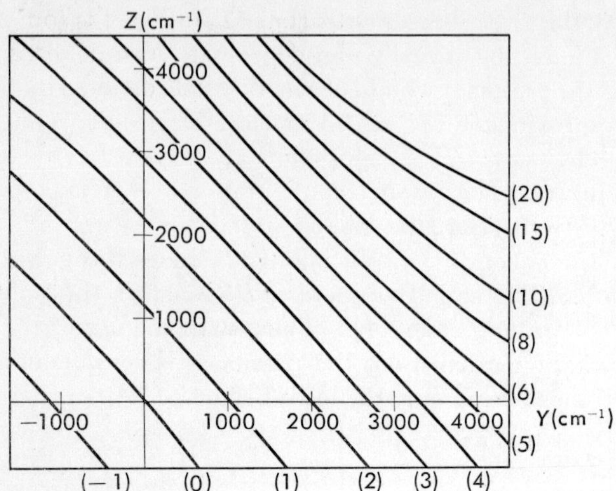

FIG. 6. D values for various Y and Z values when X is fixed at 29.0 B(*2*). D values are shown in parentheses in cm^{-1} units. The values of B and a are taken as 820 cm^{-1} and 400 cm^{-1}, respectively.

values of D for various Y and Z values are shown in Fig. 6 where the contributions from all excited $^4A_2$ and ^4E states are considered (*2*). In this figure, D values are shown in parentheses in cm^{-1} units, and the values of B and a are taken as 820 cm^{-1} and 400 cm^{-1},* respectively. The D values in other ferric hemoproteins of high spin type have later been determined, and these results are shown in Table I.

Gibson and Ingram (*10*) also made EPR measurement on ferric hemoglobin of low spin type, $Mb(Fe^{3+})N_3^-$, and obtained three principal values of g tensor as follows:

$$g_x = 1.72\,, \qquad g_y = 2.22\,, \qquad g_z = 2.80\,. \tag{2-2}$$

Griffith (*11*) derived the relation between these principal values of g tensor and the coefficients of superposition of three configurations $\xi\eta^2\zeta^2$,

* In the possible presence of weak bonding with the ligands, the values of parameters B and a may be smaller than those in free ions. The values of B used in the present report (B=820 cm^{-1} in $Hb(Fe^{3+})$ and B=917 cm^{-1} in $Hb(Fe^{2+})$) are those determined from the absorption spectra of many iron group complexes (see, Y. Tanabe and S. Sugano, *J. Phys. Soc. Japan*, **9**, 766 (1954)). The value of a is determined to be 410 cm^{-1} for free Fe^{2+} ion. For Fe^{3+} ion, the a value is undetermined, but may be almost of the same magnitude as that for Fe^{2+} ion.

TABLE I. *D* Values of Some Hemoproteins of High Spin Type

Hemoproteins	*D* value	Reference
Mb (Fe^{3+}) H_2O	10 cm^{-1}	*5, 8*
Mb (Fe^{3+}) F	7 cm^{-1}	*5*
Hb (Fe^{3+}) H_2O†	10 cm^{-1}	*5*
Catalase	12 cm^{-1}	*9*
HF-Catalase	9 cm^{-1}	*9*

† See Section 5.

$\xi^2\eta\zeta^2$ and $\xi^2\eta^2\zeta$, and obtained the energy differences of three *d*-orbitals ξ, η and ζ. The result was

$$\varepsilon(\xi)-\varepsilon(\eta)=2.26a\,, \qquad \varepsilon(\xi)-\varepsilon(\zeta)=4.45a\,. \tag{2-3}$$

If the coefficient of spin-orbit interaction a is taken to be 400 cm^{-1}, these splitting energies become 900 cm^{-1} and 1780 cm^{-1}, respectively. Since the work by Gibson and Ingram, EPR measurements have been made on many hemoproteins of low spin type and some of their results are listed in Table II. As seen from this table, it seems to be a general tendency that the anisotropy in the porphyrin plane is appreciably large in the hemoprotein of low spin type. Kamimura and Mizuhashi (*16*) explained this anisotropy in terms not only of the direct effect of the 6th ligand but also of the Jahn-Teller effect.

If the energy difference $3/2Z$ between ξ orbital (or η orbital) and ζ orbital in Mb$(Fe^{3+})H_2O$ is assummed to correspond to $\{\varepsilon(\xi)+\varepsilon(\eta)\}/2-\varepsilon(\zeta)$ in Mb$(Fe^{3+})N_3^-$, the value of Z becomes 885 cm^{-1}, and the value of Y is estimated to be 4000 cm^{-1} in Mb$(Fe^{3+})H_2O$.

TABLE II. Principal Values of *g* Tensors of Some Hemoproteins of Low Spin Type

Hemoproteins	g_x	g_y	g_z	Reference
Mb (Fe^{3+}) N_3^-	1.72	2.22	2.80	*10*
Im^-	1.55	2.29	2.93	*12*
CN^-	0.93	1.89	3.45	*12*
OH^-	1.70	2.23	2.67	*4*
Hb (Fe^{3+}) H_2O†	1.70	2.20	2.80	*13*
Beef Cyt. *C*	1.24	2.24	3.06	*14*
Bonito Cyt. *C*	1.29	2.23	3.02	*15*

† See Section 5.

For ferrous hemoproteins, the EPR method is not usually applicable, since the ferrous iron ion has an even number of d-electrons in its outer orbit. In fact, there has been no report of EPR signals of ferrous hemoproteins. However, magnetic susceptibility measurement is also applicable to the ferrous form, and the temperature dependence of the magnetic susceptibility in a very low temperature region may give information about the fine structure of the electronic state as in the case of ferric form.

Nakano *et al.* (*17*) measured the magnetic susceptibility at different temperatures ranging from 77° to 2.0°K and the magnetic field dependence of the magnetization in a strong magnetic field with a superconductive magnet up to 50 KOe on Mb(Fe^{2+}) and Hb(Fe^{2+}) at 4.2°K. The results from these two kinds of measurements were consistent with each other and the following conclusion was drawn. The fine structure of the ground quintet state is well expressed by a spin Hamiltonian DS_z^2, and the value of D is 5 cm^{-1} in both Mb(Fe^{2+}) and Hb(Fe^{2+}).

This parameter D is also related to the energy levels discussed in the preceding section as in the ferric case. In the field where d-level is split into four sub-levels as shown in Fig. 3, the lowest state $^5T_2$ in Fig. 5 is split into two levels, one of which is an orbitally non-degenerate state denoted by $^5B_2$, while the other is an orbitally twofold degenerate state denoted by ^5E. State $^5B_2$ is expressed by one configuration $e^2(^3A_2)b_2{}^2a_1b_1(^3B_1)$, and state ^5E is expressed by one configuration $e^3b_2(^3E)a_1b_1(^3B_1)$. The energy of Coulomb interaction among electrons is common to these two states, and the energy difference between these states is equal to that between orbital e and orbital b_2. The fivefold spin degeneracy of the ground state $^5B_2$ is removed by the spin-orbit interaction with the ^5E state. Thus, in this case, parameter D is expressed as

$$D=\frac{a^2}{16}\cdot\frac{1}{\varepsilon(^5E)-\varepsilon(^5B_2)}\,. \tag{2-4}$$

If the value of a is taken to be 400 cm^{-1}, the energy difference $\varepsilon(^5E)-\varepsilon(^5B_2)$ becomes 2000 cm^{-1}. As mentioned above, this energy difference is equal to $3/2Z$, and the value of Z is estimated to be 1300 cm^{-1}.

Recently, Eaton and Charney (*18*) identified the $d\rightarrow d$ transition ($^1A_1\rightarrow{}^1A_2$*) on ferrous cytochrome c from absorption and circular di-

* This state $^1A_2$ corresponds to $^1T_1$ in Fig. 5, where the values of Y and Z are taken to be zero.

chroism spectra in the wavelength region 5700–10000Å and found that the X value (in our notation) was between 17000 and 19000 cm^{-1}.

The ligand field theory predicts that the state energy of $^5T_2$ (high spin state) is equal to that of $^1A_1$ (low spin state) when the parameter X is taken to be 18.54B (see Fig. 5). Using the value of 917 cm^{-1} for Racah's parameter B, this value of X becomes about 18000 cm^{-1}. Therefore, the energy difference between the low spin and high spin states may also be very small in ferrous hemoproteins, although the phenomena of thermal equilibrium between these spin states has not been reported.

For the value of Y, there has been no experimental evidence on ferrous hemoproteins, but this may be of the order of 10^3 cm^{-1}, as inferred from the values for ferric hemoproteins.

Finally, we can say that, in myoglobin and hemoglobin, the values of splitting energies of the d-level have been determined and the energy values of various electronic states arising from d^5 and d^6 configurations can be estimated with the use of the ligand field theory. In Fig. 7, the

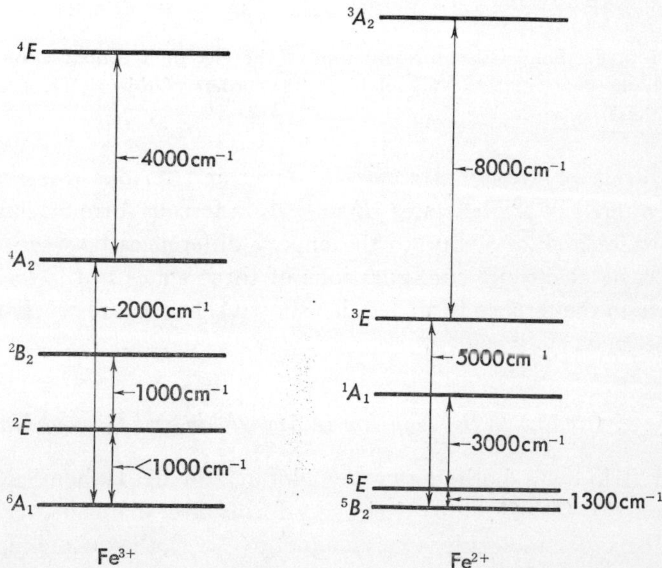

FIG. 7. Estimated energy levels of low-lying states in ferric and ferrous myoglobin. The values of parameters are taken as follows: $X=29.0B$, $Y=5B$, $Z=B$, $C/B=4.73$ and $B=820$ cm^{-1} for ferric myoglobin, $X=18.5B$, $Y=4B$, $Z=1/3B$, $C/B=4.41$ and $B=970$ cm^{-1} for ferrous myoglobin.

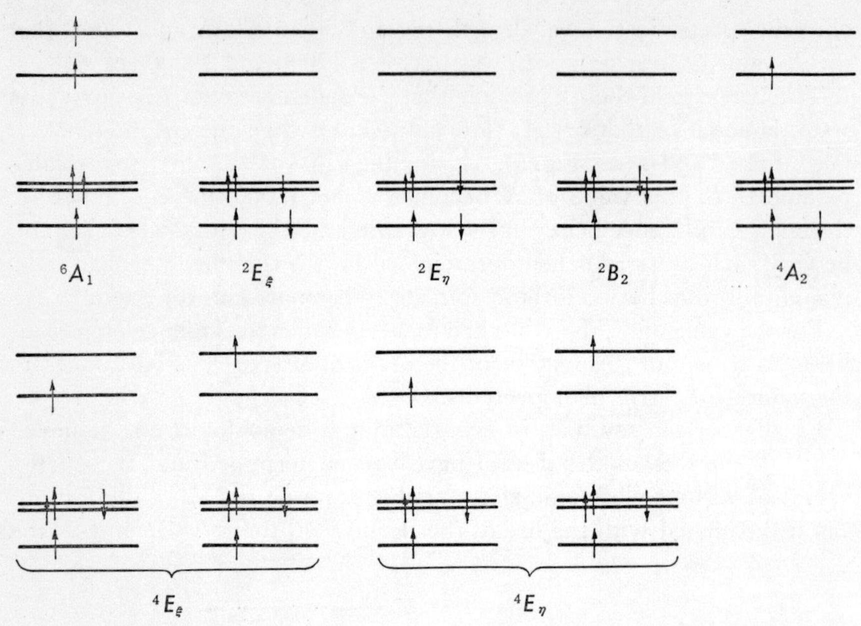

FIG. 8. Schematic representation of the electronic configurations of low-lying states in ferric myoglobin. The orders of *d*-levels are the same as those shown in Fig. 3.

energy levels of low-lying states in ferric and ferrous myoglobin are shown, where two triplet states ^3E and $^3A_2$ in ferrous form originate from the $^3T_2$ state in Fig. 5, due to the energy differences between *e* and b_2 levels. The electronic configurations of these states in the ferric form and those in the ferrous form are also shown schematically in Figs. 8 and 9, respectively.

3. *Oxygen Binding to the Iron Ion in Myoglobin and Hemoglobin**

As is well known, both deoxy-myoglobin and deoxy-hemoglobin are paramagnetic ($S=2$), but their oxy-forms are diamagnetic ($S=0$). Since the oxygen molecule is paramagnetic ($S=1$), the quenching of spin

* Throughout this section, the difference between myoglobin and hemoglobin is neglected. Therefore, the word "myoglobin" in this section is interchangeable with "hemoglobin." The difference between myoglobin and hemoglobin in their electronic states is discussed in the last section.

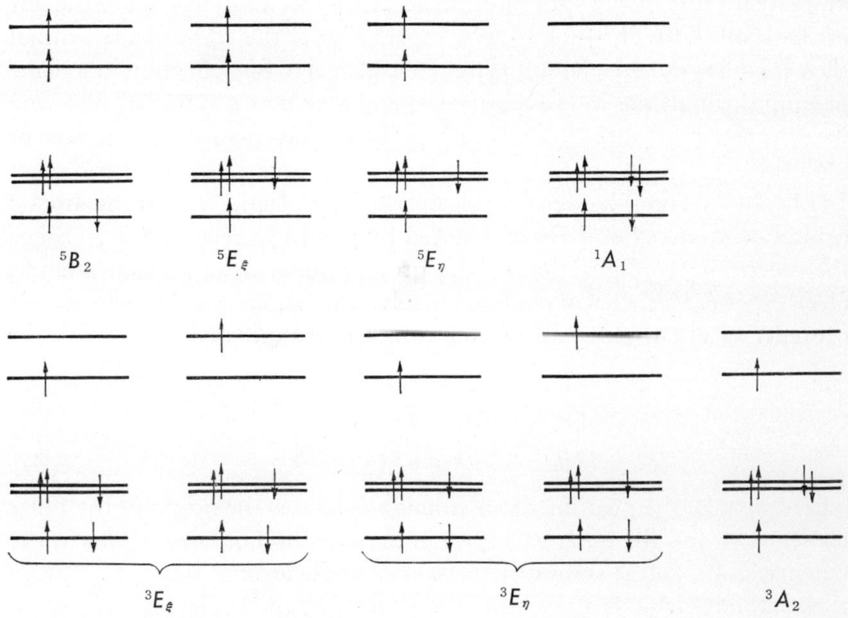

FIG. 9. Schematic representation of the electronic configurations of low-lying states in ferrous myoglobin. The orders of *d*-levels are the same as those shown in Fig. 3.

angular momentum is characteristic in their oxygenation process. The author and his collaborators (*19*) are now investigating the oxygen binding to the iron ion in these hemoproteins by quantum mechanical calculation, through which we intend to survey the biological meaning of the electronic structure of the iron ion in these hemoproteins described in Section 2.

1) Electronic structure of oxygen molecule

In a homonuclear diatomic molecule such as an oxygen molecule, molecular orbitals are classified by the absolute value $|m|$ of the orbital angular momentum along the internuclear axis and by the symmetry property for the inversion to the center of molecule. The notation for a molecular orbital or an energy level is to be marked with σ, π, δ, . . . for $|m|=0, 1, 2, \ldots$ and with g or u as the subscript showing its symmetric or antisymmetric property for the inversion. The notation for an electronic state is to be marked with Σ, Π, Δ, . . . for $\Lambda=0, 1, 2, \ldots$ where Λ shows the absolute value of the sum of m for each electron, with the numerical value of the spin multiplicity $2S+1$ as a superscript at the left of the symbol for Λ, $+$ or $-$ as a superscript at the right of the symbol showing its symmetric or antisymmetric property for reflection in a plane passing through the internuclear axis, and g or u as a subscript showing its symmetric or antisymmetric property for inversion to the center of molecule.

The two oxygen nuclei will be denoted by A and B, and the atomic orbital of atoms A and B are denoted by χ_a and χ_b, respectively. Since the electronic configuration of the ground state of an oxygen atom is $(1S)^2\,(2S)^2\,(2P)^4$, we will confine ourselves to $1S$, $2S$ and $2P$ type atomic orbitals. Let the three $2P$ atomic orbitals be expressed as

$$\left.\begin{aligned} 2P_\sigma &= cR(r)\cos \Theta \\ 2P_{\pi^+} &= c'R(r)\sin \Theta\, e^{i\varphi} \\ 2P_{\pi^-} &= c'R(r)\sin \Theta\, e^{-i\varphi} \end{aligned}\right\} \qquad (3\text{–}1)$$

where we take the position of an atomic nucleus as the origin of the polar coordinates and the polar axis $\theta=0$ in the direction of the nucleus of the partner atom, and φ shows the rotational angle around this axis. From these atomic orbitals we set up the symmetry molecular orbitals of an oxygen molecule as follows;

$$\left.\begin{aligned} (\sigma_1)_{g,u} &= 1S_a \pm 1S_b \\ (\sigma_2)_{g,u} &= 2S_a \pm 2S_b \\ (\sigma_3)_{g,u} &= 2P_{\sigma a} \pm 2P_{\sigma b} \\ (\pi^\pm)_{g,u} &= 2P_{\pi a^\pm} \mp 2P_{\pi b^\pm}\,. \end{aligned}\right\} \qquad (3\text{–}2)$$

As is usually accepted, the orbital energies of these molecular orbitals increase in the order of σ_{1g}, σ_{1u}, σ_{2g}, σ_{2u}, σ_{3g}, $\pi_u{}^\pm$, $\pi_g{}^\pm$ and σ_{3u}, so the wave functions of the ground state and of lower excited states of the oxygen molecule are composed mainly of configurations which contain two electrons in each of the σ_{1g}, σ_{1u}, σ_{2g}, σ_{2u}, σ_{3g} orbitals, and no electron in the σ_{3u} orbital. Within these limitations, it is sufficient to consider only six electrons accomodated in the $\pi_u{}^\pm$ and $\pi_g{}^\pm$ orbitals, and the molecular states compatible with this assumption are shown in Table III together with the observed vertical excitation energies relative to the ground state ${}^3\Sigma_g{}^-$ in eV,* where the observed energy values of low-lying states of $O_2{}^+$ and $O_2{}^-$ are also listed.

* $1eV = 8.0668\times10^3$ cm^{-1} $= 23.0684$ kcal/mole.

TABLE III. Electronic States of O_2, O_2^- and O_2^+

	States	π_u^+	π_u^-	π_g^+	π_g^-	Observed energy values in eV	Reference
O_2	$^3\Sigma_g^-$			1	1	0	*20*
	$^1\Delta_g$			2 	 2	1.0	*20*
	$^1\Sigma_g$			1	1	1.6	*20*
	$^1\Sigma_u^-$	 1	1 	1 	 1	7.9	*21*
	$^3\Delta_u$	1 	 1	1 	 1		
	$^3\Sigma_u^+$	 1	1 	1 	 1	7.5	*22*
	$^3\Sigma_u^-$	 1	1 	1 	 1	9.5	*20*
	$^1\Delta_u$	1 	 1	1 	 1		
	$^1\Sigma_u^+$	 1	1 	1 	 1		
O_2^-	$^2\Pi_g$			 1	1 	−0.2	*23*
O_2^+	$^2\Pi_g$			1 2	2 1	12.5	*20*
	$^4\Pi_u$	 1	1 	1 1	1 1	16.8	*20*
	$^2\Pi_u$	 1	1 	1 1	1 1	17.7	*20*

Electronic configurations are shown by specifying the unoccupied orbitals.

Since the binding energy of an oxygen molecule with the iron ion in myoglobin will be less than 7.9eV, it is sufficient to consider only three electronic states in O_2, the ground state ($^3\Sigma_g^-$) and two excited states ($^1\Sigma_g^+$ and $^1\Delta_g$), in our calculation.

2) Summary of the calculation method

The calculations were based on the method of " atoms in molecule " proposed by Moffit (*24*). In our case, we regard oxymyoglobin as a " molecule," and deoxymyoglobin and the oxygen molecule as " atoms " constituting the "molecule." That is, the eigen-functions in oxymyoglobin are assumed to be expressed linearly in terms of the antisymmetrized product-functions, each of which consists of the eigen-function of an isolated oxygen molecule and that of the iron ion in deoxymyoglobin. Only the interaction parts are calculated non-empirically with these unperturbed states.* The energy eigenvalues of these unperturbed states are estimated empirically. (In our case, those of the iron ion in myoglobin are estimated from the results of Section 2.) This treatment is based on the fact that the interaction energy (binding energy) is small relative to the total energy of the "molecule" and the electronic structures of the "atoms" are fairly conserved in the "molecule."

3) Orientation of O_2 in oxymyoglobin

Two possible simple models have been proposed for the orientation of the oxygen molecule in oxymyoglobin or oxyhemoglobin. One, in which the O-O axis is at 60° to the heme plane, was proposed by Pauling (*25*). Another, in which the oxy-form is a π-complex with the O-O axis parallel to the heme, was proposed later by Griffith (*26*). Then, the measurement of nuclear magnetic resonance on $Hb(Fe^{2+})^{17}O_2$ was performed by Maričič *et al.* (*27*) in connection with the stereo-chemistry and the nature of the oxygen binding. They obtained only one nuclear magnetic resonance line which accounted for the available $^{17}O_2$ in the sample and concluded that Griffith's model was compatible with their experimental results.** However, recent X-ray diffraction analysis (*28*) suggests that the oxygen molecule is bound to the iron ion in a manner similar to that proposed by Pauling.

With these circumstances in mind, we intend to perform the calculations on two stereo-chemical arrangements, a parallel one and an inclined one. Because of the simplicity of higher symmetry, we will take the parallel case where the O-O axis is parallel to the x axis defined in Section 1. Calculations on the other cases will be reported in the near future.

4) Electronic states in oxymyoglobin

In our parallel case, the electronic states of oxymyoglobin are classified

* For details of the calculation method, see the appendix.

** If an oxygen molecule is bound to the iron ion with the O-O axis parallel to the heme plane, the two oxygen atoms will be situated in equal distances from the iron ion and will be equivalent.

TABLE IV. Singlet States Constructed from the Low-lying States in a Ferrous Iron Ion in Myoglobin and Three Low-lying States in an Oxygen Molecule

Oxyheme	Fe^{2+}	O_2
$^1A_1$	$^1A_1$	$^1\Delta_g$
	$^1A_1$	$^1\Sigma_g{}^+$
	$^3E_\xi$	$^3\Sigma_g{}^-$
$^1A_2$	$^3E_\eta$	$^3\Sigma_g{}^-$
$^1B_1$	$^3A_2$	$^3\Sigma_g{}^-$
	$^3B_2$	$^3\Sigma_g{}^-$
$^1B_2$	$^1A_1$	$^1\Delta_g$
$^7A_1$	$^5E_\xi$	$^3\Sigma_g{}^-$
$^7A_2$	$^5E_\eta$	$^3\Sigma_g{}^-$
$^7B_1$	$^5B_2$	$^3\Sigma_g{}^-$

Septet states which are constructed from three quintet states ($^5B_2$, $^5E_\xi$ and $^5E_\eta$) and the lowest state ($^3\Sigma_g{}^-$) in the oxygen molecule are also listed. Here, the O-O axis is assumed to be parallel to the heme plane and to the x axis.

into four groups by their symmetric properties. These groups are denoted by the symbols A_1, A_2, B_1 and B_2 (the names for the irreducible representations of point group C_{2v}). Since the interaction part between the iron ion and the oxygen molecule is invariant in the symmetry operations, there are interactions only among the configurations which belong to the same group. Therefore, the energy levels in oxymyoglobin are also denoted by these symbols. Since the oxymyoglobin is diamagnetic, the configurations which constitute the ground state (bonding state) in oxymyoglobin must be singlet ($S=0$). Possible configurations of singlet states constructed from the low-lying states of the ferrous iron ion in myoglobin shown in Fig. 7 and three low-lying states in the oxygen molecule shown in Table III are listed in Table IV for each group mentioned above. In actual calculation, we have considered also all charge transfer configurations which arise from those in Table IV by transfering one electron from Fe^{2+} to O_2 ($Fe^{3+}-O_2{}^-$) or from O_2 to Fe^{2+} ($Fe^+-O_2{}^+$). In addition to these singlet states, the septet states which are constructed from the lowest state (quintet) in myoglobin, and the lowest state (triplet) in the oxygen molecule are also considered for reference, the configurations of which are included in Table IV.

5) Energy parameters of heme

The energy values relative to the ground state of O_2, which are shown in Table III, can be used as the energy eigenvalues for the unperturbed states of the oxygen molecule. As for myoglobin, however, the result of ligand field theory (Figs. 8 and 9) shows only the energy differences among the states for a definite d^N configuration. The energy differences among the states of different d^N configurations are estimated from the ionization energies of free iron ions (*29*), and the Coulomb interaction energy between *d*-electrons and the point charges of $-0.5e$ which are assumed to be distributed on the position of each nitrogen atom of the porphyrin ring. The effect of the Coulomb field due to these point charges on the oxygen molecule is also considered.

6) Results and discussion

Calculations have been carried out for the following two cases.

Case 1 The values of the parameters B, C, X, Y and Z determined for Mb(Fe^{2+}) are used in estimating the energy values of all unperturbed states of myoglobin. (The value of Y is chosen to be 2000 cm^{-1} or 4000 cm^{-1}.)

Case 2 The values of the parameters B, C, X, Y and Z determined for Mb(Fe^{2+}) are used in estimating the energy values of the unperturbed

TABLE V. Calculated Values of State Energies of Oxymyoglobin

States	Case 1		Case 2	
	$R=1.5$ Å	$R=2.0$ Å	$R=1.5$ Å	$R=2.0$ Å
$^1A_1$	−2.391	0.537	−1.917	0.558
$^1A_2$	−3.531	0.324	−3.071	0.368
$^1B_1$	−2.388	0.842	−1.670	0.877
$^1B_2$	−2.086	1.026	−1.431	1.070
$^7A_1$	−1.226	−0.203	−1.226	−0.203
$^7A_2$	−4.205	−0.331	−3.087	−0.252
$^7B_1$	−1.205	−0.160	−1.300	−0.122

The O–O axis is taken to be parallel to the heme plane and the x axis. The energy values are given in eV, and the sum of the unperturbed state energy of the $^5B_2$ state in Mb (Fe^{2+}) and that of $^3\Sigma_g^-$ in O_2 is taken to be zero. R shows the distance between the iron ion and the center of the oxygen molecule. The value of the undetermined parameter Y in Mb (Fe^{2+}) is chosen to be 2000 cm^{-1} in these calculations.

state of Fe^{2+} and Fe^{+} configurations.* The energy values, however, of the unperturbed states of Fe^{3+} configurations are estimated with the use of the parameter values determined for Mb(Fe^{+3}).

The calculated values of state energies are shown in Table V where the sum of the unperturbed state energy of $^5B_2$ in Mb(Fe^{2+}) and that of $^3\Sigma_g^-$ in O_2 is taken to be zero. The main features of the results are

a) The energy values of all $S\Gamma$ states are negative at 1.5Å. b) The state energy of $^1A_2$ is lowest among the singlet states. c) In Case 2, the state energy of $^1A_2$ is almost equal to that of $^7A_2$, although the energy of $^1A_2$ is slightly larger than that of $^7A_2$ in Case 1. But, if the value of parameter Y in Fe^{2+} configurations is chosen to be 4000 cm^{-1} in the treatment of Case 2, $^1A_2$ becomes lowest. That is, $E(^1A_2) = -3.460eV$ and $E(^7A_2) = -3.260eV$.

The reasons for a) are

1) The configuration interaction with the charge transfer configurations becomes effective at this distance. 2) The Coulomb interaction term becomes negative mainly due to the interaction of the electrons accomodated into the bonding orbitals of the oxygen molecule with the core of the iron ion. 3) The values of exchange repulsions are fairly small at this distance.

The reasons for b) are

1) The values of the overlap integrals $(e|\pi_g^+)$ and $(e|\pi_g^-)$ are largest, and the configuration interaction with charge transfer configurations ($e \rightarrow \pi_g^+$ and $e \rightarrow \pi_g^-$), which appear in $^1A_2$, are most effective. 2) The exchange integrals $(e\pi_g^+|e\pi_g^+)$ and $(e\pi_g^-|e\pi_g^-)$ appear with negative values in the diagonal element only of the configuration ($^3E_\eta \leftrightarrow {}^3\Sigma_g^-$) in the $^1A_2$ state. Such an exchange effect was used first by Heitler and London (*30*) in order to successfully explain the problem of chemical bonds for the hydrogen molecule.

The results of c)** indicate that the values of parameters X and Y are sensitive to the explanation of the diamagnetism of oxymyoglobin.

In conclusion, the triplet state ^3E which is an excited state in ferrous myoglobin mainly contributes to the bonding state. The results of calculation in the inclined case cannot yet be obtained, but the combina-

* We cannot obtain Mb(Fe^{+}) and cannot use the experimental data for Fe^{+} configurations. However, the states of O_2^+ are much higher in energy, and the effect of Fe^{+} configurations on bonding state is negligible.

** The reason for c) is rather physical, and will be explained elsewhere.

tion of the triplet state in myoglobin and the triplet state (ground state) in the oxygen molecule seems to construct the bonding state also in this case, for both exchange effect and the configuration interaction with charge transfer configurations seem to be most effective in this combination. As noticed in Sections 1 and 2, this triplet state ($^3E_\eta$) is anomalously low in energy compared with that of free Fe^{2+} ions.

4. Thermal Equilibrium between High-spin and Low-spin States in Ferrihemoproteins

George *et al.* (*6*, *31*, *32*) measured carefully the temperature dependences (in the range from 30° to 0°C) of magnetic susceptibilities and optical absorption spectra of various derivatives of ferrimyoglobin and ferrihemoglobin, which showed the intermediate values of magnetic susceptibilities at room temperature. They analyzed their data by assuming that these hemoproteins were a mixture of high spin compounds and low spin compounds. That is, the observed magnetic susceptibility $\chi_{\text{ob.}}$ was expressed by

$$\chi_{\text{ob.}} = (1-\alpha)\chi_H + \alpha\chi_L\,, \qquad (4\text{–}1)$$

where α is the ratio of the amount of low spin compounds to the total amount of compounds of the two spin types, and χ_H and χ_L are the values of magnetic susceptibilities expected from purely high spin compounds and low spin compounds, respectively. The logarithms of the equilibrium constants $K \equiv \alpha/(1-\alpha)$ between the compounds thus obtained were plotted for various temperatures. Since these values of K were almost on a straight line when plotted against inverse temperature $1/T$, they claimed that the compounds which showed the intermediate values of magnetic susceptibility were in the thermal equilibrium between the high and low spin states. Furthermore, they determined the difference of enthalpy and entropy, ΔH and ΔS, between these two spin states from the slope of the line and the extrapolated value of ln K at $1/T=0$. The entropy change ΔS thus obtained was much larger than that expected from the ratio of the spin state number of these two spin states.

Afterwards, Iizuka and Kotani (*33–35*) measured more accurately the temperature dependences of magnetic susceptibilities in the range from room temperature to liquid nitrogen temperature. According to George

TABLE VI. Experimental Values of ε and γ for Various Ferrihemoproteins (*33–35*)

Hemoproteins	6th ligand	Ground state	ε (cm^{-1})	γ	T_0 (°K)
Mb (Fe^{3+})	H_2O	high	−1270	3.05×10^{-3}	388
	N_3^-	low	1310	4.01×10	391
	I_m^-	low	3783	4.12×10^6	331
Hb (Fe^{3+})	OH^-	low	655	4.20	372
	N_3^-	low	1770	3.02×10^2	374
CCP (Fe^{3+})	H_2O	high	−1230	5.59×10^{-4}	274
	OH^-	low	1830	2.58×10^4	232
	N_3^-	low	1500	4.64×10^2	298
	OCN^-	low	390	2.84	264

T_0 is the temperature defined by the relation $\ln 3\gamma-\varepsilon/kT_0=0$ for each hemoprotein.

et al., they expressed the observed magnetic susceptibility $\chi_{ob.}$ as follows:

$$\chi_{ob.}=\frac{\chi_L+3\gamma e^{-\varepsilon/kT}\chi_H}{1+3\gamma e^{-\varepsilon/kT}}. \tag{4–2}$$

All experimental data were well explained by assuming that both γ and ε were temperature independent parameters. It may be considered that ε and $\ln 3\gamma$ correspond to the changes of enthalpy and entropy, respectively, in the thermodynamical language. The values of ε and γ for various ferrihemoproteins which they determined are shown in Table VI. In this table, T_0 is the characteristic temperature at which the difference of free energy between high and low spin states, or $\ln 3\gamma-\varepsilon/kT_0$, becomes zero.

The main features of their results were as follows;

i) The characteristic temperature T_0 is almost independent of the type of the sixth ligand, but depends on the type of hemoprotein. Thus, T_0 seems to be characteristic of the type of apoprotein. ii) Whether the ground state is of high spin or not, the statistical weight of the state of higher level seems to be much larger than that of the ground state. iii) Although the energy difference or enthalpy change ε between the two spin states is much larger than the energy corresponding to room temperature, large entropy change seems to cancel this large enthalpy change at the temperature T_0, and the coexistence of the two spin states can be observed near room temperature.

The author (*36*) has proposed a model which explains the character of this phenomenon by considering the temperature dependence of the position of the porphyrin ring relative to the iron ion due to the cooperative breaking of the van der Waals contacts between porphyrin and apoprotein. In this original paper, the statistical probabilities of two spin states are calculated with the use of the Ising model. The picture of the proposed model is described qualitatively in the present report.

At the first glance, one may have an idea that the spin change will be coupled with the low-frequency degree of freedom of internal rotations or oscillations in the apoprotein part. However, this idea seems unable to explain the main feature ii), in what way this coupling occurs. Another possibility is that the energy difference between the two spin states depends on temperature due to the shifts of the ligands around the iron ion. This energy difference may depend almost linearly on temperatures in a narrow range. Let us consider the main features mentioned above in terms of the latter possibility.

Since the characteristic temperature T_0 seems to be characteristic of the type of apoprotein, and to be almost independent of the type of the sixth ligand, it may be reasonable to consider this phenomenon not by the shift of the sixth ligand but by the relations among the iron ion, porphyrin ring and apoprotein. The molecular structures of myoglobin (*37*) and hemoglobin (*38*) have been determined by high-resolution X-ray diffraction analysis. According to these results, the porphyrin ring in these hemoproteins has a number of van der Waals contacts with the hydrophobic residues of apoprotein in the heme crevice. The iron ion in the heme is situated nearly at the center of the porphyrin ring and is bound to four nitrogen atoms of the porphyrin by weak bonds. This iron ion has also relatively rigid connection with the nitrogen atom of the imidazol residue of histidine (F 8) in apoprotein. This histidine is in the helical region, and the conformation of the neighbors of this amino acid will be also rather rigid. Therefore, it seems that the van der Waals contacts between the porphyrin and apoprotein are most sensitive to the temperature. The higher the temperature, the more contacts may be broken. When a contact is broken, the contacts in the neighborhood of this contact will tend to be broken, and the broken contacts will appear together in one portion of the porphyrin. Thus, this portion of porphyrin shifts away from the apoprotein, and the distances between the iron ion and the four nitrogen atoms of porphyrin will vary.

Next, in order to explain the main feature ii), let us assume that the apoprotein portion which supports the porphyrin is situated on the opposite side of the imidazol residue of histidine (F 8) to the heme plane. Then, the portion of the porphyrin ring where the contacts are broken becomes closer to the iron ion, and thus the low spin state will appear at higher temperatures, if the iron ion is situated out of the porphyrin plane and to the side of the imidazol residue in the compound which is of high spin at low temperatures. On the other hand, if the iron ion is situated nearly in the porphyrin plane or slightly out of it to the opposite side of the imidazole residue in the compound which is of low spin at low temperatures, then the distance between the iron ion and the porphyrin ring becomes longer and thus the high spin state will appear at higher temperatures. It has been suggested for hemin that the in-plane and out-of-plane configurations of the heme iron favor low- and high-spin states, respectively (*39*). On the author's model, Perutz (*40*) also recognizes such a tendency in hemoglobin by X-ray diffraction analysis.

A large value of entropy change can also be explained quantitatively with probable values of parameters, according to this model. (For details on this point, see the original paper.)

Recently, according to the author's suggestion, the measurements of the temperature dependence of the optical spectrum and paramagnetic susceptibilities of reconstituted mesohemoproteins have been carried out in the temperature range from 77° to 296°K in order to examine the effect of chemical modification of the heme group on the heme-apoprotein interaction (*41*). The results of these experiments are shown in Table VII. The T_0 values of mesohemoproteins are 50 to 100 degrees higher than those of the corresponding protohemoproteins. (Compare Table VII

TABLE VII. Experimental Values of ε and γ for Various Meso-ferrihemoproteins (*41*)

Hemoproteins	6th ligand	Ground state	ε (cm^{-1})	γ	T_0 (°K)
Mb (Fe^{3+})	H_2O	high	-1010	1.68×10^{-2}	487
	N_3^-	low	800	4.12	455
Hb (Fe^{3+})	N_3^-	low	1050	1.27×10	415
CCP (Fe^{3+})	N_3^-	low	1220	3.89×10	370
	OCN^-	low	500	2.97	326

The value of T_0 of meso-ferrihemoproteins are 50 to 100 degrees higher than those of the corresponding proto-ferrihemoproteins.

with Table VI.) Thus, the author's model is experimentally confirmed by this study, at least concerning the point that the contacts between the porphyrin and apoprotein play an important role in the thermal equilibrium of the two spin states in ferric hemoproteins.

5. *Discussion*

As is well known, the hemoglobin molecule is composed of four subunits, each of which is similar to myoglobin in its molecular structure. These four subunits behave in the binding of the oxygen molecule as if there is interaction among them. However, the difference of electronic structure between ferrous myoglobin and ferrous hemoglobin could not be detected from the magnetic measurements, as mentioned in Section 2. Therefore, we did not distinguish between myoglobin and hemoglobin in the calculation in Section 3.

But actually, acid ferri-hemoglobin $Hb(Fe^{3+})H_2O$ shows an unusual magnetic property which is different from that of acid ferri-myoglobin $Mb(Fe^{3+})H_2O$ (*42*). $Mb(Fe^{3+})H_2O$ is of a purely high-spin state in the low temperature region (from 77° to 2°K) and is in thermal equilibrium between the high-spin state and the low-spin state in the higher temperature region. On the other hand, $Hb(Fe^{3+})H_2O$ is of a purely high-spin state in the temperature region from 30° to 0°C, but the ratio of low-spin state increases suddenly at 0°C, and then this ratio decreases gradually as the temperature becomes lower. In the low temperature region (from 77° to 2°K), the ratio of low-spin state to high-spin state is quite independent of temperature. D value of the high-spin state is given in Table I, and the principal values of g tensor of the low spin state are given in Table II. $Hb(Fe^{3+})OH^-$ and $Hb(Fe^{3+})N_3^-$ show the usual thermal equilibrium between the two spin states as myoglobins and cytochrome C peroxidases. This behavior characteristic of $Hb(Fe^{3+})H_2O$ is unexplainable at the present stage. Thus, we cannot yet satisfactorily interprete the difference in electronic structures of the iron ion between myoglobin and hemoglobin.

However, it must certainly be the characteristic feature of hemoprotein that there are many low-lying states of the iron ion and the energy differences among these states are changeable in response to the conformation change of apoprotein as indicated by the phenomenon of thermal equilibrium. In myoglobin and hemoglobin, the low-lying excited state, probably triplet state, seems to contribute mainly to the binding with the oxygen molecule, and the energy change of this triplet state may strongly influence the affinity of the iron ion for the oxygen molecule. It is also interesting that this triplet state is not the ground state but a low-lying excited state in myoglobin and hemoglobin. These problems will be the subject of future studies by the author.

Summary

Energy levels of various electronic states of the iron ion in hemoproteins are estimated on the basis of the ligand field theory with the use of the experimental data of magnetic and optical properties. This work makes clear that there are many low-lying states in both ferric and ferrous hemoproteins. Quantum mechanical calculation is then carried out on the binding of oxygen molecules to the iron ion in myoglobin or hemoglobin, with the purpose of relating the electronic structure to the biological function. The result of this calculation shows that the triplet, low-lying excited state in ferrous myoglobin contributes mainly to the bonding state in oxygenated myoglobin. In a free Fe^{2+} ion, the exciting energy of this triplet state is much larger than the binding energy obtained in this calculation. Finally, one interpretation is given for the phenomenon of the thermal equilibrium between the two electronic states of ferric hemoprotein and it is indicated that the electronic structures of the iron ion seem to be changeable in response to the conformation change of apoprotein.

Appendix

We expand the eigenfunction Φ, representing a stationary state of oxymyoglobin, linearly in terms of the members of product-function Ψ_i

$$\Phi = \sum_i c_i \Psi_i \,. \tag{A-1}$$

The finite expansion (A-1) is to represent a solution of the eigenvalue problem of oxymyoglobin, namely,

$$H\Phi = E\Phi \,, \tag{A-2}$$

where H is the total Hamiltonian operator and E is the energy of the stationary state Φ. Substituting (A-1) in (A-2), we find

$$\sum_i c_i H\Psi_i = E\sum_i c_i\Psi_i . \tag{A-3}$$

Multiplying both sides of this equation by $\Psi_j{}^*$ and integrating over all space and spin coordinates, we then have

$$\sum_i H_{ji}c_i = E\sum_i S_{ji}c_i , \tag{A-4}$$

where we use the abbreviations

$$\left.\begin{aligned} S_{ji} &= \int \Psi_j{}^*\Psi_i dV = S_{ij}{}^* \\ H_{ji} &= \int \Psi_j{}^* H\Psi_i dV = H_{ij}{}^* . \end{aligned}\right\} \tag{A-5}$$

The relation (A-5) may be said to define the energy matrix H and the overlap matrix S. Eliminating c_i from this system of equations, we obtain the determining relation

$$| H_{ji} - S_{ji}E | = 0 . \tag{A-6}$$

Regarded as an equation in E, (A-6) may be solved to give the energy eigenvalues of oxymyoglobin.

Next, let us consider the calculation of a matrix element of total Hamiltonian H_{ji}. The explicit forms of Ψ_i and Ψ_j are;

$$\left.\begin{aligned} \Psi_i &= A\{\psi_{iI}(q_1q_2\cdots q_n)\cdot\phi_{iII}(q_{n+1}q_{n+2}\cdots q_N)\} \\ \Psi_j &= A\{\psi_{jI}(q_1q_2\cdots q_{n'})\cdot\phi_{jII}(q_{n'+1}q_{n'+2}\cdots q_N)\} \\ &\qquad (\text{generally } n \neq n') \end{aligned}\right\} \tag{A-7}$$

where ψ shows the wave function of the iron ion, ϕ the wave function of the oxygen molecule, q_k the coordinate of the k-th electron including spin, and A is the antisymmetrizing operator for N electrons. The Hamiltonian of the total system may be resolved into the forms according to the electronic configuration of $\psi_{iI}\cdot\phi_{iII}$.

$$\left.\begin{aligned} H &= H_{\rm I} + H_{\rm II} + V + U \ (\text{core}\cdot\text{core}) \\ H_{\rm I} &= \sum_{k=1}^{n}\left\{-\frac{1}{2}\Delta_k + v_1(k)\right\} + \sum_{l>k=1}^{n}\frac{1}{r_{kl}} \\ H_{\rm II} &= \sum_{k=n+1}^{N}\left\{-\frac{1}{2}\Delta_k + v_2(k)\right\} + \sum_{l>k=n+1}^{n}\frac{1}{r_{kl}} \\ V &= \sum_{k=1}^{n} v_2(k) + \sum_{k=n+1}^{N} v_1(k) + \sum_{k=1}^{n}\sum_{l=n+1}^{N}\frac{1}{r_{kl}} , \end{aligned}\right\} \tag{A-8}$$

where H_{I} and H_{II} show the Hamiltonian of the iron ion in myoglobin and the oxygen molecule respectively, and V shows the interaction part between the iron ion and the oxygen molecule. U (core•core) is the interaction between the core of the heme and that of the oxygen molecule, and v_1 and v_2 are core potentials of the heme and oxygen molecule, respectively. The matrix element of H calculated for the Hamiltonian resolved as in (A-8) will be denoted by $\langle\Psi_j|H\Psi_i\rangle$ Since $\psi_{i\mathrm{I}}$ and $\phi_{i\mathrm{II}}$ are the eigenfunctions of H_{I} and H_{II}, respectively, these functions satisfy the following equations,

$$\left.\begin{aligned} H_{\mathrm{I}}\psi_{i\mathrm{I}} &= E_{i\mathrm{I}}\psi_{i\mathrm{I}} \\ H_{\mathrm{II}}\phi_{i\mathrm{II}} &= E_{i\mathrm{II}}\phi_{i\mathrm{II}} \end{aligned}\right\} \qquad \text{(A-9)}$$

where $E_{i\mathrm{I}}$ and $E_{i\mathrm{II}}$ are the energy eigenvalues of unperturbed states of heme the and oxygen molecule, respectively. With the use of these relations, $\langle\Psi_j|H\Psi_i\rangle$ may be written as

$$\langle\Psi_j\,|\,H\Psi_i\rangle = (E_{i\mathrm{I}}+E_{i\mathrm{II}})\langle\Psi_j\,|\,\Psi_i\rangle + \langle\Psi_j\,|\,V\Psi_i\rangle + U\langle\Psi_j\,|\,\Psi_i\rangle\,. \qquad \text{(A-10)}$$

On the other hand, $\langle\Psi_i|H\Psi_j\rangle$ may be written as

$$\langle\Psi_i\,|\,H\Psi_j\rangle = (E_{j\mathrm{I}}+E_{j\mathrm{II}})\langle\Psi_i\,|\,\Psi_j\rangle + \langle\Psi_i\,|\,V\Psi_j\rangle + U\langle\Psi_i\,|\,\Psi_j\rangle\,. \qquad \text{(A-11)}$$

Then, $\langle\Psi_j|H\Psi_i\rangle$ is not always equal to $\langle\Psi_i|H\Psi_j\rangle^*$. This shows that the matrix of H constituted from the matrix element calculated in this method is not Hermitian and the eigenvalues of this matrix are not necessarily real. To avoid this difficulty, we approximate that the matrix element of H between j state and i state, H_{ji}, may be written as

$$H_{ji} = H_{ij}^* \approx \frac{1}{2}\{\langle\Psi_j\,|\,H\Psi_i\rangle + \langle\Psi_i\,|\,H\Psi_j\rangle\}\,. \qquad \text{(A-12)}$$

Finally, the following expression is obtained

$$\begin{aligned} H_{ji} &= \frac{1}{2}\{(E_{i\mathrm{I}}+E_{i\mathrm{II}})+(E_{j\mathrm{I}}+E_{j\mathrm{II}})\}\langle\Psi_j\,|\,\Psi_i\rangle \\ &\quad + \frac{1}{2}\{\langle\Psi_j\,|\,V\Psi_i\rangle + \langle\Psi_i\,|\,V\Psi_j\rangle\} + U\langle\Psi_j\,|\,\Psi_i\rangle\,. \end{aligned} \qquad \text{(A-13)}$$

The overlap integral $\langle\Psi_i|\Psi_j\rangle$ and the interaction part $\langle\Psi_j|V\Psi_i\rangle$ or $\langle\Psi_i|V\Psi_j\rangle$ are calculated non-empirically up to the second order of overlap integral between one electron orbitals, where we use the analytical Hartree Fock wave functions in the ^5D state of free Fe^{2+} ion for d-orbitals of the iron ion and the Schmidt orthogonalized molecular orbitals constructed from the Gaussian type atomic orbitals determined by Fujinaga for the wave functions of the oxygen molecule.

References

1 J. Ōtsuka, *J. Phys. Soc. Japan*, **21**, 596 (1966).
2 J. Ōtsuka, *J. Phys. Soc. Japan*, **24**, 885 (1968).
3 L. Pauling and C. D. Coryell, *Proc. Natl. Acad. Sci. U.S.*, **22**, 159 (1936).
4 J. F. Gibson, D. J. E. Ingram and D. Schonland, *Discussions Faraday Soc.*, **26**, 72 (1958).
5 A. Tasaki, J. Ōtsuka and M. Kotani, *Biochim. Biophys. Acta*, **140**, 284 (1967).
6 P. George, J. Beetlestone and J. S. Griffith, *in* "Hematin Enzymes," ed. by J. E. Falk, R. Lemberg and R. K. Morton, Pergamon Press, New York, p. 106 (1961).
7 M. Kotani, *Rev. Mod. Phys.*, **35**, 717 (1963).
8 H. Uenoyama, T. Iizuka, H. Morimoto and M. Kotani, *Biochim. Biophys. Acta*, **160**, 159 (1968).
9 K. Torii, Y. Ogura, J. Ōtsuka and A. Tasaki, *J. Biochem.*, **66**, 791 (1969).
10 J. F. Gibson and D. J. E. Ingram, *Nature*, **180**, 29 (1957).
11 J. S. Griffith, *Nature*, **180**, 30 (1957).
12 H. Hori and H. Morimoto, private communication.
13 N. Nakano, private communication.
14 I. Salmeen and G. Palmer, *J. Chem. Phys.*, **48**, 2049 (1968).
15 H. Hori and H. Morimoto, *Biochim. Biophys. Acta*, **200**, 581 (1970).
16 H. Kamimura and S. Mizuhashi, *J. Appl. Phys. Suppl.*, **39**, 684 (1968).
17 N. Nakano, J. Ōtsuka and A. Tasaki, *Biochim. Biophys. Acta*, **236**, 222 (1971).
18 W. A. Eaton and E. Charney, *J. Chem. Phys.*, **51**, 4502 (1969).
19 J. Ōtsuka, Y. Seno, N. Fuchigami and O. Matsuoka, unpublished.
20 G. Herzberg, "Molecular Spectra and Molecular Structure (I. Spectra of diatomic molecules)," D. Van Nostrand Co. Inc., New York (1950).
21 G. Herzberg, *Canad. J. Phys.*, **31**, 657 (1953).
22 G. Herzberg, *Canad. J. Phys.*, **30**, 185 (1952).
23 K. Yoshino and Y. Tanaka, *J. Chem. Phys.*, **48**, 4859 (1968).
24 W. Moffit, *Proc. Roy. Soc.*, **A210**, 245 (1951).
25 L. Pauling, *in* "Haemoglobin (Sir Joseph Barcroft Memorial Symposium)," ed. by F. J. V. Roughton and J. C. Kendrew, Butterworth, London, p. 57 (1949).
26 J. S. Griffith, *Proc. Roy. Soc.*, **A235**, 23 (1956).

27 S. Maričič, J. S. Leigh, Jr., and D. E. Sunko, *Nature*, **214**, 462 (1967).
28 H. C. Watson and C. L. Nobbs, " 19 Colloquim der Gesellschaft für Biologische Chemie," Springer-Verlag, Berlin-New York, p. 37 (1968).
29 C. E. Moore, "Atomic Energy Levels, Circular of the National Bureau of Standards," U. S. Government Printing Office, Washington, Vol. II, p. 49 (1949).
30 W. Heitler and F. London, *Z. Phys.*, **44**, 455 (1927).
31 J. Beetlestone and P. George, *Biochemistry*, **3**, 707 (1961).
32 P. George, J. Beetlestone and J. S. Griffith, *Rev. Mod. Phys.*, **36**, 441 (1964).
33 T. Iizuka and M. Kotani, *Biochim. Biophys. Acta*, **154**, 417 (1968).
34 T. Iizuka, M. Kotani and T. Yonetani, *Biochim. Biophys. Acta*, **167**, 257 (1968).
35 T. Iizuka and M. Kotani, *Biochim. Biophys. Acta*, **181**, 275 (1969).
36 J. Ōtsuka, *Biochim. Biophys. Acta*, **214**, 233 (1970).
37 J. C. Kendrew, *Science*, **139**, 1259 (1963).
38 M. F. Perutz, H. Muirhead, J. M. Cox, L. C. G. Goaman, F. S. Mathew, E. L. McGrady and L. E. Webbs, *Nature*, **219**, 29 (1968).
39 J. L. Hoard, *in* " Hemes and Hemoproteins," ed. by B. Chance, R. W. Estabrook and T. Yonetani, Academic Press, New York, p. 9 (1966).
40 M. P. Perutz, *Nature*, **288**, 726 (1970).
41 T. Yonetani, T. Iizuka, T. Asakura, J. Ōtsuka and M. Kotani, submitted to *J. Biol. Chem.*
42 T. Iizuka and M. Kotani, *Biochim. Biophys. Acta*, **194**, 351 (1969).

Received for publication July 12, 1971.

REACTION MECHANISMS OF FLAVIN-CONTAINING OXYGENASES

Shigeki Takemori, Kenzi Suzuki, Masayuki Katagiri and Takao Nakamura
Department of Chemistry, Faculty of Science, Kanazawa University, Ishikawa, and Department of Biology, Faculty of Science, Osaka University, Osaka

During recent years, the roles of flavins as " co-oxygenases " have received considerable attention in the study of biological oxidation (*1–3*). The first evidence that flavoprotein catalyzes the incorporation of molecular oxygen into an organic substance was provided in 1957 by Hayaishi and Sutton (*4*) who demonstrated by using heavy oxygen as a tracer that an FMN-containing flavoprotein, " lactate oxidative decarboxylase " from *Mycobacterium phlei*, was an oxygenase. In 1962 Katagiri *et al.* (*5–7*) obtained salicylate hydroxylase from *Pseudomonas putida* as the first purified and well characterized NAD(P)H oxygenase. The prosthetic group, a novel co-oxygenase, was identified as FAD. Since then an increasing number of oxygenases containing flavins as their prosthetic groups have been found in many metabolic reactions, and many of these have now been crystallized (*8–14*), permitting detailed analysis of the mechanism of their actions.

Some oxygenases require an external electron donor, such as reduced pyridine nucleotides, while other oxygenases consume internal hydrogen atoms of a substrate and do not require any additional reductant. On

TABLE I. Flavin-containing Oxygenases

Electron flow system	Product	Example	Prosthetic group
PN ⟶ FP ⟶ O_2	SO+H_2O	Salicylate hydroxylase	FAD
		p-Hydroxybenzoate hydroxylase	FAD
		Imidazoleacetate hydroxylase	FAD
PN ⟶ FP ⟶ O_2	$S(OH)_2$	3-Hydroxypyridine ring oxygenase	FAD
S ⟶ FP ⟶ O_2	SO+H_2O	L-Lactate oxygenase	FMN
		L-Lysine oxygenase	FAD
		L-Arginine oxygenase	FAD

PN, reduced pyridine nucleotides such as NADH or NADPH; FP, flavoprotein; S, substrate.

the basis of these characteristics, the over-all stoichiometry catalyzed by these oxygenases may be schematically represented by either of the following equations.

$$SH_2+O_2 \longrightarrow SO+H_2O \text{ or } S(OH)_2$$
$$AH_2+S+O_2 \longrightarrow A+SO+H_2O \text{ or } A+S(OH)_2,$$

where S and SH_2 denote the substrate molecule and AH_2 an external electron donor.

Table I summarizes various types of flavin-containing oxygenase. All the oxygenases listed in the table, except for arginine oxygenase, have so far been crystallized and shown to have characteristics of flavoproteins. In the studies on the mechanism of these oxygenase reactions interest is now focused on the role played by flavins in the " activation " of molecular oxygen coupled with the electron flow from substrate or external electron donor to oxygen.

Analyses of the action mechanisms of two representative types of these oxygenases, lactate oxygenase and salicylate hydroxylase, have been carried out in our laboratories (*15–21*). In this paper it is intended to present the functional characteristics of the flavin-containing oxygenases on the basis of observations made in these recent investigations.

Salicylate Hydroxylase

1. Molecular characteristics

Salicylate hydroxylase was purified from the cells of *Pseudomonas putida* which had been grown with salicylate as the sole source of carbon and energy. The purified enzyme was bright yellow in color and was obtained in a crystalline form (Fig. 1). This enzyme, which contains FAD as the prosthetic group, catalyzes in the presence of molecular oxygen, the oxidation of NADH resulting in decarboxylation-hydroxylation of salicylate to catechol as follows:

$$\text{C}_6\text{H}_4(\text{OH})(\text{COOH}) + \text{NADH} + \text{H}^+ + \text{O}_2 \longrightarrow \text{C}_6\text{H}_4(\text{OH})(\text{OH}) + \text{NAD}^+ + \text{H}_2\text{O} + \text{CO}_2.$$

No transition metal was detected in the purified preparation of the enzyme (*22*). Salicylate hydroxylase was reversibly separated into FAD and apoenzyme moieties by ammonium sulfate precipitation at low pH. Of the flavin derivatives tested, only FAD was able to restore the activity of the apoenzyme (*6*). In order to analyze the formation of the holoen-

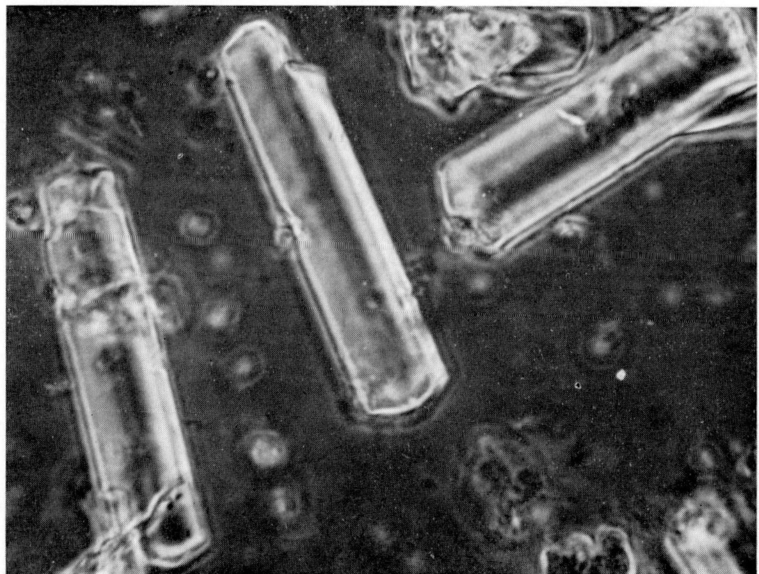

FIG. 1. Crystals of salicylate hydroxylase (*16*).

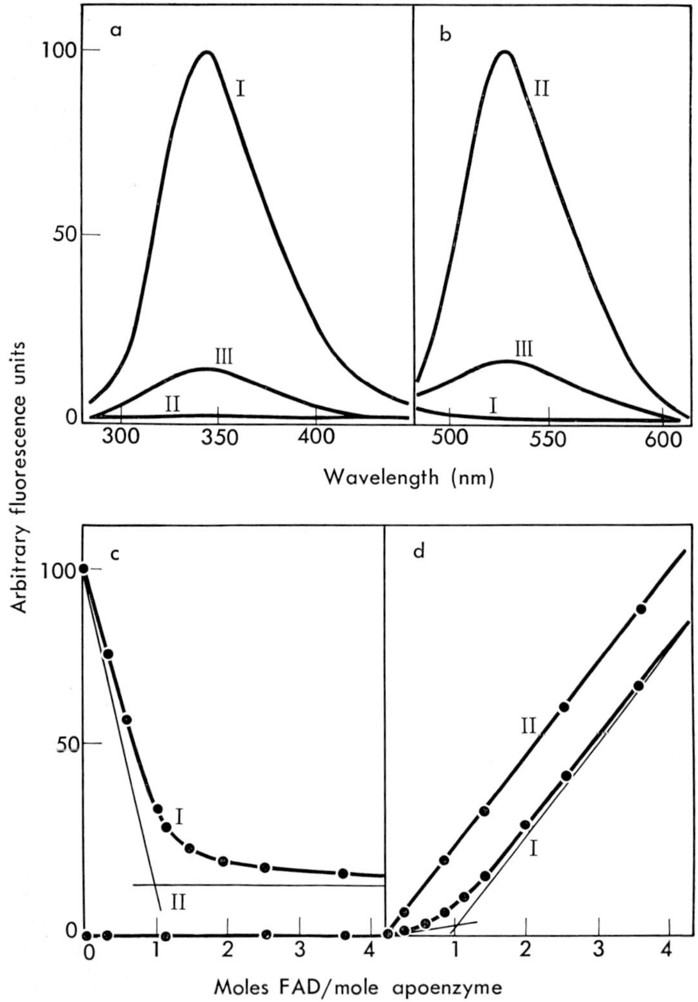

FIG. 2. Interaction between apo-salicylate hydroxylase and FAD (*18*). (a and b): Emission spectra of the apoenzyme, FAD and the mixture activated at 292 nm(a) and at 450 nm(b). I, 7.0 μM apoenzyme; II, 7.4 μM FAD; III, apoenzyme+FAD. (c and d): Fluorometric titration of the apoenzyme with FAD. Fluorescence activated at 292 nm was measured at 342 nm(c) and at 525 nm(d), respectively. Titrations were carried out with 0.5 μM apoenzyme (solid circles in Curve I). The thick line in Curve I was calculated theoretically by using the equation: [apoenzyme] [FAD]/[holoenzyme]=45 nM. Curve II shows the changes of fluorescence caused by FAD alone.

zyme from the apoenzyme and FAD, the fluorescent behavior of FAD and the aromatic amino acid residues of salicylate hydroxylase were examined (*18*). As shown in Figs. 2-a and -b, when FAD was added to the apoenzyme solution, the protein fluorescence at 342 nm markedly decreased. The FAD fluorescence at 525 nm was also quenched on adding the apoenzyme. It was graphically deduced from the titration curves of Figs. 2 -c and -d that one mole of FAD was bound to one mole of the apoenzyme. This value was also in accordance with the result of equilibrium dialysis (*6*). The dissociation constant of FAD in the holoenzyme was calculated to be 45 nM, which agreed with the K_m value for FAD determined from the changes of hydroxylase activity with variation of FAD and apoenzyme concentrations in the assay system. The molecular weight was estimated to be 57,000 on the basis of the sedimentation constant of 3.40 S, the diffusion constant of 5.80×10^{-7} cm^2 per sec and the partial specific volume of 0.75 ml/g (*6*). This value was in agreement with a minimum molecular weight per flavin calculated from protein concentration and 450-nm absorbance. In the presence of 5 M guanidine hydrochloride-0.1 M mercaptoethanol, the enzyme was not decomposed into smaller subunits (*23*). Thus, salicylate hydroxylase exists as a monodisperse monomer containing one molecule of FAD. Recently, interesting results concerning the structure of a salicylate hydroxylase from a soil bacterium were reported by White-Stevens and Kamin (*24*). Their enzyme has a molecular weight of 91,000 and consists of two flavins and two subunits of similar size.

2. *Formation of the enzyme-substrate complex*

Optical absorption spectra of salicylate hydroxylase were of great practical usefulness to demonstrate the formation of the enzyme-substrate complex and its enzymatic function (*16*). The visible spectrum of the holoenzyme was very similar to that of free FAD except that the absorption intensity of bound FAD was slightly lower. On adding salicylate to the holoenzyme, the absorption peaks at 450 and 375 nm were shifted to 455 and 385 nm, respectively, and a marked shoulder appeared at around 480 nm (Fig. 3). Similar spectral shifts were also brought about by other substrates for the hydroxylase reaction, namely aromatic compounds such as 2,3-, 2,4-, 2,5- and 2,6-dihydroxybenzoates, *p*-aminosalicylate, 1-hydroxy-2-naphthoate and 3-methylsalicylate. Since the complex formation was manifested by the absorption increase at the shoulder at

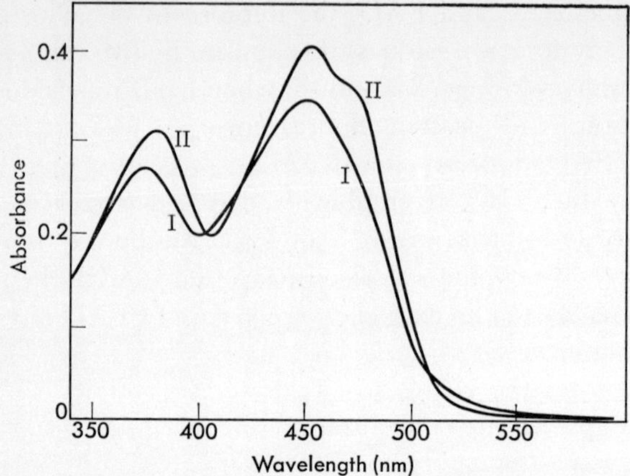

FIG. 3. Optical absorption spectra of salicylate hydroxylase (I) and its complex with salicylate (II) (*15*). Holoenzyme, 31 μM; salicylate, 0.3 mM. Measured in 33 mM potassium phosphate buffer (pH 7.0).

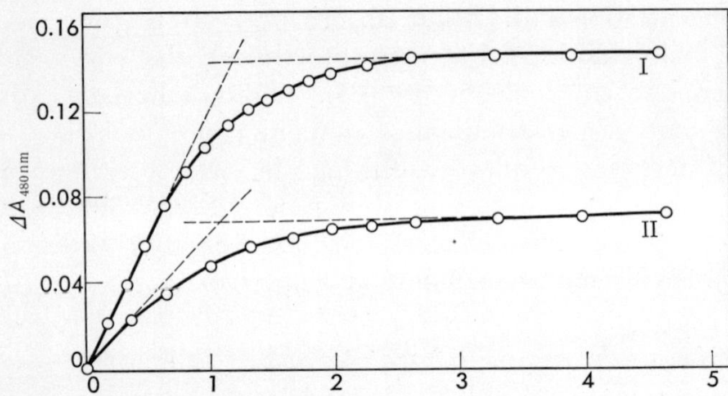

FIG. 4. Spectrophotometric titration of salicylate hydroxylase with salicylate (*16*). Holoenzyme, 153 nmoles (Curve I) or 76 nmoles (Curve II). Measured in 3.0 ml of 33 mM potassium phosphate buffer (pH 7.0).

480 nm, the stoichiometry was determined by titration of the holoenzyme with salicylate. This is illustrated in Fig. 4 where the increase in absorbance at 480 nm is plotted against the amount of salicylate added. It is obvious that the initial slope of the titration curve intercepts the maximum value observed at 1.0 mole salicylate per mole enzyme-bound FAD. From the results of Fig. 4, the dissociation constant of salicylate in its complex with the holoenzyme was calculated to be 3.5 μM.

The formation of a complex between salicylate hydroxylase and the substrate was evidenced by various facts. The enzyme-substrate complex formation was confirmed by a fluorometric analysis, in which the enzyme combined specifically with the substrate to form a fluorescent complex; the molar ratio of the components in the complex was 1: 1 (*18*). The circular dichroism spectrum of salicylate hydroxylase differed somewhat from that in the enzyme-substrate complex (*25*). The holoenzyme showed a negative CD centered at approximately 450 nm and a positive band at 375 nm. Addition of salicylate to the holoenzyme resulted in an

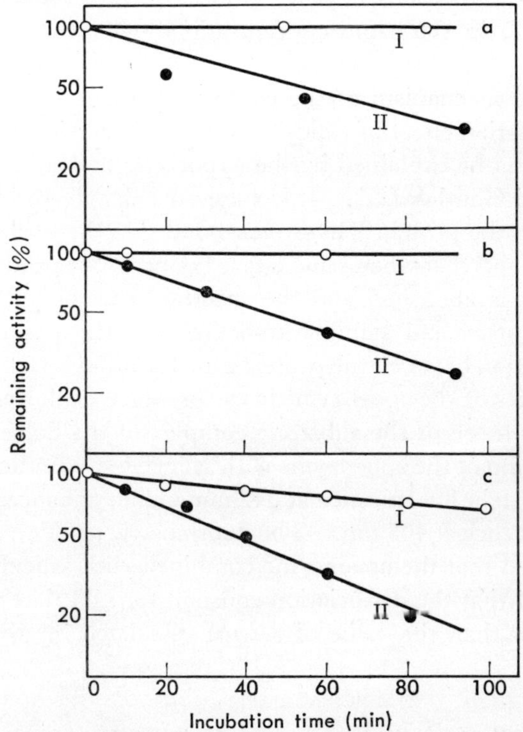

FIG. 5. Effect of salicylate on acid-(a), heat-(b) and Nagarse-(c) inactivations of salicylate hydroxylase (*16*). (a) 30 mM sodium acetate buffer (pH 4.2) at 20°C; (b) 30 mM potassium phosphate buffer (pH 7.0) at 41°C; (c) 16 mM Tris-HCl buffer (pH 8.0) in the presence of Nagarse (3.3 μg/ml) at 29°C. Curve I, with 400 μM salicylate; Curve II, without salicylate. Holoenzyme, 2.9 μM.

enhancement of the 450-nm CD band, together with a decrease in the 375-nm band.

The protective effect of the substrate against the stability of salicylate hydroxylase was additional evidence that the holoenzyme combined with salicylate to form the substrate complex (*16*). The enzyme-substrate complex was much more stable than the holoenzyme against various denaturing conditions. The stability against acid and heat treatments was remarkably enhanced by the complex formation (Figs. 5-a and b) as well as by the proteinase treatment (Fig. 5-c). Similar stabilizing effects were also brought about by incubation with other substrates such as 2,5-dihydroxybenzoate or *p*-aminosalicylate, while non-substrates such as catechol or benzoate, in the same concentrations, gave no detectable protection.

Details of the real mechanism of the complex formation remain obscure. The perturbation effect of salicylate on the absorption spectrum of the holoenzyme can be explained by the hypothesis of Harbury *et al.* (*26*) and Massey and Ganther (*27*). It is suggested that in the holoenzyme a group within the protein is hydrogen-bonded to the isoalloxazine moiety of the flavin, and that when salicylate is bound to the protein moiety, this interaction is abolished and the spectral perturbation is produced. The fluorimetry and stability tests (*16*, *18*) also point to the possibility of interaction between the protein moiety and salicylate. The stability characteristics of the apoenzyme in the presence of the substrate were very similar to those of the substrate complex of the holoenzyme. The complex formation of the apoenzyme with salicylate was indicated by quenching of the protein fluorescence at 342 nm and an enhancement of the salicylate fluorescence at 405 nm. The fluorometric titration at these wavelengths indicated that the apoenzyme combined with salicylate at a molar ratio of 1, and that the dissociation constant for salicylate was 1.8 μM, somewhat lower than the value of 3.2 μM calculated for the holoenzyme complex.

3. *Reactivity of the flavin in the enzyme-substrate complex*

In most of the oxygenases, which contain flavin as the prosthetic group and require the reduced pyridine nucleotide as the source of an external electron donor, the electron transfer from the donor to flavin and flavin to molecular oxygen can be demonstrated by the observation of the flavin absorption spectrum. The interaction of the flavin moiety with NADH

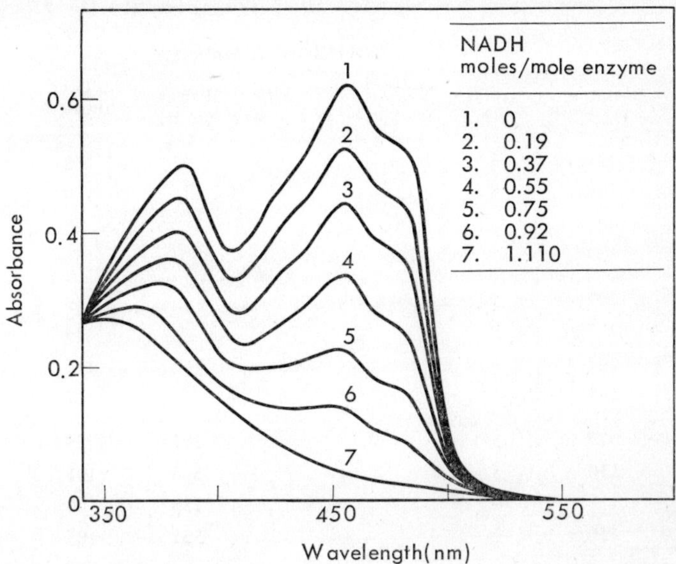

FIG. 6. Reductive titration of the enzyme-salicylate complex with NADH (*16*). Holoenzyme, 47 μM; salicylate, 1.7 mM. Measured with the use of a Thunberg-type cuvette in 33 mM potassium phosphate buffer (pH 7.0).

was first studied (*16*) to determine whether or not the flavin moiety serves as the most important functional part of the electron transfer reaction. Figure 6 shows the reductive titration of the enzyme-substrate complex with NADH. Upon addition of appropriate amounts of NADH to a solution of the complex, the absorption over the whole spectral range diminished instantaneously. During the course of titration the stoichiometric relation between added NADH and formed $FADH_2$ was demonstrated. When the holoenzyme, without salicylate, was also reduced with NADH, the situation was quite similar to that of its substrate complex; there was no long wavelength absorption or any other spectrally distinct intermediate which could be attributed to a one-electron reduced form of the enzyme-flavin. Upon introduction of air into the reduced form of the complex, the reduced flavin moiety was rapidly reoxidized, and the absorption spectrum of the oxidized form of the enzyme reappeared. Under these conditions, the reaction product, catechol, was stoichiometrically produced in the reaction mixture (Table II, I-A and-B). When the holoenzyme was reduced with NADH in the

TABLE II. Stoichiometry in the Salicylate Hydroxylase Reaction (*16, 17*)

Expt No.	Enzymes (nmoles)	Salicylate (μmoles)	External reductants: NADH (nmoles)	External reductants: $Na_2S_2O_4$ (nmoles)	External reductants: Light (min)	Enzymes reduced (nmoles)	Catechol formed (nmoles)
I–A	151	5	80			82	78
I–B	149	5	123			130	125
I–C[a]	140	—	82			90	74
I–D[b]	153	—	210			153	0
II–A	147	5		205		149	174
II–B	147	5		76		79	64
II–C[a]	147	—		103		115	107
II–D[b]	147	—		206		147	0
III–A	77	5			181	58	41
III–B	110	5			568	101	98
III–C[a]	110	—			138	60	48
III–D[b]	110	—			551	105	0

Each reaction was carried out in a Thunberg-type cuvette with a 3-ml assay system containing 100 μmoles of potassium phosphate buffer (pH 7.0) and the components indicated. Illumination was performed with the use of a 100-W tungsten lamp in the presence of 150 μmoles of EDTA.

[a] After the enzyme was reduced, 5 μmoles of salicylate were added from the side arm, and the reaction mixture was then exposed to air. [b] The reduced enzyme was exposed to air before the addition of salicylate.

absence of salicylate and was then allowed to react with air after addition of salicylate, a stoichiometric amount of catechol was also detected as observed in the complex (Table II, I-C). The titration data given above indicate that the enzyme-substrate complex is stoichiometrically converted by NADH to a stable two-electron reduced intermediate, and this reduced species of the enzyme is an active intermediate capable of supporting the aerobic hydroxylation of the substrate bound to the enzyme. These findings are consistent with the view that the enzyme-bound $FADH_2$ formed by NADH serves not only as an ultimate electron donor in the hydroxylation reaction, but also functions for the activation of molecular oxygen. This hypothesis could be further strengthened since the non-specific reducing system other than NADH led to the same results (*17*). Figure 7-a shows the results of anaerobic titration of salicylate hydroxylase with dithionite in the presence of salicylate. Reductive

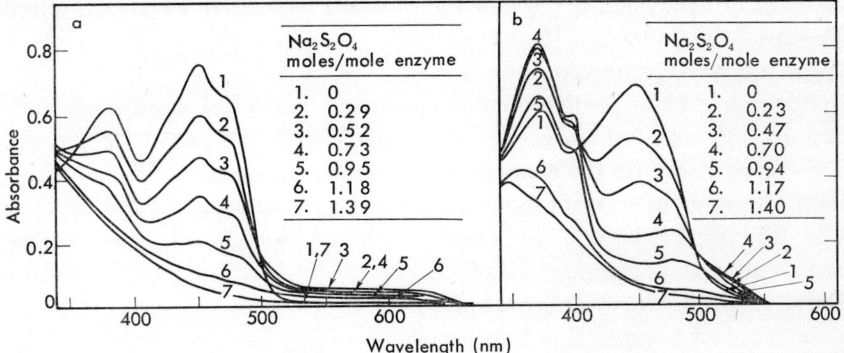

FIG. 7. Sodium dithionite titration of salicylate hydroxylase in the presence (a) and absence (b) of 1.7 mM salicylate (*17*). Holoenzyme, 49 μM. Measured using a Thunberg-type cuvette in 33 mM potassium phosphate buffer (pH 7.0).

titration by dithionite led to formation of an intermediate with a long wavelength absorption at above 500 nm. The full reduction with complete elimination of the long wavelength absorption was obtained with one mole of dithionite per mole of the bound-flavin. On admitting air into the reaction mixture, the reduced complex was rapidly reoxidized, and one mole of the hydroxylated product, catechol, was produced from one mole of the reduced complex (Table II, II-A and-B). In the absence of salicylate, the reductive titration of salicylate hydroxylase with dithionite gave a reddish intermediate, whose absorption maxima were situated at 375, 400 and 480 nm as illustrated in Fig. 7-b. Finally, the complete reduction of the enzyme was obtained with one mole of dithionite per mole flavin. When salicylate was added to the reduced holoenzyme anaerobically and the reaction mixture was then exposed to air, the product, catechol, was formed in an amount stoichiometric to the reduced flavin (Table II, II-C).

Spectral changes closely similar to those observed in dithionite reduction were obtained in light irradiation of salicylate hydroxylase. On illumination in the presence of EDTA, a long wavelength-absorbing intermediate was seen with the enzyme in the presence of salicylate and prolonged illumination gave a fully reduced spectrum. Under these conditions, catechol in an amount stoichiometric with the reduced flavin was also produced upon addition of air (Table II, III-A and-B). In the absence of salicylate, a reddish intermediate appeared which exhibited absorption maxima at 375, 400 and 480 nm, and finally a fully reduced spectrum was produced. After anaerobic addition of salicylate, the reoxidation of the reduced enzyme with air produced a stoichiometric amount of catechol as observed in the complex (Table II, III-C). Therefore, based on the experimental data described above, the stoichiometry among the enzyme-bound flavin, the reductant, and the product was well established. The reduced complex produced, whether the reductant was dithionite, EDTA-light or NADH, was enzymatically active in the hydroxylation reaction.

It should be noted that two spectroscopically distinguishable species are produced depending on the presence or absence of salicylate during the reductive titration of the enzyme with a nonspecific reductant. The reddish intermediate in the absence of salicylate gave a typical $g=2.00$ semiquinone EPR signal. Recently, Massey and Palmer (*28*) demonstrated two different spectral types of flavoprotein semiquinone which they called " red " and " blue." On the basis of the data on both absorption and EPR spectra, the reddish intermediate of salicylate hydroxylase appeared to be identical with the red radical described by them. The reddish intermediate was easily converted into the long wavelength-absorbing intermediate by mixing with salicylate under anaerobic conditions. However, the latter intermediate gave no EPR signal. The long wavelength-absorbing intermediate arising in the presence of salicylate is somewhat more difficult to interpret. From the EPR results this appeared to be due not to the blue radical, although it had a similar absorption spectrum. More detailed evidence is required to elucidate the nature of this species.

4. *Activation effect of salicylate on reaction of NADH with holoenzyme*

NADH at relatively low concentrations was not oxidized to an appreciable extent in the absence of salicylate. However, when the concentration of NADH was raised, salicylate hydroxylase showed an NADH oxidase activity even in the absence of salicylate. Under these conditions, the K_m value of the holoenzyme for NADH was estimated to be 3.2 mM (*29*). In the presence of salicylate, the K_m value for NADH in the salicylate hydroxylase reaction was determined to be 2.6 μM (*6*). As will be described in the following section more in detail, the rate of reduction of the enzyme flavin by NADH was remarkably increased by the presence of salicylate (*cf.* Section 5-E). These results strongly suggest that the binding of salicylate to salicylate hydroxylase facilitates both the binding of NADH and the reduction of the enzyme flavin by NADH.

5. *Kinetic analysis of the salicylate hydroxylase reaction*

Since salicylate hydroxylase interacts with three kinds of substrates, *i.e.*, salicylate, NADH and oxygen, there are *a priori* a total of six possible sequences by which hydroxylation can be accomplished through supposed intermediates. The hypothetical reaction mechanism is illustrated in Fig. 8, in which intermediate complexes conceivably produced during

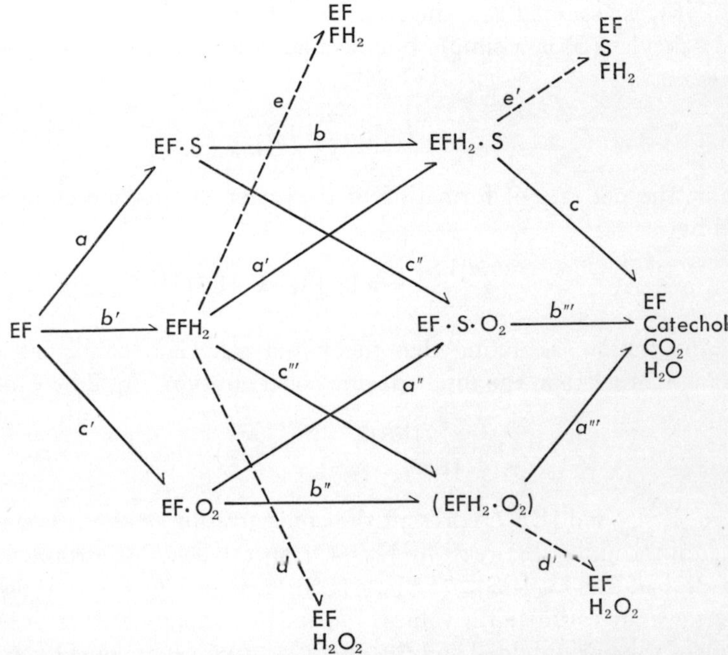

FIG. 8. Representation of possible intermediates during salicylate hydroxylase reaction. EF and S denote salicylate hydroxylase and salicylate, respectively. EF·S and EFH_2·S denote the oxidized and reduced enzyme-substrate complexes, respectively. Other complexes are also defined in similar ways. Steps *d*, *d′* and *e*, *e′* represent side reactions; either "oxidase reaction" or "diaphorase reaction" in the presence of externally added FAD.

the reactions are included. As described in the preceding section, a complex of the holoenzyme with salicylate has been identified as the predominant species. Based on the stoichiometric relation demonstrated by using substrate level amounts of the enzyme, it is suggested that the reduction-reoxidation of this complex leads to hydroxylation of the bound salicylate. This hypothetical reaction mechanism illustrated as steps *a*, *b* and *c* in Fig. 8 could be more completely substantiated by rapid and sensitive kinetic measurements (*19*, *30*). The kinetics of the salicylate hydroxylase reaction was measured in a temperature-controlled flow system which was essentially the same as that designed by Chance and Legallais (*31*).

A. *Salicylate binding with oxidized and reduced holoenzymes* (*steps a and a′*): If it is assumed that the reaction between salicylate hydroxylase(E) and salicylate(S) is a simple bimolecular reaction, it is given by the expression

$$E+S \underset{k_{-s}}{\overset{k_s}{\rightleftharpoons}} ES \tag{1}$$

Thus, the net rate of formation of the enzyme-salicylate complex (ES) will be

$$\frac{d[ES]}{dt} = k_s[E][S]-k_{-s}[ES] \tag{2}$$

On integration, assuming that the initial salicylate concentration $[S]_0$ is much larger than the total enzyme concentration, Eq. 2 becomes

$$2.3 \log \frac{[ES]_{eq}}{[ES]_{eq}-[ES]_t} = (k_s[S]_0+k_{-s})t \tag{3}$$

Here $[ES]_{eq}$ and $[ES]_t$ represent the concentration of the ES complex at an equilibrium and at reaction time t, respectively. According to Eq. 3, the plot of 2.3 $\log [ES]_{eq}/[ES]_{eq}-[ES]_t$ against t yields a straight line. Thus, we can estimate the values of k_s and k_{-s} from the values of both the slope of the line obtained and the dissociation constant, $K_s=k_{-s}/k_s$, of the ES complex. The experiments were carried out following the increase of absorbance at 480 nm due to the formation of the enzyme-salicylate complex. Figure 9-a shows a typical example of a flow trace for measuring the rate constant. Combining the value of the slope obtained by Eq. 3 and a dissociation constant of 3.5 μM, the k_s value of salicylate binding was calculated to be 1.8×10^7 M^{-1} sec^{-1}, and k_{-s}, 62 sec^{-1}.

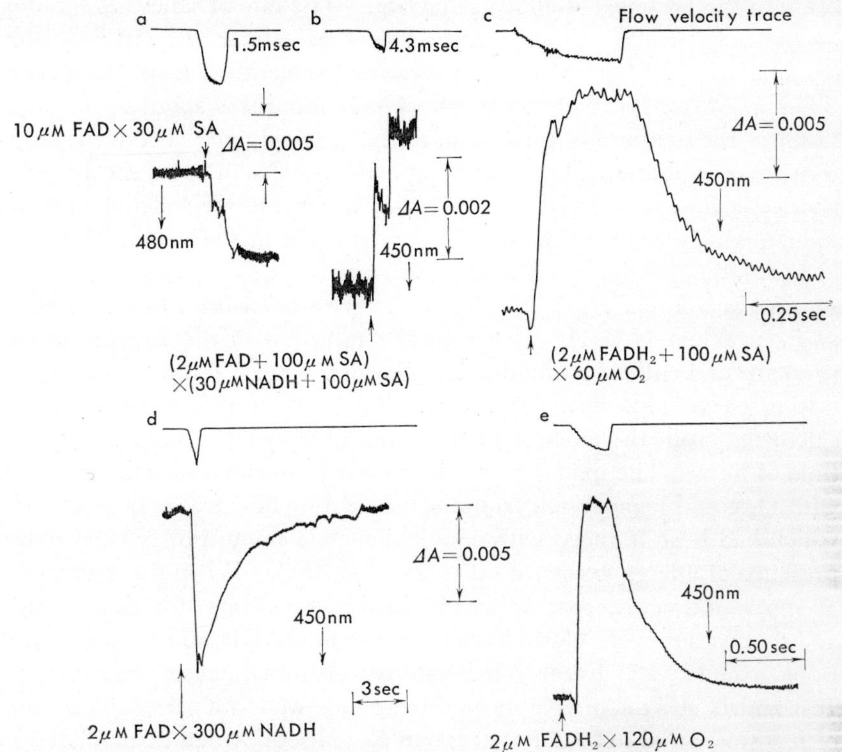

FIG. 9. Flow kinetic measurements of salicylate hydroxylase reaction (*19, 30*). Measured in 33 mM potassium phosphate buffer (pH 7.0) at 25°C. (a) Changes in absorbance at 480 nm on mixing salicylate hydroxylase and salicylate. (b and c) Time course of reduction and reoxidation of the enzyme-salicylate complex. (d and e) Time course of reduction and reoxidation of the enzyme in the absence of salicylate.

Very recently, kinetic experiments on the salicylate binding to the holoenzyme in the reduced form were performed (*32*). The data obtained, though still incomplete, seem to allow us to compare the relevant state of affairs with that of the oxidized enzyme. No spectral change in absorption in the visible range was observed with reduced holoenzyme when salicylate was added. However, a rapid partial quenching of fluorescence of salicylate with a maximum at 405 nm caused by the addition of the reduced holoenzyme was detectable under anaerobic conditions. A fluorescence-quenching experiment with salicylate gave a rate of 0.29

sec^{-1} in the presence of 30 μM salicylate. The rate of salicylate binding by the holoenzyme in the oxidized form was estimated from the value of k_s to be 540 sec^{-1} under the same concentration of salicylate. Thus, the extent of the rate of salicylate binding was about 1,900 times faster in the oxidized enzyme than in the reduced one. From the titration curve in fluorescence-quenching experiments, an apparent dissociation constant for salicylate in the reduced enzyme was calculated to be 17 μM, which was considerably larger than that in the oxidized enzyme (3.5 μM).

B. Reduction and reoxidation of the enzyme-salicylate complex (steps b and c): When NADH was aerobically mixed with the enzyme in the presence of a sufficient amount of salicylate, a rapid reduction occurred during continuous flow (Fig. 9-b). The rate of reduction, k_{red}, was calculated from the amount of the reduced enzyme formed during the time of flow. The oxidation of the reduced enzyme-salicylate complex with O_2 could be analyzed from the reoxidation flow trace of the enzyme which had been reduced with a stoichiometric amount of NADH in the presence of a large excess of salicylate (Fig. 9-c). When the reciprocals of apparent first order rates thus obtained for reactions of both reduction and reoxidation were plotted against those of NADH and O_2 concentrations, respectively, linear relationships were obtained as seen in Figs. 10-a and-b, and the intercepts on the ordinate were not zero. The value of k_{red} was found to be 230 sec^{-1}, and k_{ox}, 21 sec^{-1} when extrapolated to infinite NADH and O_2 concentrations, respectively. In these cases, the following two-step reactions *via* an intermediate X_1 or X_2 are considered to be involved.

$$EF.S + NADH + H^+ \longrightarrow X_1 \xrightarrow{230\ sec^{-1}} EFH_2.S + NAD^+$$

$$EFH_2.S + O_2 \longrightarrow X_2 \xrightarrow{21\ sec^{-1}} EF + Catechol + CO_2 + H_2O$$

Although we proposed the reaction intermediate as a kinetic assumption, no intermediate stage other than the steady transition stage from the oxidized to the fully reduced form was observed. It should be noted regarding *p*-hydroxybenzoate hydroxylase reported by Howell and Massey (*33*) that spectroscopically observable intermediates are formed in the course of the complete reduction of the enzyme by NADPH in the presence of *p*-hydroxybenzoate.

C. Reduction and reoxidation of the holoenzyme in the absence of salicylate

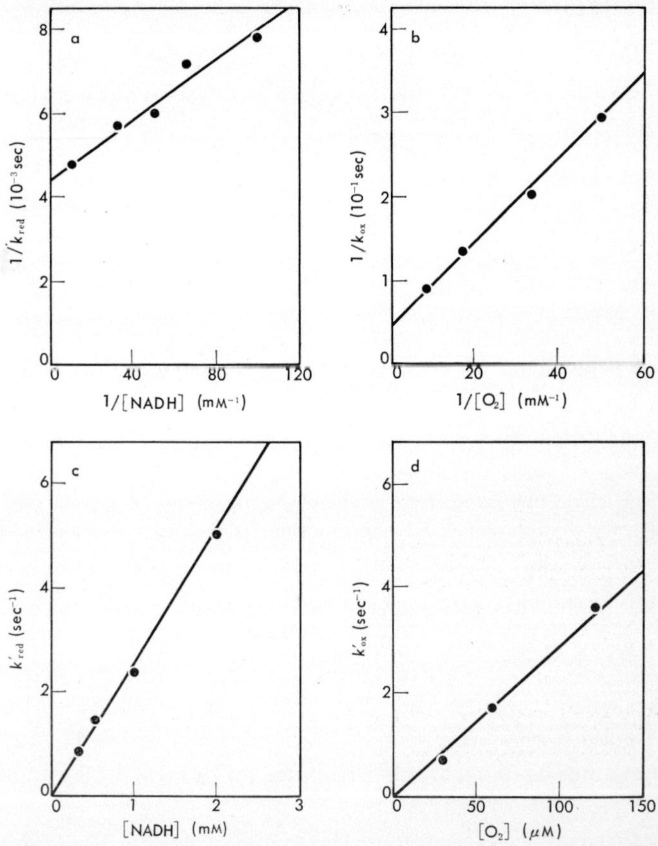

FIG. 10. Lineweaver-Burk plot of flow kinetic data (*29*). a and b, in the presence of 100 μM salicylate; c and d, in the absence of salicylate. Conditions were the same as described for Fig. 9 except that varying concentrations of NADH and oxygen were used.

(*steps b′ and c‴, d′*): The reduction of the FAD moiety of the enzyme was analyzed by mixing anaerobically NADH with the enzyme in the absence of salicylate. The reoxidation of the $FADH_2$ moiety was analyzed by mixing O_2 with the enzyme which had been reduced by prior addition of a stoichiometric amount of NADH. The values of k'_{red} and k'_{ox} in the reduction-reoxidation of the FAD moiety were computed from the flow traces shown in Figs. 9-d and -e. Figures 10-c and -d show reciprocal plots showing the dependence of the rates of reduction and

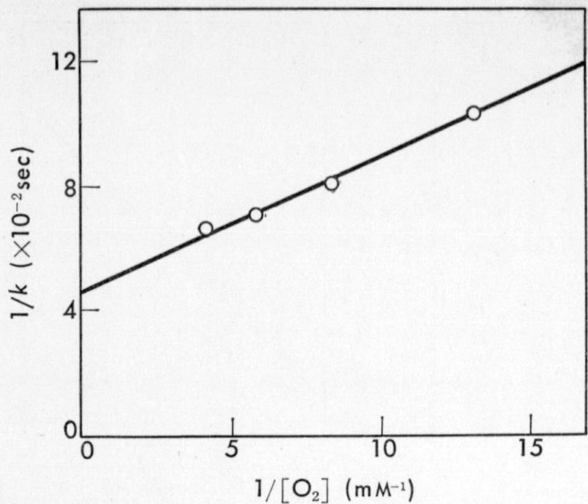

FIG. 11. Plot of $1/k$ against $1/[O_2]$ (*29*). Salicylate, 100 μM. Measured in 33 mM potassium phosphate buffer (pH 7.0) at 25°C with a Clark oxygen electrode. The value of k stands for the maximum rate obtained by extrapolating to infinite NADH concentration at various fixed concentrations of oxygen.

reoxidation on the concentrations of NADH and O_2 ,respectively. Both plots gave straight lines, with intercepts at the zero point on the ordinate. The second-order rate constant of the reaction was calculated to be $k'_{red}=2.5\times10^3\,M^{-1}\,sec^{-1}$, k'_{ox}, $2.9\times10^4\,M^{-1}\,sec^{-1}$, showing that the reduction-reoxidation reactions of the holoenzyme in the absence of salicylate are one-step reactions involving on intermediate unlike in the case of the presence of salicylate.

D. Overall reaction: The rate of the overall reaction was measured by oxygen uptake at various concentrations of NADH and oxygen in the presence of a large excess of salicylate. On the basis of the Lineweaver-Burk plot for NADH concentration at various fixed concentrations of oxygen, the maximum values of the rate extrapolated to infinite NADH concentration were obtained. The reciprocal values of the rates thus obtained were replotted against those of oxygen concentrations (Fig. 11). From the intercept, the maximum value of the molecular activity to be obtained at infinite NADH and oxygen concentrations was estimated to be 21 sec^{-1} at 25°C.

E. Reaction sequence in salicylate hydroxylase reaction: When the reduction rate of the flavin moiety of the holoenzyme was compared with that of the enzyme-salicylate complex, the values of the free enzyme and the complex were 0.08 and 180 sec^{-1}, respectively, with 30 μM NADH concentration in the enzyme assay. Thus the rate of reduction of the flavin moiety is more than 2,000 times larger in the presence of salicylate than in its absence. The rate of salicylate binding to the oxidized holoenzyme was far larger than that with the reduced holoenzyme (*cf.* Section 5-A). Since the molecular activity for the salicylate hydroxylase reaction is 21 sec^{-1}, the results described here lead to the conclusion that step *c* with k_{ox}= 21 sec^{-1} is the rate-limiting step in the overall reaction, and steps b' and a' do not essentially participate in the hydroxylation reaction. The other alternatives are sequences involving steps c', c'' or c'''. The rates of binding of substrate to the holoenzyme and the reduction of flavin moiety of the enzyme-salicylate complex were essentially the same, either in the presence of oxygen or in its absence. These observations seem to be at variance with the involvement of these steps in the hydroxylation reaction. At present, all pieces of experimental evidence favor the hypothesis that steps *a*, *b*, and *c* should be predominant in the overall catalytic process of salicylate hydroxylase reaction. In this proposed mechanism, it should be noted that the binding of the substrate to the enzyme in the reaction sequence causes an enhancement of the affinity for NADH in the protein.

Lactate Oxygenase

1. Preparation

Lactate oxygenase was first crystallized from cells of *Mycobacterium phlei* by Sutton in 1957 (*34*). In 1968 Takemori *et al.* (*20*) devised an improved method for purifying the enzyme to obtain crystalline preparations with an excellent yield. Lactate oxygenase could be extracted with water from cells ground mechanically with aluminum oxide. The extracted enzyme was concentrated by salting out and was readily crystallized by adding ammonium sulfate from the enzyme solution of 0.1 M phosphate buffer, pH 6.0. Electrophoretically and ultracentrifugally homogeneous crystalline preparations were obtained after several recrystallizations (Fig. 12). The yield from 200 g cells was about 150 to 200 mg of four-times crystallized enzyme. When stored as a crystal suspension, the enzyme was practically indefinitely stable.

2. Molecular properties

Lactate oxygenase, which is a typical flavoprotein with FMN as its

FIG. 12. Crystals of lactate oxygenase (*20*).

prosthetic group, catalyzes the reaction: $CH_3CHOHCOOH+O_2 \longrightarrow CH_3COOH+CO_2+H_2O$. FMN was released from lactate oxygenase upon treatment of the enzyme with acidic ammonium sulfate. The activity of the apoenzyme could be restored by addition of FMN and riboflavin. FAD had no effect. No transition metal was detected in the crystalline preparations of the enzymes. A very small amount of sugar (hexose, $0.158 \pm 0.004\%$ and hexosamine, $0.120 \pm 0.005\%$) was found in the enzyme preparation. However, it is not clear whether or not the trace sugars are genuine components of the enzyme (*20*).

The visible spectrum of the enzyme has absorption peaks at 375 and 454 nm, and marked shoulders occur at 430 and 480 nm. The enzyme catalyzes the oxidation of only L-lactate (K_m, 2.5×10^{-2} M) and α-hydroxy-*n*-butyrate, and is inhibited competitively by D-lactate (K_i, 5.2×10^{-3} M) (*20*).

The molecular weight values of lactate oxygenase determined by various methods are summarized in Table III. The minimum molecular weight from the analysis of flavin and protein content was 56,000. Quantitative N-terminal amino acid analysis by the DNP method showed the presence of about 1 mole of serine/56,000 g of protein. These re-

TABLE III. Summary of Molecular Weight Determinations of Lactate Oxygenase (*20, 35*)

Methods	Molecular weight
Sedimentation and diffusion	370,000
Archibald method	352,000
Gel filtration	340,000
Electron microscopic analysis (N×V×ρ)†	58,600×6 (352,000)
Flavin analysis	56,000×6 (336,000)

† N, Avogadro's number; V, calculated volume of the subunit sphere by using r= 25.6Å; and ρ, density of the enzyme, 1.39.

TABLE IV. Molecular Weights of the Subunits Obtained from Lactate Oxygenase (*35*)

Lactate oxygenase	$s^0_{20,w}$	$D^0_{20,w}$	Molecular weight
5 M Guanidine-0.1 M mercapto-ethanol-treated	2.25 S	3.55×10^{-7}	57,100
Carboxymethylated	2.44 S	4.10×10^{-7}	53,600

sults indicate that lactate oxygenase exists in an oligomer structure composed of six subunits (*35*). The bonds responsible for holding these subunits together appear to be sufficiently strong, *i.e.*, the oligomer does not dissociate into subunits under a wide range of conditions such as in the presence of 8 M urea-0.1 M mercaptoethanol. The enzyme was dissociated into smaller units by 5 M guanidine hydrochloride-0.1 M mercaptoethanol. A sedimentation velocity experiment performed in this solvent revealed a single schlieren peak. The preparation which had been carboxymethylated in the presence of 5 M guanidine hydrochloride -0.1 M mercaptoethanol was also homogeneous ultracentrifugally, and only one protein band was observed when disc electrophoresis was performed on the dissociated enzyme at pH 9.5 in 8 M urea. As summarized in Table IV, the value of the subunit molecular weight was calculated to be 54,000 to 57,000, which was in good agreement with that of the minimum molecular weight obtained from flavin content.

Electron microscope studies of lactate oxygenase, which were performed in colaboration with Prof. T. Oda of Okayama University, provided further evidence for the presence of six subunits in the molecule of

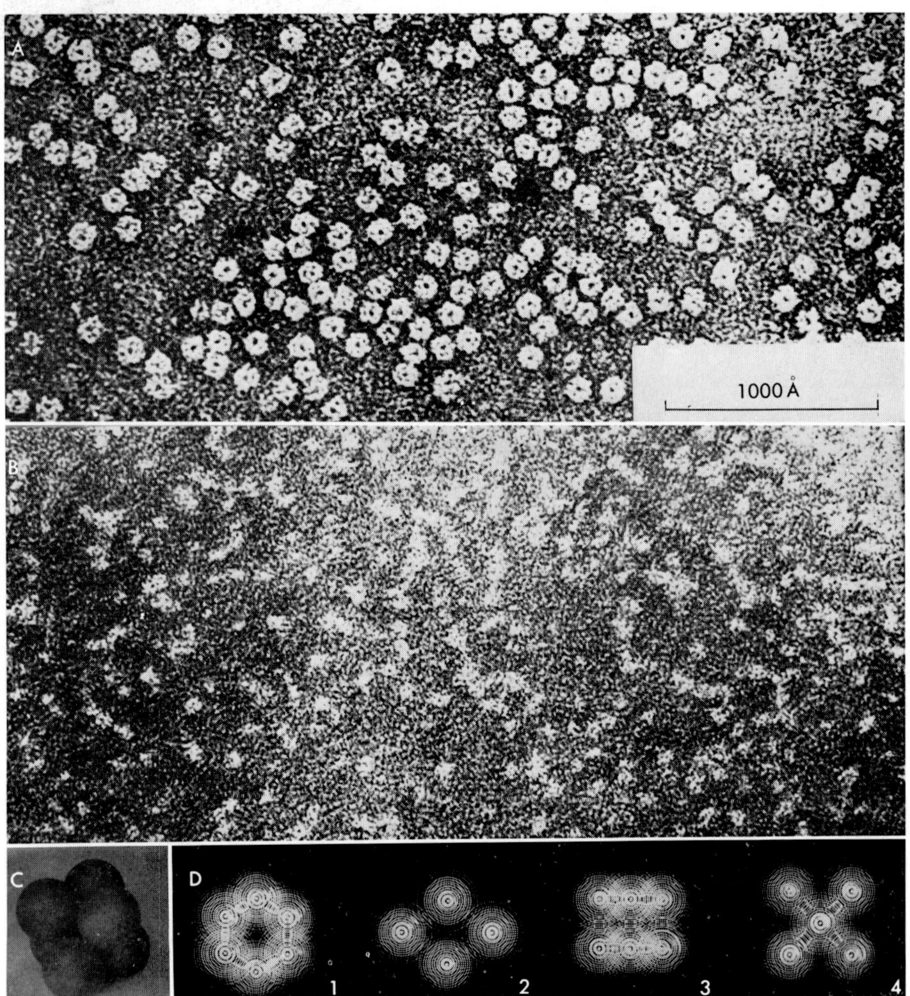

FIG. 13. Electron micrographs of lactate oxygenase negatively stained with uranyl acetate, pH 4.3 (*35*). (A) native enzyme; (B) 5 M guanidine-treated enzyme; (C) the hexameric model of lactate oxygenase; (D) the superimposed appearance of a structural model of the enzyme molecule viewed perpendicularly at four typical orientations: 1, three point attachment; 2, two point attachment; 3, one point attachment; and 4, one point attachment on the supporting film, respectively. The rough images of density distributions are represented by the rings drawn in the circles.

this enzyme (*35*). Typical negatively stained electron microscopic images of the lactate oxygenase molecules are shown in Fig. 13-A. A tentative model of the molecule (Fig. 13-C) is presented on the basis of the assumption that the enzyme molecule is composed of six identical protomers, each of which occupies exactly equivalent positions at the corners of an octahedron. The drawings in Fig. 13-D represent the superimposed appearances of the model viewed from four typical directions. As shown in Table III, the value of the molecular weight estimated from the electron microscopic analysis was consistent with the values obtained from other measurements. Figure 13-B is an electron micrograph taken after the enzyme was incubated in the presence of 5 M guanidine hydrochloride-0.1 M mercaptoethanol. It is evident that dissociation of the subunits is complete under these conditions. The dissociated subunits appear as random structures on the granular background.

3. *Reaction mechanism*

In order to elucidate the reaction mechanism of lactate oxygenase, in which lactate functions not only as substrate but also as internal electron donor, it may be worthwhile to study the stoichiometry of the reaction under aerobic or anaerobic conditions. As shown in Table V, when a substrate amount of lactate oxygenase was incubated aerobically with a limited amount of L-lactate, practically all the substrate was converted stoichiometrically to acetate. When a substrate amount of the enzyme was titrated anaerobically with varying amounts of L-lactate by using a Thunberg-type cuvette, the absorption over the whole spectral range diminished instaneously. During the reductive titration of the enzyme, no stable semiquinone-like intermediate could be detected. The full reduction was observed on the addition of 1 mole of L-lactate per mole

TABLE V. Stoichiometry of the Aerobic Reaction of Lactate Oxygenase (nmoles) (*36*)

Enzyme-FMN	L-Lactate added	Acetate formed	Pyruvate formed
150	150	168	0
149	75	95	0

Experiments were performed in a 3-ml assay system containing 60 μmoles of potassium phosphate buffer (pH 7.0) and the components indicated.

TABLE VI. Stoichiometry of the Reaction When Lactate Oxygenase Was Reduced Anaerobically and Then Exposed to Air (nmoles) (*36*)

Enzyme-FMN	L-Lactate added	Enzyme-$FMNH_2$ formed	Pyruvate formed
148	148	144	144
147	120	120	121
148	74	72	83
150	40	40	40

Experiments were performed using a Thunberg-type cuvette in a 3-ml assay system containing 60 μmoles of potassium phosphate buffer (pH 7.0) and the components indicated.

of the enzyme-bound FMN. Upon exposure of the reduced enzyme to air, the FMN moiety was rapidly oxidized and absorption in the visible region increased to the level of that of the untreated preparations. As shown in Table VI, under these conditions, all the added L-lactate was found to be converted to pyruvate corresponding to the amount of reduced flavin. When the reaction mixture was acidified before oxygen was admitted, essentially similar results were obtained. These investigations indicate that enzyme-bound $FMNH_2$ and pyruvate corresponding to the dehydrogenated L-lactate were formed anaerobically. However, the oxygenated product, acetate, was not produced on admitting air to the reduced enzyme; thus, the $FMNH_2$ moiety formed under anaerobic conditions was essentially not functional as an electron donor capable of supporting the oxygenative decarboxylation of the substrate. Clearly these results were different from those observed with salicylate hydroxylase in which the enzyme-bound $FADH_2$ formed by NADH could react enzymatically with molecular oxygen to convert salicylate into catechol. From these results it seems likely that the flavin of lactate oxygenase does not function in a manner similar to that proposed for salicylate hydroxylase. Whether or not such a dehydrogenated substrate, pyruvate, is a real intermediate in the process of oxygenative decarboxylation remains unanswered. Addition of catalase to the aerobic reaction did not affect oxygen consumption. These data indicate that hydrogen peroxide produced by the oxidation of reduced flavin is not involved as a transient intermediate of active oxygen in the oxygenative decarboxylation reaction, although it is familiar as a classic Holleman reaction that pyruvate is non-enzymatically converted to acetate in the presence of H_2O_2.

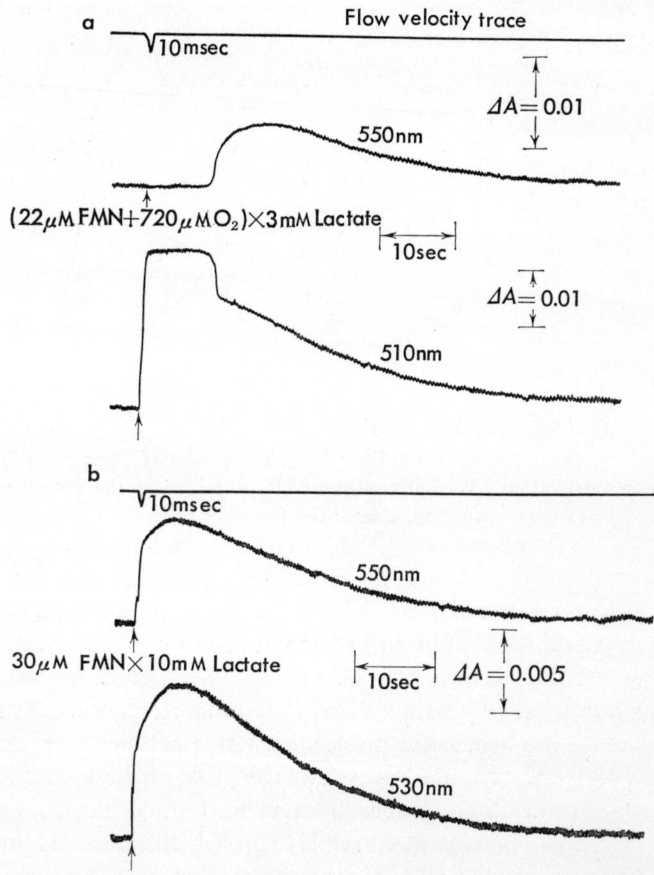

FIG. 14. Absorbance changes under aerobic (a) and anaerobic (b) conditions (*21*). Measured in 50 mM potassium phosphate buffer, pH 7.0, at 25°C.

This is also the case with lysine oxygenase which catalyzes the oxygenative decarboxylation of L-lysine (*37*). The enzyme-bound FAD of lysine oxygenase was reduced by L-lysine anaerobically, but the dehydrogenated form of lysine, α-keto-ε-aminocaproic acid, was found only after oxygen was admitted to the reaction mixture.

In an effort to obtain further information on the reaction mechanism, stopped-flow techniques were applied to aerobic and anaerobic reactions of lactate oxygenase (*21*). As shown in Fig. 14-a, the steady-state levels were recorded by aerobic mixing of the enzyme with L-lactate. Absorb-

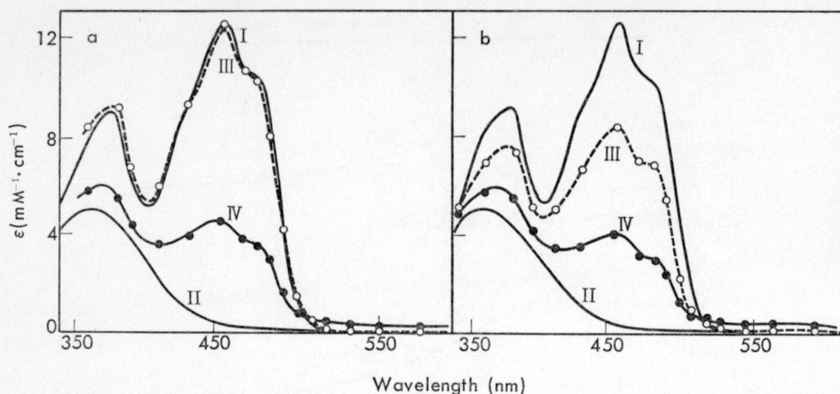

FIG. 15. Spectra obtained at various times in the experiments described in Fig. 14 (*21*). Curve I, oxidized enzyme; Curve II, enzyme reduced fully with L-lactate anaerobically; Curve III, obtained at the steady-state level (a) or obtained immediately after the flow stopped (b); Curve IV, obtained when the absorbance value at 550 nm was maximal.

ance changes at 550 and 510 nm which occurred on depletion of oxygen were grossly dissimilar. The absorption at 550 nm increased immediately, and reached the maximum in 16.5 sec. The absorption then disappeared much more slowly with a decay half-time of 9.5 sec. At 510 nm, a rapid decrease in absorbance proceeded with a parallel increase in absorbance at 550 nm. Figure 15-a shows the plots of the extinction coefficient of the enzyme obtained at each wavelength under the same reaction conditions. As can be seen in Curve III (broken line), the flavin moiety of the enzyme proceeded in a highly oxidized state during the steady state. Here, no spectral intermediate attributable to a flavin semiquinone was seen. When the absorbance value at 550 nm attained the maximum, a long wavelength-absorbing intermediate was observed (Curve IV). The transformation to the fully reduced enzyme resulted in loss of the long wavelength absorption.

When L-lactate was anaerobically mixed with the enzyme, the absorptions at 550 and 530 nm increased simultaneously (Fig. 14-b). Under these conditions, it is apparent that the development of these wavelength absorptions on addition of L-lactate is a biphasic event. The half-time of the absorbance change occurring in the fast phase of the reaction was 15 msec in the presence of 10 mM L-lactate. The absorbance then fell

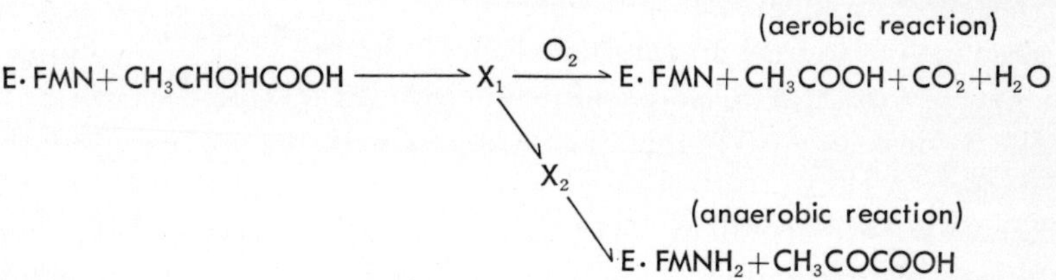

Fig. 16. Proposed mechanism of lactate oxygenase reaction. E·FMN denotes the enzyme.

very slowly until the enzyme was completely reduced. When the absorbance changes occurring in the fast phase of the reaction were plotted against wavelength, an absorption spectrum as represented by Curve III of Fig. 15-b was obtained. As shown by Curve IV of Fig. 15-b, the long wavelength-absorbing intermediate was again observed when the absorbance value at 550 nm was maximal. Although spectral and kinetic evaluations of the lactate oxygenase reaction are very complex, we attempted to present a hypothetical reaction sequence. This is summarized in Fig. 16. Here, an intermediate, X_1, corresponds to the species that appeared at the fast phase of the reaction and is actually common to both aerobic and anaerobic reactions. When the enzyme was mixed with a final concentration of 10 mM L-lactate, the pseudo-first-order rate constant for the fast phase was 46 sec^{-1}, which was considerably greater than the turnover number in the overall aerobic reaction (9.5 sec^{-1}). Therefore, the intermediate X_1 does not participate in a rate-limiting step in the overall reaction. The rate-determining step might be some other step, *e.g.*, the dissociation of the product from the enzyme. The intermediate expressed by X_2 corresponds to the long wavelength-absorbing intermediate. This species might be the transient intermediate in the production of the fully reduced enzyme and pyruvate under anaerobic conditions.

Summary

A spectroscopically well-defined complex in the presence of a substrate has been demonstrated for salicylate hydroxylase and *p*-hydroxybenzoate hydroxylase (*8*, *38*), and the functional properties of their flavin moiety have been clarified. One of the interesting observations was that the coupling of the substrate to the enzyme induces the reactivity of the enzyme toward an external electron donor. Both the affinity of the enzyme toward reduced pyridine nucleotides and the flow of electrons to the flavin moiety were remarkably augmented by the presence of the substrate. A similar activation phenomenon was recently observed with another flavin-dependent oxygenase. In 3-hydroxypyridine ring oxygenase, the reduction of the flavin moiety by NADH was accelerated by the addition of a substrate, 2-methyl-3-hydroxypyridine-5-carboxylate (*13*). It is likely that the coupling of the substrate to the enzyme induces the conformational change of the protein moiety which is usually common to all allosteric phenomena. The CD spectrum and visible absorption spectrum of the enzyme-substrate complex were very different from those of the free enzyme. The stability of the enzyme was also remarkably enhanced by the complex formation. These results speak in favor of a conformational change accompanying the binding of the substrate to the enzyme.

Similar phenomena have been recently observed for the benzoate-stimulated oxidation of NADH by salicylate hydroxylase (*24*, *39*). A substrate analogue with a free carboxyl group was not hydroxylated in the reoxidation of the enzyme by oxygen, but enhanced the reactivity of the enzyme toward NADH. The rate of reduction of the FAD moiety by NADH was also remarkably increased by the presence of benzoate. The enzyme-benzoate complex was detectable by a change of the absorption spectrum in the presence of the excess amounts of benzoate. These phenomena were not initiated by substances such as methyl salicylate and phenol which do not have a free carboxyl group in the molecule. These data suggest that the free carboxyl group of the substrate molecule is a functional group which may initiate the activation of the enzyme.

A quite similar activation mechanism was reported for the camphor hydroxylase system from *Pseudomonas putida* in which cytochrome P-450 functions as the terminal oxidase (*40*, *41*). In this system the coupling of camphor to cytochrome P-450 was assumed to be the primary reaction in the hydroxylation reaction. Since the enzyme-substrate complex is a much more efficient system for accepting electrons from reduced pyridine nucleotides than the free enzyme, an activation mechanism coupled by a substrate may be common to many flavo- and hemo-oxygenases.

Acknowledgments

Our warm thanks are due to Messrs. H. Yasuda, K. Mihara, K. Nakazawa, Y. Nakai, M. Nakamura and H. Tajima for their cooperation and help we have received throughout the work.

References

1 D. Wellner, *Ann. Rev. Biochem.*, **36**, 669 (1967).
2 O. Hayaishi, *Ann. Rev. Biochem.*, **38**, 21 (1969).
3 M. Katagiri and S. Takemori, *in* " Flavins and Flavoproteins," ed. by H. Kamin, University Park Press, Baltimore, p. 447 (1971).
4 O. Hayaishi and W. B. Sutton, *J. Am. Chem. Soc.*, **79**, 4809 (1957).
5 M. Katagiri, S. Yamamoto and O. Hayaishi, *J. Biol. Chem.*, **273**, PC 2413 (1962).
6 S. Yamamoto, M. Katagiri, H. Maeno and O. Hayaishi, *J. Biol. Chem.*, **240**, 3408 (1965).
7 M. Katagiri, H. Maeno, S. Yamamoto, O. Hayaishi, T. Kitao and S. Oae, *J. Biol. Chem.*, **240**, 3414 (1965).
8 K. Hosokawa and R. Y. Stainer, *J. Biol. Chem.*, **241**, 2453 (1966).
9 K. Yano, M. Morimoto, N. Higashi and K. Arima, *in* " Biological and Chemical Aspects of Oxygenases," ed. by K. Bloch and O. Hayaishi, Maruzen Co., Tokyo, p. 329 (1966).
10 Y. Maki, S. Yamamoto, M. Nozaki and O. Hayaishi, *J. Biol. Chem.*, **244**, 2942 (1969).
11 H. Takeda, S. Yamamoto, Y. Kojima and O. Hayaishi. *J. Biol. Chem.*, **244**, 2939 (1969).
12 D. B. Pho, A. Olomucki and N. V. Thoai, *Biochim. Biophys. Acta*, **118**, 311 (1966).
13 L. G. Sparrow, P. P. K. Ho, T. K. Sundaram, D. Zach, E. J. Nyns and E. Snell, *J. Biol. Chem.*, **244**, 2590 (1969).
14 C. C. Levy and P. Frost, *J. Biol. Chem.*, **241**, 997 (1966).
15 M. Katagiri, S. Takemori, K. Suzuki and H. Yasuda, *J. Biol. Chem.*, **241**, 5675 (1966).
16 S. Takemori, H. Yasuda, K. Mihara, K. Suzuki and M. Katagiri, *Biochim. Biophys. Acta*, **191**, 58 (1969).
17 S. Takemori, H. Yasuda, K. Mihara, K. Suzuki and M. Katagiri, *Biochim. Biophys. Acta*, **191**, 69 (1969).
18 K. Suzuki, S. Takemori and M. Katagiri, *Biochim. Biophys. Acta*, **191**, 77 (1969).
19 S. Takemori, M. Nakamura, M. Katagiri and T. Nakamura, *in* " Flavins

and Flavoproteins," ed. by H. Kamin, University Park Press, Baltimore, p. 463 (1971).
20 S. Takemori, K. Nakazawa, Y. Nakai, K. Suzuki and M. Katagiri, *J. Biol. Chem.*, **243**, 313 (1968).
21 S. Takemori, Y. Nakai, M. Katagiri and T. Nakamura, *FEBS Letters*, **3**, 214 (1969).
22 S. Yamamoto, H. Takeda, Y. Maki and O. Hayaishi, *J. Biol. Chem.*, **244**, 2951 (1969).
23 K. Honnami, S. Takemori and M. Katagiri, unpublished data.
24 R. H. White-Stevens and H. Kamin, *Biochem. Biophys. Res. Commun.*, **38**, 882 (1970).
25 K. Suzuki and M. Katagiri, unpublished data.
26 H. A. Harbury, K. F. Lanoue, P. A. Loach and R. M. Amick, *Proc. Natl. Acad. Sci. U.S.*, **45**, 1708 (1959).
27 V. Massey and H. Ganther, *Biochemistry*, **4**, 1161 (1965).
28 V. Massey and G. Palmer, *Biochemistry*, **5**, 3181 (1966).
29 M. Nakamura, S. Takemori, M. Katagiri and T. Nakamura, unpublished results reported at the 43rd Japanese Biochemical Society Meeting, Tokyo, October 1970 (in Japanese).
30 S. Takemori, M. Nakamura, K. Suzuki, M. Katagiri and T. Nakamura, *FEBS Letters*, **6**, 305 (1970).
31 B. Chance and V. Legallais, *Rev. Sci. Instrum.*, **22**, 6727 (1951).
32 M. Nakamura, S. Takemori and M. Katagiri, unpublished data.
33 L. G. Howell and V. Massey, *in* " Flavins and Flavoproteins," ed. by H. Kamin, University Park Press, Baltimore, p. 499 (1971).
34 W. B. Sutton, *J. Biol. Chem.*, **216**, 749 (1955).
35 M. Katagiri, S. Takemori, H. Tajima, T. Oda and T. Matsumoto, Abstracts, the 20th Symposium on the Protein Structure, Osaka, p. 10 (1969) (in Japanese).
36 Y. Nakai, K. Nakazawa, K. Suzuki, S. Takemori and M. Katagiri, Abstracts, the 19th Symposium on Enzyme Chemistry, Kanazawa, p. 225 (1968) (in Japanese).
37 T. Nakazawa, S. Yamamoto, Y. Maki, H. Takeda, Y. Kajita, M. Nozaki and O. Hayaishi, *in* " Flavins and Flavoproteins," ed. by K. Yagi, University of Tokyo Press, Tokyo, p. 214 (1968).
38 S. Nakamura, Y. Ogura, K. Yano, N. Higashi and K. Arima, *in* " Flavins and Flavoproteins," ed. by H. Kamin, University Park Press, Baltimore, p. 475 (1971).
39 K. Suzuki and M. Katagiri, *Seikagaku*, **42**, 563 (1970) (in Japanese).
40 M. Katagiri, B. N. Ganguli and I. C. Gunsalus, *J. Biol. Chem.*, **243**, 3543 (1968).

41 M. Katagiri, S. Takemori, M. Nakamura and T. Nakamura, Proceedings of the 2nd International Symposium on Oxidases and Related Redox Systems, University Park Press, Baltimore, in press.

Received for publication September 24, 1971.

STRUCTURE AND ACTION MECHANISM OF α-KETO ACID DEHYDROGENASE MULTIENZYME COMPLEX

Masahiko Koike, Minoru Hamada, Kichiko Koike, Nobuyuki Tanaka, Kin-Ichi Otsuka and Tadashi Suematsu

Department of Pathological Biochemistry, Atomic Disease Institute, Nagasaki University School of Medicine, Nagasaki

The living cells are the fundamental units of which all organisms are composed. Anatomic and chemical studies of the cells have shown that they are not merely droplets of protoplasm but rather highly organized molecular factories. It has become apparent that enzymes, *i.e.*, metabolic workers in these factories, are not randomly distributed in intact cells but instead are specifically organized into significant functional assemblages which are normally found to be bound to or embedded in the membranes or the matrices. At least a few of the enzyme aggregates are isolated from the cells in soluble forms. As previously reviewed by Reed and Cox (*1*), such multienzyme complexes in the cell may exist to increase efficiency in the metabolic processes and may also provide the basis for the regulatory mechanisms of the metabolism to maintain the homeostasis of living bodies. It seems worthwhile, therefore, to scrutinize the structures and action mechanisms of the mammalian α-keto acid dehydrogenase complexes as examples of organized multienzyme complexes. The present report is a summary of the current knowledge of the pig heart α-keto acid dehydrogenase complexes in terms of their structures and functions.

Properties of the Pig Heart α-Keto Acid Dehydrogenase Complexes

Two species of the multienzyme complex catalyzing a coenzyme A and nicotinamide adenine dinucleotide-linked oxidative decarboxylation of pyruvate and 2-oxoglutarate (Eq. 1), respectively, were isolated as giant functional units from the Keilin-Hartree preparation of the pig heart muscle (*2–6*) as well as from the cell-free extract of *Escherichia coli* (*7, 8*).

$$\begin{aligned} &RCOCO_2H + CoA\text{-}SH + NAD^+ \\ &\longrightarrow RCO\text{-}S\text{-}CoA + CO_2 + NADH + H^+ \end{aligned} \qquad (1)$$

The molecular weights of the highly purified pig heart pyruvate and 2-oxoglutarate dehydrogenase complexes (PDC and OGDC) are about 7.4 million and 2.7 million, respectively (*5, 6, 9, 10*), as summarized in Table I. Upon ultracentrifugation and Tiselius electrophoresis these complexes show themselves to be homogeneous. Both complexes exhibited not only the overall reaction (Eq. 1) activity but also the activities of ferricyanide-linked α-keto acid dehydrogenase, lipoate acyltransferase and lipoamide dehydrogenase (*cf.* Tables II and IV). For the degradation of α-keto acids including branched chain α-keto acids, thiamine pyrophosphate (TPP), bivalent cations such as Mg^{2+} or Ca^{2+}, enzyme-bound lipoic acid (LiA) and FAD are essential cofactors in addition to CoA and NAD. Both complexes contain enzyme-bound coenzymes which are associated with enzymatic activities, such as thiamine pyrophosphate, lipoic acid and FAD with the exception that thiamine pyrophosphate is lacking in the pyruvate dehydrogenase complex, as presented in Table I. These data indicate that both complexes are

TABLE I. Hydrodynamic Parameters and Enzyme-bound Coenzyme Contents of Pig Heart α-Keto Acid Dehydrogenase Complexes

Complex	$s^\circ_{20,w}$	$D_{20,w}$ ($\times 10^{-7}$ cm^2 $\times$ sec^{-1})	Mol. Wt.[a] ($\times 10^6$)	Coenzyme contents		
				TPP	LiA	FAD
				(mole/mole of the complex)		
PDC	68.5 S	1.35	7.4	0[b]	24	12–14
OGDC	35.7 S	1.18	2.7	6	8	8–12

[a] Determined by the sedimentation equilibrium described by Yphantis (*11*).
[b] $K_m = 4.2 \times 10^{-6}$ M, determined by the overall reaction assay.

composed of three individual constituent enzymes which could be separated from each other.

Resolution and Reconstitution of the Pig Heart α-Keto Acid Dehydrogenase Complexes

The pig heart pyruvate dehydrogenase complex was separated into three constituent enzymes—pyruvate dehydrogenase (PDH), lipoate acetyltransferase (LAT) and lipoamide dehydrogenase (PDC-Fp) (*9*), in a manner similar to those comprising the corresponding *E. coli* complex (*12*). This complex was first dissociated into lipoamide dehydrogenase and a colorless subcomplex in the presence of 4 M urea on a calcium phosphate gel cellulose column. The latter subcomplex which exhibited pyruvate dehydrogenase and lipoate acetyltransferase activities was further separated into pyruvate dehydrogenase and partially purified lipoate acetyltransferase by ammonium sulfate precipitation in the presence of 0.3 M potassium iodide. Partially purified lipoate acetyltransferase was highly purified by Sepharose 6B gel filtration in the presence of potassium iodide followed by ultracentrifugation in a sucrose density gradient (*13*). The enzyme complex resembling the original complex was reassembled from these isolated enzymes simply by mixing and was obtained as the yellow pellet by successive ultracentrifugation. Comparative data on the properties of the original complex, its constituent enzymes and the reconstituted complex are shown in Table II.

TABLE II. Properties of the Pyruvate Dehydrogenase Complex, Its Constituent Enzymes and Reconstituted Complex

Enzyme	$s_{20,w}$	Specific activities (μmole/hr/mg protein)			
		Overall reaction	Dehydrogenase	Acetyltransferase	Lipoamide dehydrogenase
PDC	67.5	120	6.5	122	378
PDH	7.5	0	12.8	0	0
LAT	28.2	3	0.4	282	45
PDC-Fp	5.6	0	0	0	4630
Reconstituted PDC	65.0	110	7.8	125	462

TABLE III. Physical and Chemical Properties and Coenzyme Contents of the Pyruvate Dehydrogenase Complex, Constituent Enzymes and Reconstituted Complex

Enzyme	Mol. Wt.[a]	Subunit[b] (mol. wt. & No.)	Assumed No. of molecules per mole of the complex	Total No. of coenzymes per mole of the complex
PDC	7.43×10^6			
PDH	15.3×10^4	$36{,}000 \times 2$ $41{,}000 \times 2$	30	TPP = 0
LAT	1.98×10^6	$36{,}000 \times 48$	1	LiA = 24
PDC-Fp	10.8×10^4	$55{,}000 \times 2$	6	FAD = 12
Reconstituted PDC	7.45×10^6			

[a] Determined by the sedimentation equilibrium described by Yphantis (*11*).
[b] Determined by sodium dodecyl sulfate-polyacrylamide gel electrophoresis (*14*, *15*).

Studies on the properties of the isolated constituent enzymes have been carried out and the results are summarized in Table III (*9*). It appears that the lipoate acetyltransferase (mol. wt. about 1.98 million) consists of 48 identical polypeptide chains with a molecular weight of about 36,000 (*13*), which was determined by sodium dodecyl sulfate-polyacrylamide gel electrophoresis (*14*, *15*), and that these chains are linked by noncovalent bonds. It appears that one of the two polypeptide chains contains one molecule of covalently bound lipoic acid (*16*). Pyruvate dehydrogenase (mol. wt. about 153,000) consists of two pairs of two different polypeptide chains with molecular weights of about 36,000 and 41,000, respectively (*17*). Lipoamide dehydrogenase (mol. wt. about 108,000) consists of two identical polypeptide chains with a molecular weight of about 55,000. This enzyme is a flavoprotein and contains two molecules of FAD per mole of enzyme (*17*). It is apparent that the individual constituent enzymes are linked in the complex by noncovalent bonds to construct the enzyme complex. Binding experiments have shown that pyruvate dehydrogenase and lipoamide dehydrogenase do not combine with each other, but each of these constituent enzymes combines with lipoate acetyltransferase, which displays a dual function, that is, catalytic and structural. These data indicate that the pyruvate dehydrogenase complex consists of about 30 molecules of pyruvate dehydrogenase, 6 molecules of lipoamide dehydrogenase and 1 molecule of lipoate acetyltransferase.

TABLE IV. Properties of the 2-Oxoglutarate Dehydrogenase Complex, Its Constituent Enzymes and Reconstituted Complex

Enzyme	$s_{20,w}$	Specific activities (μmoles/hr/mg protein)			
		Overall reaction	Dehydrogenase	Succinyltransferase	Lipoamide dehydrogenase
OGDC	35.7	315	110	36	681
OGDH	7.2	0	276	16	0
LST	14.9	0	29	135	0
OGDC-Fp	5.7	0	0	0	5170
Reconstituted OGDC	31.5	335	120	39	860

The pig heart 2-oxoglutarate dehydrogenase complex is first dissociated into lipoamide dehydrogenase (OGDC-Fp) and a colorless subcomplex in the presence of 2.5M urea by fractionation on a calcium phosphate gel cellulose column. The latter subcomplex exhibited 2-oxoglutarate dehydrogenase and lipoate succinyltransferase activities and could be separated into lipoate succinyltransferase (LST) and 2-oxoglutarate dehydrogenase (OGDH) by gel filtration on Sepharose 6B in the presence of 0.7 M guanidine hydrochloride and 2 mM dithiothreitol at pH 7 (*18*). From these isolated enzymes the large enzyme complex resembling the original complex was reconstituted. Comparisons of the properties of the original complex, its constituent enzymes and the reconstituted complex are summarized in Table IV.

Preliminary studies on the properties of the isolated constituent enzymes have been attempted, and the results are summarized in Table V (*10*). It appears that the lipoate succinyltransferase (mol. wt. about 1.02 million) consists of 16 identical polypeptide chains with a molecular weight of about 53,000 which was determined by sodium dodecyl sulfate-polyacrylamide gel electrophoresis, and that these chains are linked by noncovalent bonds. This enzyme contains about 8 molecules of lipoic acid per mole of the enzyme. Therefore, one of the two polypeptide chains apparently contains one molecule of covalently bound lipoic acid (*16*). 2-Oxoglutarate dehydrogenase (mol. wt. about 223,000) appears to consist of two identical polypeptide chains with a molecular weight of about 120,000. This enzyme contains one molecule of thiamine pyrophosphate per mole of the enzyme. Lipoamide dehydrogenase is very

TABLE V. Physical and Chemical Properties and Coenzyme Contents of the 2-Oxoglutarate Dehydrogenase Complex, Constituent Enzymes and Reconstituted Complex

Enzyme	Mol. Wt.[a]	Subunit[b] (mol. wt. & No.)	Assumed No. of molecules per mole of the complex	Total No. of coenzymes per mole of the complex
OGDC	2.70×10^{6}			
OGDH	22.3×10^{4}	$120{,}000\times2$	6	TPP = 6
LST	1.02×10^{6}	$53{,}000\times16$	1	LiA = 8
OGDC-Fp	10.8×10^{4}	$55{,}000\times2$	4–6	FAD = 8–12
Reconstituted complex	2.50×10^{6}			

[a] Determined by the sedimentation equilibrium described by Yphantis (*11*).
[b] Determined by sodium dodecyl sulfate-polyacrylamide gel electrophoresis (*14*, *15*).

similar to that from the pyruvate dehydrogenase complex as reported recently (*19*). Binding experiments indicate that 2-oxoglutarate dehydrogenase and lipoamide dehydrogenase do not combine with each other, but each of these enzymes combines with lipoate succinyltransferase. Lipoate succinyltransferase also possesses two functions similar to those of lipoate acetyltransferase. Lipoamide dehydrogenase from these two complexes are interchangeable to reconstitute the two complexes, but the other constituent enzymes are not interchangeable (*19*). These data indicate that the 2-oxoglutarate dehydrogenase complex consists of about 6 molecules of 2-oxoglutarate dehydrogenase, 4–6 molecules of lipoamide dehydrogenase and 1 molecule of lipoate succinyltransferase.

Macromolecular Structure of the Pig Heart α-Keto Acid Dehydrogenase Complexes

It is apparent from these biochemical studies that the pyruvate and 2-oxoglutarate dehydrogenase multienzyme complexes are, respectively, a mosaic of three constituent enzymes linked by noncovalent bonds to permit efficient coupling of the individual reactions catalyzed by these constituent enzymes (*cf.* Fig. 5). This has further been confirmed by the correlative electron microscopy (*9*, *18*, *20*).

Electron micrographs of isolated lipoate acetyltransferase negatively stained with 0.25% sodium phosphotungstate (pH 7.2) by the cross-

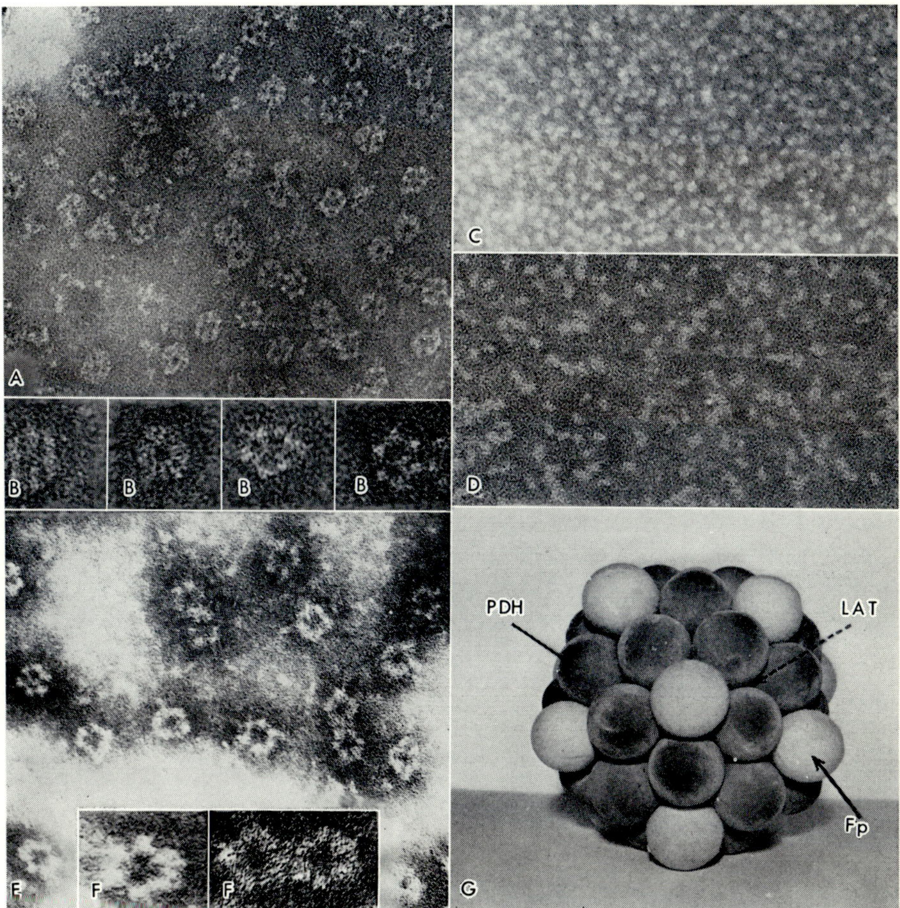

FIG. 1. Electron micrographs of the pyruvate dehydrogenase complex and its constituent enzymes negatively stained with sodium phosphotungstate, and a tentative model of the complex. A, lipoate acetyltransferase (×300,000); B, its selected individual images (×600,000); C, pyruvate dehydrogenase (×300,000); D, lipoamide dehydrogenase (×300,000); E, the pyruvate dehydrogenase complex (×300,000); F, its selected images (×600,000); and G, tentative model of the pyruvate dehydrogenase complex (a fivefold axis)—PDH refers to pyruvate dehydrogenase, LAT to lipoate acetyltransferase and Fp to lipoamide dehydrogenase.

spraying technique (*21*) showed a polyhedral structure with a gross diameter of approximately 170 Å as shown in Fig. 1-A and-B. The appearance of this enzyme suggests that its morphological subunits are situated at the vertices of a pentagonal dodecahedron. Electron micrographs of the pyruvate dehydrogenase and lipoamide dehydrogenase with a gross diameter of about 50–70 Å are shown in Fig. 1-C and-D, respectively. The gross appearance of the pyruvate dehydrogenase complex, which seemed to be composed of three enzymes, shows various aspects of a polyhedron with a gross diameter of about 210 to 250 Å with no clearly determined internal structure, as shown in Fig. 1-E and-F. The appearance of both the pyruvate dehydrogenase-lipoate acetyltransferase subcomplex and the lipoamide dehydrogenase-lipoate acetyltransferase subcomplex showed images similar to those of lipoate acetyltransferase because of dissociation of the constituent enzymes during negative staining. The appearance of the pyruvate dehydrogenase complex shadow-casted with Pt-C showed a globular structure with a diameter of about 284 Å and a height of about 260 Å. The basic geometry was strikingly different from that of the bacterial enzyme complex (*22*); however, it was similar in structure to the beef kidney enzyme complex (*23*). Our present view is that about 30 molecules of pyruvate dehydrogenase component and about 12 molecules of monomer of the lipoamide dehydrogenase component are symmetrically distributed and bound through non-covalent linkages to the lipoate acetyltransferase molecule. A model which accounts for the biochemical and electron microscopic data is shown in Fig. 1-G. In this model, pyruvate dehydrogenase and a monomer of lipoamide dehydrogenase molecules are represented as spheres.

The gross appearances of the 2-oxoglutarate dehydrogenase complex and lipoate succinyltransferase negatively stained with sodium phosphotungstate were remarkably similar to that observed in the *E. coli* enzymes (*10*, *18*, *22*). The image of lipoate succinyltransferase showed a cube-like or tetramer structure with a side of about 117 Å as shown in Fig. 2-A and-B. The appearance of the lipoate succinyltransferase suggests that its morphological subunits are situated at the eight vertices of a cube. The appearance of the complex showed a polyhedral structure with a diameter of about 260 Å as shown in Fig. 2-C. In this photograph, the lipoate succinyltransferase molecule seems to occupy the central portion of the polyhedron, and the molecules of 2-oxoglutarate dehydrogenase and lipoamide dehydrogenase presumably correspond to the

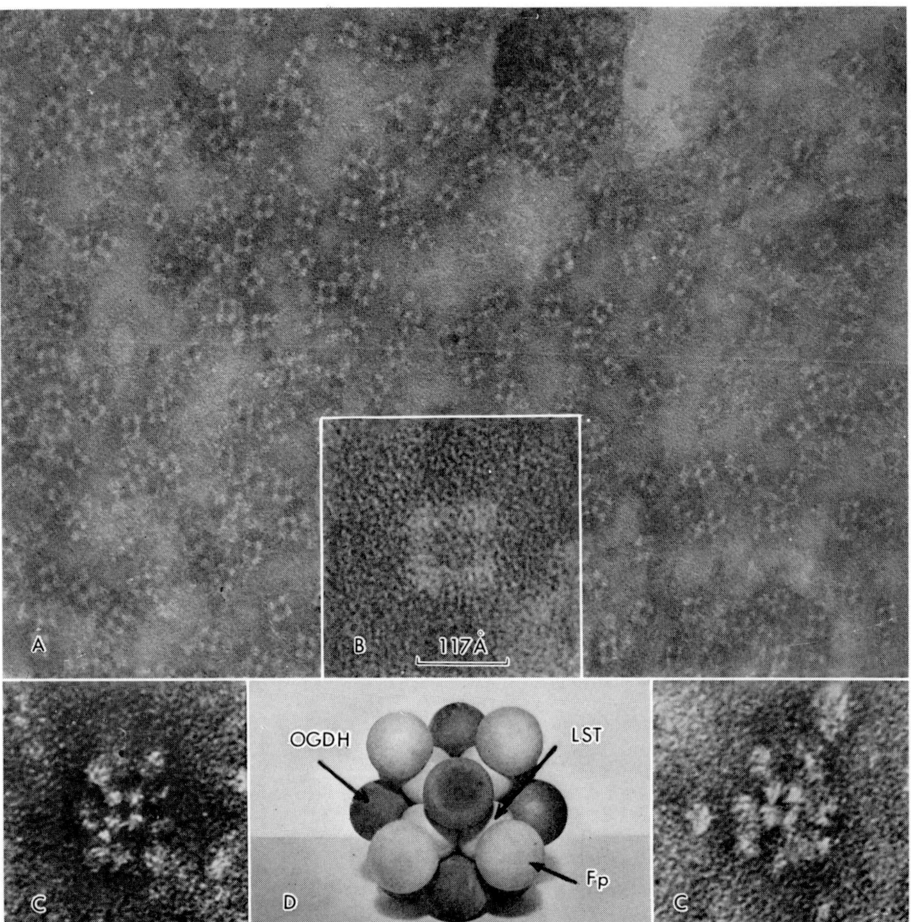

FIG. 2. Electron micrographs of the 2-oxoglutarate dehydrogenase complex and its constituent enzymes negatively stained with phosphotungstate, and a tentative model of the complex. A, lipoate succinyltransferase (×300,000); B, its selected individual image (×1,200,000); C, selected individual image of the 2-oxoglutarate dehydrogenase complex (×600,000); a tentative model of the 2-oxoglutarate dehydrogenase complex (a fourfold axis)—OGDH refers to 2-oxoglutarate dehydrogenase, LST to lipoate succinyltransferase and Fp to lipoamide dehydrogenase.

peripheral subunits. The data obtained thus far suggest that 6 molecules of the 2-oxoglutarate dehydrogenase component could be situated on the six faces of lipoate succinyltransferase and 4 or 6 molecules of the lipoamide dehydrogenase component on four or six faces of lipoate succinyltransferase. A model which accounts for these data is shown in Fig. 2-D. In this model, 2-oxoglutarate dehydrogenase and lipoamide dehydrogenase molecules and morphological subunits of lipoate succinyltransferase are represented as spheres.

It is still necessary to establish further in detail the macromolecular structures of these two multienzyme complexes by the new technique described by De Rosier and Klug (*24*) for the reconstitution of their three dimensions from their electron micrographs.

Action Mechanism of the Oxidative Decarboxylation of α-Keto Acid

The available evidence obtained suggested that several enzymes participate in a CoA- and NAD-linked multistage oxidative decarboxylation of α-keto acids, such as pyruvate, 2-oxoglutarate and branched α-keto acids, which are essential intermediates of the citric acid cycle and amino acid metabolism. This reaction is a main pathway of α-keto acid metabolism in animal tissues. The overall reaction (Eq. 1) occurs within the multienzyme complexes *via* a coordinated sequence of decarboxylation, acyl-generation, acyl-transfer and electron-transfer reactions shown in Eqs. 2–6, where the brackets indicate enzyme-bound intermediates (*cf.* Fig. 5).

$$\mathrm{RCOCO_2H+TPP\text{-}E_1 \longrightarrow [RCHO\text{-}TPP]\text{-}E_1+CO_2} \tag{2}$$

$$\mathrm{[RCHO\text{-}TPP]\text{-}E_1+LipS_2\text{-}E_2 \longrightarrow [RCO\text{-}S\text{-}LipSH]\text{-}E_2+TPP\text{-}E_1} \tag{3}$$

$$\mathrm{[RCO\text{-}S\text{-}LipSH]\text{-}E_2+CoA\text{-}SH \longrightarrow [Lip(SH)_2]\text{-}E_2+RCO\text{-}S\text{-}CoA} \tag{4}$$

$$\mathrm{[Lip(SH)_2]\text{-}E_2+FAD\text{-}E_3\Big\langle{S \atop S}\Big| \longrightarrow LipS_2\text{-}E_2+[H\dot{F}AD]\text{-}E_3\Big\langle{\dot{S} \atop SH}} \tag{5}$$

$$\mathrm{[H\dot{F}AD]\text{-}E_3\Big\langle{\dot{S} \atop SH} +NAD^+ \longrightarrow FAD\text{-}E_3\Big\langle{S \atop S}\Big| +NADH+H^+} \tag{6}$$

$\mathrm{R=CH_3\text{-}, HOOC(CH_2)_2\text{-}}$, TPP = thiamine pyrophosphate,
$\mathrm{LipS_2}$ = lipoic acid, $\mathrm{Lip(SH)_2}$ = dihydrolipoic acid,
E_1 = α-keto acid dehydrogenase, E_2 = lipoate acyltransferase,
E_3 = lipoamide dehydrogenase.

The decarboxylation reaction (Eq. 2), which is catalyzed by α-keto acid dehydrogenase (E_1), is thought to be a cleavage of the α-keto acid to produce CO_2 and an α-keto acid dehydrogenase-bound "aldehyde-thiamine pyrophosphate" compound, *i.e.*, "active aldehyde." It is generally accepted that thiamine pyrophosphate is ionized at the 2-position of the thiazolium ring, and that the resulting zwitterion reacts with the carbonyl group of the α-keto acid, such as pyruvate, to form an intermediate compound (2-lactyl-thiamine pyrophosphate) which undergoes decarboxylation to produce "active aldehyde (2-hydroxyethyl-thiamine pyrophosphate)." Oxidation of α-keto acid with potassium ferricyanide as an electron acceptor (Eq. 7) is catalyzed by the mammalian α-keto acid dehydrogenase complexes or isolated α-keto acid dehydrogenases.

$$RCOCO_2H + 2Fe(CN)_6^{3-} + H_2O \longrightarrow RCO_2H + CO_2 + 2Fe(CN)_6^{4-} + 2H^+ \qquad (7)$$

Ferricyanide apparently oxidizes the "active aldehyde" to RCO_2H. When the ferricyanide-linked oxidation of pyruvate (Eq. 7) proceeds in the presence of inorganic phosphate and *E. coli* pyruvate dehydrogenase, pyruvate is converted to acetyl phosphate, suggesting the production of an energy-rich intermediate compound (*25*). In contrast with these observations, only a small amount of acetyl phosphate was formed by the pig heart pyruvate dehydrogenase complex in the presence of dibasic phosphate ion, and a high concentration of phosphate ion inhibited the oxidation of pyruvate (*26*). The reason is not yet clear, but the postulated intermediate is presumably 2-acetyl-thiamine pyrophosphate. The occurrence of this intermediate compound can be inferred from the following observations. Synthetic 2-acetyl-thiazolium salts have been shown to be unstable, and they undergo nucleophilic attack by water to give acetic acid, by hydroxylamine to give acetylhydroxamate, and by mercaptide ions to give thiolacetates (*27*).

The acyl-generation reaction (Eq. 3) may be regarded as being a reductive acylation of enzyme-bound lipoic acid. In particular, "active aldehyde" is believed to attack the dithiolane ring of enzyme-bound lipoic acid in a nucleophilic displacement reaction, followed by a reverse condensation to form 6-acyldihydrolipoyl-lipoate acyltransferase (E_2). There is no conclusive evidence that this reaction is catalyzed either by α-keto acid dehydrogenase or lipoate acyltransferase.

The acyl-transfer reaction (Eq. 4) is a nucleophilic displacement of dihydrolipoyl-lipoate acyltransferase by CoA to produce acyl-S-CoA. Equations 3 and 4 are apparently catalyzed by lipoate acyltransferase which contains enzyme-bound lipoic acid. Evidence for Eq. 4 is based largely on a model reaction carried out with substrate amounts of lipoic acid and *E. coli* lipoate acetyltransferase in the presence of phosphotransacetylase with acetyl phosphate and a catalytic amount of CoA. The thioester produced in this reaction has been isolated and characterized as (+)-6-S-acetyldihydrolipoic acid (*28*). Reversal of this model reaction appears to be a process represented by Eq. 4. Mono-S-succinyldihydrolipoic acid is produced enzymatically by coupling succinic thiokinase with *E. coli* lipoate succinyltransferase in the presence of dihydrolipoic acid, succinate, ATP and Mg^{2+}. These model reactions carried out with a substrate amount of dihydrolipoic acid or its derivatives do not involve the catalytically active enzyme-bound lipoic acid. It is apparent that acylation of enzyme-bound lipoic acid occurs at one active site on lipoate acyltransferase (Eq. 3), and acyl-transfer to CoA occurs at some other active site (Eq. 4). Evidence for Eq. 4 can also be adduced from inhibition studies of overall reaction (Eq. 1) with arsenite in the presence of 2-oxoglutarate, CoA, NAD and the pig heart 2-oxoglutarate dehydrogenase complex (*29*). A possible reaction sequence (Eq. 3) in physiological oxidation of pyruvate is illustrated as follows;

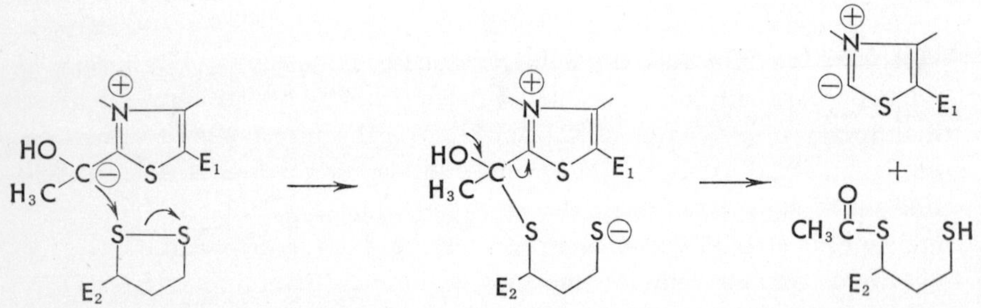

E_1 = pyruvate dehydrogenase, E_2 = lipoate acetyltransferase

The postulated intermediate is a semi-thioketal of 2-acetylthiamine pyrophosphate. It is presumed that the acetyl group is attached to the secondary thiol group (at C_6 in dihydrolipoyl-E_2) in view of evidence obtained in the model reaction with substrate amounts of dihydrolipoic acid, and subsequently is transferred to CoA which possibly attaches itself to an unidentified site on lipoate acetyltransferase.

The electron-transfer reaction (Eqs. 5 and 6) is an oxidation of enzyme-bound dihydrolipoic acid followed by reduction of NAD. These reactions are catalyzed by lipoamide dehydrogenase (E_3) which is a flavoprotein. This enzyme contains a reactive disulfide group which apparently participates in the catalytic mechanism of the enzyme. A biradical form of the enzyme, comprising a sulfur radical and flavin semiquinone, is believed to be an intermediate in the overall two-electron transfer from enzyme-bound dihydrolipoic acid to NAD. The evidence for Eqs. 5 and 6 is based on model reactions carried out with substrate amounts of dihydrolipoic acid and its derivatives. The mechanism of this reaction is discussed in detail in a review by Massey (*30*). Lipoamide dehydrogenase isolated from the pyruvate and 2-oxoglutarate dehydrogenase complexes was examined in detail for its various properties. It was found that the only differences in electrophoretic mobility of these two enzymes seemed to be associated with the conformational changes around the active site, including the chromophoric group, *i.e.*, the disulfide group and FAD (*19*).

Interpretation of the Relation between the Macromolecular Structure and the Catalytic Properties

It is apparent that the lipoyl moiety, which is covalently bound to lipoate acyltransferase, undergoes a cycle of transformation, such as acyl-generation, acyl-transfer and electron-transfer (*cf.* Fig. 5) and interacts with " aldehyde-thiamine pyrophosphate " which is bound to α-keto acid dehydrogenase, with FAD which is bound to lipoamide dehydrogenase and with CoA which is presumably bound to lipoate acyltransferase. These interactions occur within a complex in which the movement of the constituent enzymes is limited, and from which the intermediates do not appear to be dissociated during the reaction. Therefore, it is assumed that these three constituent enzymes must have a special position in the complex permitting efficient interactions of the three coenzymes to achieve multistage reaction sequences. Two possibilities have been considered: a) one or more of the constituent enzymes undergoes a conformational change during coordinated reaction sequences, and b) a flexible arm of the lipoyllysine moiety (14 Å) with a reactive dithiolane ring (*16*, *31*) rotates as shown in Fig. 3. Thus, the preliminary attempts to detect the conformational changes of the complex during the overall reaction

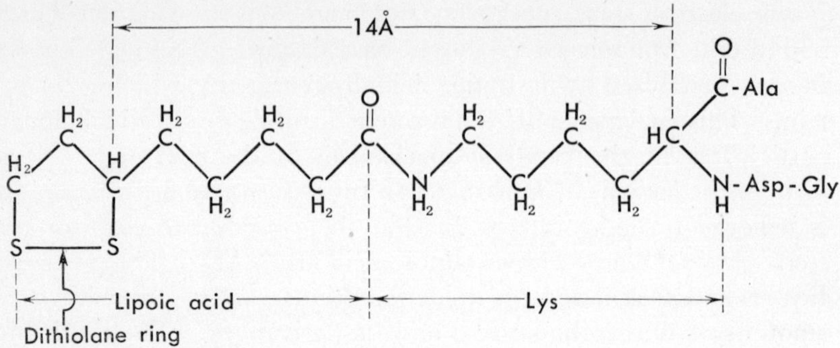

FIG. 3. The lipoyllysine moiety in the *E. coli* pyruvate dehydrogenase complex.

by bringing the enzyme-bound coenzymes into proper positions have met a limited success by measuring the optical rotatory dispersion spectra and circular dichroism spectra. Further investigations are still in progress along with the resolution and reconstitution experiment of the complex. Thus, the second possibility is presumed to be stronger for the achievement of the efficient interaction within a complex among the

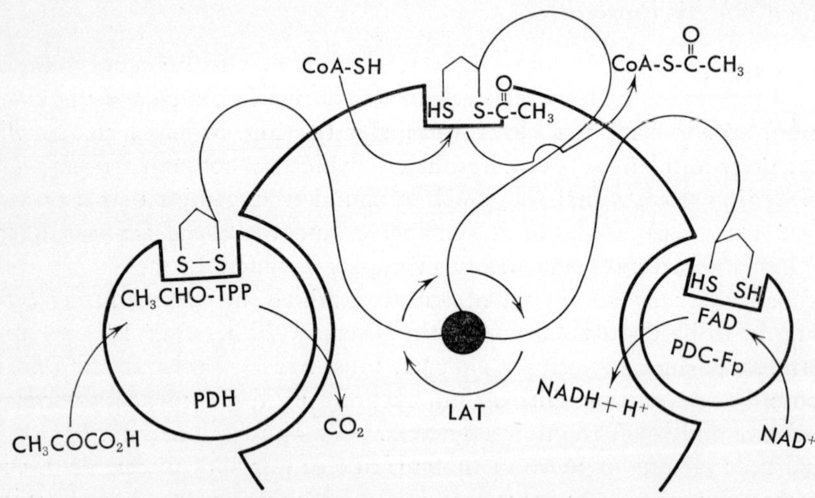

FIG. 4. Hypothetical scheme for inter-coenzyme translocation of the flexible lipoyllysine moiety of lipoate acetyltransferase (LAT) within the pyruvate dehydrogenase complex. PDH refers to pyruvate dehydrogenase and PDC-Fp to lipoamide dehydrogenase.

enzyme-bound coenzymes. The net charge on the lipoyl moiety (thiolane ring) during the coordinate reaction sequence may be 0, minus 1, or minus 2. Probably this change in the net charge provides the driving force for displacement of the lipoyl moiety from one site to the next within the complex as shown in Fig. 4.

Regulation of α-Keto Acid Dehydrogenase Complexes

Since the pyruvate dehydrogenase complex occupies an important branch point in glycolysis, the citric acid cycle and the β-oxidation system, the complex would be a possible candidate for a metabolic regulator. Previous studies (*32–37*) have revealed that the overall activity of the pyruvate dehydrogenase complex from mammals and bacteria is competitively inhibited by acetyl CoA and NADH which are the oxidation products of this reaction. The sites of the inhibition by acetyl CoA and NADH are, respectively, the pyruvate dehydrogenase and lipoamide dehydrogenase components of the complex, and these inhibitions are abolished by CoA and NAD. The overall oxidation activities of the pig heart pyruvate dehydrogenase complex with pyruvate, α-ketobutyrate and α-ketoisocaproate are competitively inhibited by α-ketovalerate and α-keto-β-methylvalerate (*34*). These results indicate that α-keto acid such as pyruvate competes with other α-keto acids for the same active site on the pyruvate dehydrogenase complex, especially in the case of the disorder of α-keto acid metabolism, such as in the maple syrup urine disease. Occurrence of this type of regulation would be a favorable situation *in vivo*. As was already mentioned, thiamine pyrophosphate in pyruvate dehydrogenase can be completely dissociated, and the activity of the overall oxidation of pyruvate is completely dependent on added thiamine pyrophosphate. This activity is inhibited by various nucleotides, *i.e.*, ATP, ADP, AMP, GTP, GDP, GMP, TMP, 3′, 5′-cyclic AMP and others, and this inhibition is reversed by an excess of thiamine pyrophosphate (*38*). These results suggest that various nucleotides compete for a binding site of thiamine pyrophosphate on pyruvate dehydrogenase in the multienzyme complex. In contrast to the pyruvate dehydrogenase complex, this type of inhibition is not observed in the 2-oxoglutarate dehydrogenase complex which contains enzyme-bound thiamine pyrophosphate. The enzyme which hydrolyzes free thiamine pyrophosphate to an extent similar to that observed with nucleotides diphosphates

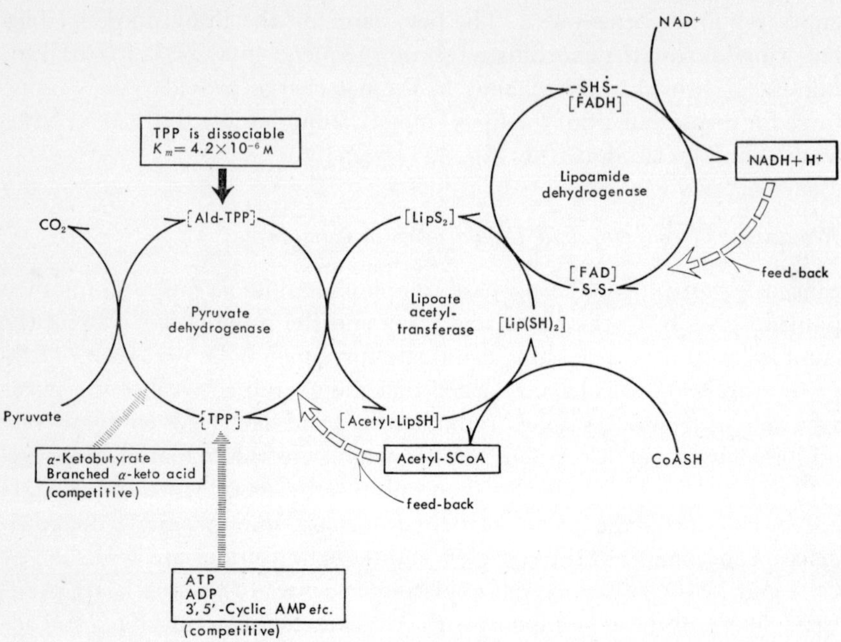

FIG. 5. Multistage reaction sequences in the oxidative decarboxylation of pyruvate and the sites of its regulation.

has been purified from bovine liver microsomes (*39*, *40*). Both activities were markedly enhanced by the presence of ATP as an allosteric effector. From this finding, it may be inferred that when ATP accumulates in the cells as a result of over-production it would stimulate the hydrolysis of thiamine pyrophosphate dissociated from the α-keto acid dehydrogenase complex. The decrease in concentration of thiamine pyrophosphate would subsequently affect the flow rate of the citric acid cycle *via* the regulation of α-keto acid oxidation. Presumably, the ATP level in the cell might be controlled through the regulation of the level of thiamine pyrophosphate by its hydrolyzing enzyme. Recently it has been reported that the activities of the pyruvate dehydrogenase complexes from bovine tissues (*41–43*) and pig heart muscle (*44*) are regulated by phosphorylation and dephosphorylation of the complex, especially the pyruvate dehydrogenase component. Phosphorylation and the concomitant inactivation of the complex are catalyzed by an ATP-specific kinase, and dephosphorylation and the concomitant reactivation are catalyzed by phosphatase. We regret to state that it could not yet be duplicated by the pyruvate dehydrogenase complex isolated from the pig heart particles.

SUMMARY

Two kinds of enzyme systems which catalyze an oxidative decarboxylation of α-keto acid have been isolated from the Keilin-Hartree preparation of the pig heart muscle as well as from the cell-free extract of *E. coli* cells as the soluble and organized multienzyme complexes with distinct morphology, and each of them was separated into three constituent enzymes. The reconstitution of the complexes was achieved. Interpretation of the relation between the macromolecular structure and the action mechanism of the α-keto acid dehydrogenase multienzyme complexes was given in detail. The regulation properties of the multienzyme compelxes were also summarized briefly.

REFERENCES

1 L. J. Reed and D. J. Cox, *Ann. Rev. Biochem.*, **35**, 57 (1966).
2 D. R. Sanadi, J. W. Littlefield and R. M. Bock, *J. Biol. Chem.*, **197**, 851 (1952).
3 V. Massey, *Biochim. Biophys. Acta*, **38**, 447 (1960).
4 T. Hayakawa, H. Muta, M. Hirashima, S. Ide, K. Okabe and M. Koike, *Biochem. Biophys. Res. Commun.*, **17**, 51 (1964).
5 T. Hayakawa, M. Hirashima, S. Ide, M. Hamada, K. Okabe and M. Koike, *J. Biol. Chem.*, **241**, 4694 (1966).
6 M. Hirashima, T. Hayakawa and M. Koike, *J. Biol. Chem.*, **242**, 902 (1967).
7 M. Koike, L. J. Reed and W. R. Carroll, *J. Biol. Chem.*, **235**, 1924 (1960).
8 G. Dennert and S. Höglund, *Eur. J. Biochem.*, **12**, 502 (1970).
9 T. Hayakawa, T. Kanzaki, T. Kitamura, Y. Fukuyoshi, Y. Sakurai, K. Koike, T. Suematsu and M. Koike, *J. Biol. Chem.*, **244**, 3660 (1969).
10 N. Tanaka, K. Koike, M. Hamada, K.-I. Otsuka, T. Suematsu and M. Koike, *J. Biol. Chem.*, in press.
11 D. A. Yphantis, *Biochemistry*, **3**, 297 (1964).
12 M. Koike, L. J. Reed, and W. R. Carroll, *J. Biol. Chem.*, **238**, 30 (1963).
13 K.-I. Otsuka, unpublished data.
14 A. L. Shapiro, E. Viñuela and J. V. Maizel, Jr., *Biochem. Biophys. Res. Commun.*, **28**, 815 (1967).

15 K. Weber and M. Osborn, *J. Biol. Chem.*, **244**, 4406 (1969).
16 H. Nawa, W. T. Brady, M. Koike and L. J. Reed, *J. Am. Chem. Soc.*, **82**, 896 (1960).
17 M. Hamada and T. Kanzaki, unpublished data.
18 K. Koike, N. Tanaka, M. Hamada, K.-I. Otsuka, T. Suematsu and M. Koike, *J. Biochem.*, **69**, 1143 (1971).
19 Y. Sakurai, Y. Fukuyoshi, M. Hamada, T. Hayakawa and M. Koike, *J. Biol. Chem.*, **245**, 4453 (1970).
20 H. Muta, *J. Electron Microscopy*, **19**, 32 (1970).
21 H. Fernández-Morán, *Res. Publ. Ass. Nerv. Ment. Dis.*, **40**, 235 (1962).
22 L. J. Reed and R. M. Oliver, *Brookhaven Symp. Biol.*, **21**, 397 (1969).
23 E. Ishikawa, R. M. Oliver and L. J. Reed, *Proc. Natl. Acad. Sci. U.S.*, **56**, 534 (1966).
24 D. J. DeRosier and A. Klug, *Nature*, **217**, 130 (1968).
25 M. L. Das, M. Koike and L. J. Reed, *Proc. Natl. Acad. Sci. U.S.*, **47**, 753 (1961).
26 T. Hayakawa and M. Koike, *J. Biochem.*, **65**, 645 (1969).
27 K. Daigo and L. J. Reed, *J. Am. Chem. Soc.*, **84**, 659 (1962).
28 I. C. Gunsalus, L. S. Barton and W. Gruber, *J. Am. Chem. Soc.*, **78**, 1763 (1956).
29 D. R. Sanadi, M. Langley and F. White, *J. Biol. Chem.*, **235**, 1924 (1960).
30 V. Massey, *in* "The Enzymes," ed. by P. Boyer, H. Lardy and K. Myrbäck, Academic Press, New York, Vol. VII, p. 275 (1963).
31 K. Daigo and L. J. Reed, *J. Am. Chem. Soc.*, **84**, 666 (1962).
32 J. Bremer, *Eur. J. Biochem.*, **8**, 535 (1969).
33 O. Wieland, B. von Jagow-Westermann and B. Stukowski, *Hoppe-Seyler's Z. Physiol. Chem.*, **350**, 329 (1969).
34 T. Kanzaki, T. Hayakawa, M. Hamada, Y. Fukuyoshi and M. Koike, *J. Biol. Chem.*, **244**, 1183 (1969).
35 R. G. Hansen and U. Henning, *Biochim. Biophys. Acta*, **122**, 355 (1966).
36 E. R. Schwartz, L. O. Old and L. J. Reed, *Biochem. Biophys. Res. Commun.*, **31**, 495 (1968).
37 E. R. Schwartz and L. J. Reed, *Biochemistry*, **9**, 1434 (1970).
38 M. Hamada, unpublished data.
39 M. Yamazaki, *Biochem. Biophys. Res. Commun.*, **16**, 416 (1964).
40 M. Yamazaki and O. Hayaishi, *J. Biol. Chem.*, **243**, 2934 (1968).
41 T. C. Linn, F. H. Pettit and L. J. Reed, *Proc. Natl. Acad. Sci. U.S.*, **62**, 235 (1969).
42 T. C. Linn, F. H. Pettit, F. Hucho and L. J. Reed, *Proc. Natl. Acad. Sci. U.S.*, **64**, 227 (1969).
43 L. J. Reed, T. C. Linn, F. H. Pettit, F. Hucho and G. Namihira,

The abstract of papers, 8th Meeting of International Congress of Biochemistry, Switzerland, Sept. 3–9, p. 235 (1970).
44 O. Wieland and E. Siess, *Proc. Natl. Acad. Sci. U.S.*, **65**, 947 (1970).

Received for publication June 22, 1971.

CATALYTIC MECHANISM OF LYSOZYME REACTION

Katsuya Hayashi
The Laboratory of Biochemistry, Faculty of Agriculture, Kyushu University, Fukuoka

The mechanism of lysozyme catalyzed reaction or the mode of lysozyme catalysis has become increasingly clear since the three-dimensional structure of the crystals of the lysozyme-oligosaccharide complex as the first enzyme-substrate complex was demonstrated by X-ray crystallographic method (*1*, *2*). Later, chemical studies on lysozyme action in solution have produced abundant information on the catalytic mechanism of lysozyme action.

The most fundamental reaction pathway and the mode of action of catalytic groups in a lysozyme catalyzed reaction may be represented by Scheme I. The catalytic group, the protonated γ-carboxyl group of Glu-35, acts as a general acid catalyzer to donate protons to the oxygen atom of β-1, 4-glucosaminide linkage, forming the partial oxonium ion. The sigma electrons between the C_1 atom of the pyranose ring and the oxygen atom of glycoside linkage then shift toward the oxonium ion and consequently bond-cleavage takes place, forming a carbonium ion on the sugar residue at subsite D. The carbonium ion intermediate which may be stabilized in certain structures for an interval necessary for the

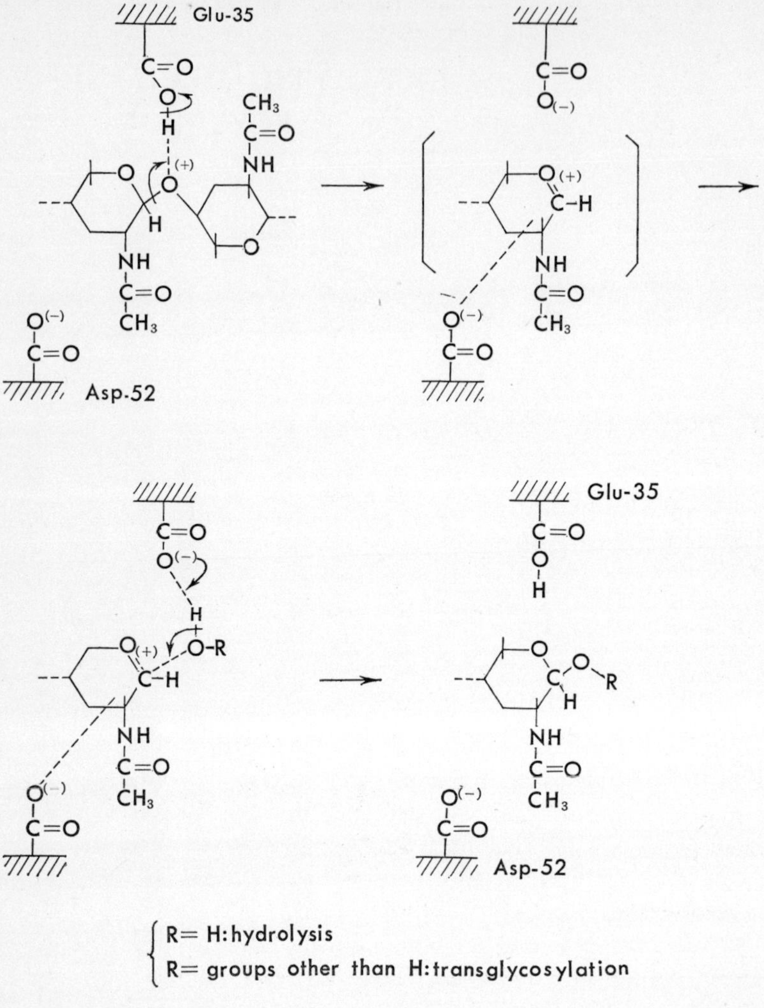

Glu-35

Asp-52

R= H:hydrolysis
R= groups other than H:transglycosylation

Scheme I

attack of nucleophiles on the intermediate. In the first step, Glu-35 loses its proton to form a conjugate base, carboxylate anion. The attack of nucleophiles on the intermediate is assisted by the general base catalysis of Glu-35 in a conjugate base form. In the case of hydrolysis, Glu-35 in the conjugate base form withdraws proton from the water molecule,

Some Possible Structures of Intermediates in Lysozyme Catalyzed Reactions (continued)

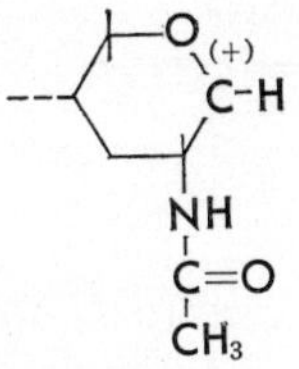

C1-Carbonium ion

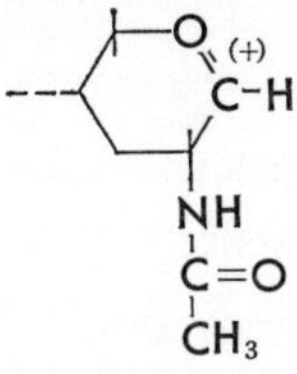

Pyranooxonium ion

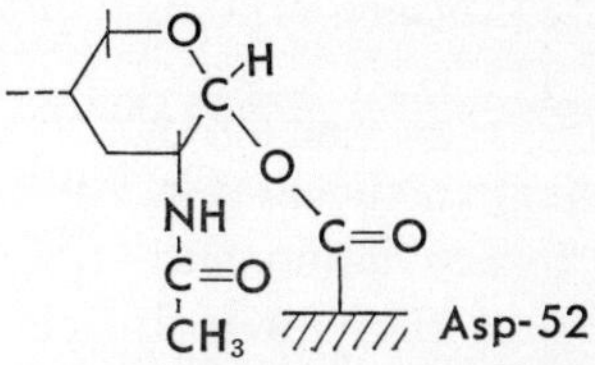

Glycosyl enzyme

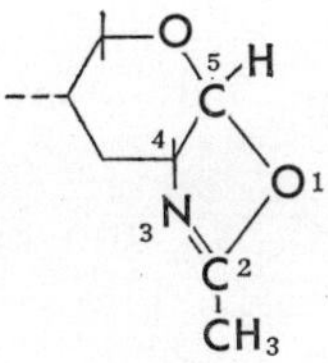

Oxazoline

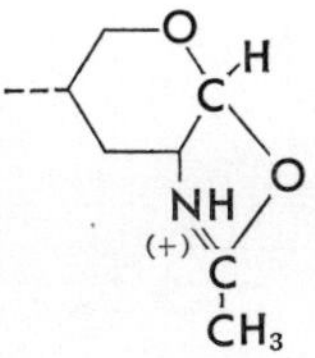

Protonated oxazoline

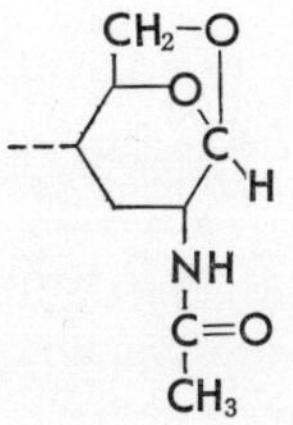

1,6-Anhydrosugar

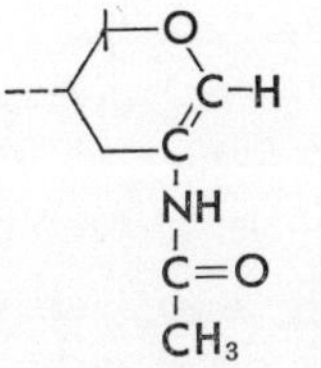

Glycal

and the resulting hydroxide ion attacks the intermediate. The attack of nucleophiles other than water causes a transglycosylation (*3–5*). This reaction is also assisted in the same way by Glu-35 as a general base catalyzer.

The retention of β-anomeric structure (*6*) in the hydrolyzed or transglycosylated product may be explained by proper orientation of three groups, the intermediate, the attacking nucleophile and the catalyzer Glu-35, that may permit the attack of the nucleophile from a fixed direction only.

At the present time, however, some problems concerning the detailed mechanism have remained unsolved; those are the precise structure and stabilization mechanism of the intermediate, the stabilization mechanism of Glu-35 in the conjugate base form, the role of Asp-52, and so on.

It is well known that lysozyme catalyzes the transglycosylation with high efficiency in addition to the hydrolysis. For instance, the transglycosylation with the acceptor, oligo-N-acetylglucosaminide, proceeds at a rate about 10^3 times larger than that of the hydrolysis in terms of molar concentration. For the occurrence of transglycosylation, some time interval is needed for the splitting out of the product from subsites E and F, and for the binding of the acceptor nucleophile at the same subsites. Consequently, the intermediate and the conjugate base of Glu-35 should be stable during such an interval. The stabilization of these structures needs specific devices provided by the active site, or their own specific, stable structures.

It was generally recognized that the reactions involving the C_1 atom of the pyranose ring proceed *via* an intermediate with the C_1 carbonium ion or pyranooxonium ion which is believed to have a half-chair structure. The intermediate is thought to have, most probably, these cationic structures (*7*, *8*). It was reported, on the contrary, that the bond-cleavage of β-1, 4-glycoside linkage of the substrate, which has 2-acetamido-2-deoxy-D-glucopyranose residue at subsite D, was assisted by the intramolecular nucleophile of the 2-acetamido group to form the oxazoline ring as an intermediate. Thus, 2-methyl-Δ^2-4,5-(glucopyrano)-oxazoline residue was considered to be an intermediate of the lysozyme catalyzed reaction (*9–11*). Furthermore, other forms such as 1,6-anhydrosugar and glycal structures seem not to be neglected as intermediates.

The other catalytic group, Asp-52 in carboxylate anion form, was believed to play a role in stabilization of the intermediate, which is attributable to the electrostatic interaction arising from the anion of Asp-52 and the cation of the intermediate. The other type of role, such as the formation of an ester bond between the C_1 carbonium ion and carboxylate anion of Asp-52, has been thought to be implausible because the distance between both groups is too far to form the covalent bonding if the sugar residue at subsite D is fixed at the same position before and after the cleavage of β-1,4-glycoside linkage. These observations, however, do not mean that the participation of Asp-52 in the lysozyme catalyzed reaction as a general acid-base catalysis is precluded.

On the other hand, the mode of the catalytic group of enzymes has recently been attracting the interest of investigators in various fields and the outline of general features of the mode of catalysis has been established by extensive studies on the model systems of enzyme catalysis (*12*, *13*). The results thus obtained indicate that for displaying a high efficiency an enzyme should fulfill, at least, the following prerequisites. 1) Catalytic reaction should proceed as an intramolecular catalysis that is attained by the formation of the enzyme-substrate complex; 2) catalytic reaction should proceed under the cooperation of two or more catalytic groups. In fact, it was known that the enzymes contain two or more catalytic groups in an active site.

When catalytic reaction takes place in a mode of intramolecular catalysis, several specific features may be seen, which are not observable in the case of intermolecular catalysis. The most important of such features may be the change in the mode of action of the catalytic group which is closely related to the high efficiency of intramolecular catalysis. For example, a carboxyl group scarcely shows catalytic capability as a general acid or general base in intermolecular catalysis. In contrast, a carboxyl group in intramolecular catalysis quite often exhibits considerable capability as a general acid or general base catalyzer. Another feature to be emphasized is that a catalytic group in intramolecular catalysis can control the direction of attack of the reagent on the substrate to form a product of a certain structure. This may be connected to the so-called substrate or product specificity. The latter refers to the fact that the product of enzyme catalyzed reaction consists of only one isomer of optical isomerism or diastereomerism.

This knowledge may permit some theorizing about general characteristics of the real enzyme catalyzed reactions, although a mechanism of enzyme catalysis involves complicated patterns that cannot be fully determined merely on the basis of results of a simple model experiment.

It will be meaningful to show a few examples to explain the above. The most characteristic feature of the enzyme catalysis, which differs markedly from that of a simple model catalysis, may be that the enzymatic reaction takes place in a heterogeneous environment which would bring about a remarkable acceleration of the reaction velocity. The active site in which the catalytic reaction proceeds is formed by various types of side chain of amino acid residues attached to a fixed polypeptide chain. Consequently, it is reasonable to assume that the active site would contain a mosaic-like phase consisting of a relatively hydrophobic region and a hydrophilic region. Furthermore, the binding of the substrate to the active site may, in some cases, expel water molecules originally surrounding the catalytic groups, forming a nonaqueous environment in the reaction site. The relationship between such an environment and the reaction velocity should be established in the near future for more understanding of the mechanism of enzyme catalysis. Such a specific environment would also induce formation of a specific water structure in the active site, which seems to control the rate and mechanism of catalytic reaction. Another characteristic of enzymatic reaction is that the activity of enzymes is reversibly altered by many factors, physical conditions of the solution, coexistence of substances other than the substrate, and so on. This is very important in connection with the controlled metabolism in organisms.

It is our purpose to clarify the general mechanism or the mode of action of catalytic groups of enzyme molecules by dealing with problems of the intermediate of lysozyme catalyzed reaction. The author is especially interested in the mechanism of a general acid-base catalysis found in this enzymatic reaction. For example, the protonated carboxyl group of Glu-35 of lysozyme donates protons to the glycoside oxygen atom of the substrate, forming a partial oxonium ion in a transition state in the cleavage of the glycoside linkage. It is believed that Glu-35 has a the pK value of 6 and the glycoside oxygen as acid has the pK value of about -3. It cannot be expected in a test tube reaction that both groups with such pK values exchange protons to a great extent at a pH range between both pK values. In contrast, the exchange of protons in the lysozyme catalyzed hydrolysis of the substrate must reach a considerable extent, because lysozyme exhibits activity in a pH range of 3 to 6. Such an exchange may be attributable partly to the firm contact between both groups owing to the formation of an enzyme-substrate complex. However, some questions remain. For instance, what are the factors stimulating such a great exchange? Where is the limit of such an exchange? How does the limiting difference in the pK values of the two groups bring about such a great extent of exchange? These questions seem to relate to the fundamental principle of the catalytic mechanism of enzyme action.

Although questions on the mechanism of catalysis are easily postulated by careful examination of the facts so far observed with lysozyme action, the methods or procedures adoptable for elucidating these questions are very difficult to establish. Considering the factors described above, the author investigated the behavior of intermediated analogues in the hope of finding some methods to answer the above questions.

Hydrolysis of 1-Benzoyl-2-acetamido-2-deoxy-β-D-glucopyranoside

1-Benzoyl-2-acetamido-2-deoxy-β-D-glucopyranoside (BNAG) was synthesized essentially according to the method of Fletcher, which was applied to the synthesis of ester of D-glucose (*14*). Since sodium 2-acetamido-2-deoxy-4,6-benzylidene-D-glucopyranose is soluble in ordinary organic solvents, coupling with benzoyl chloride was conducted without the isolation of the sodium compound.

The non-enzymatic hydrolysis of BNAG was followed by difference spectrophotometry. BNAG exhibited an absorption spectrum with a peak at 233 nm. After complete hydrolysis, the reaction mixture showed a peak at 230 nm, which was identical with that of benzoic acid. The difference spectrum between BNAG and the reaction mixture showed a negative peak at 240 nm (Fig. 1), the optical density of which was measured to estimate the extent of the hydrolysis. The hydrolysis was carried out at 50°C in the pH range from 0.3 to 10.5.

The rate constant k_{obs} was calculated from the data by assuming that the hydrolysis was a pseudo-first order reaction. The pH dependence of calculated $\log k_{obs}$ is shown in Fig. 2. The pH-$\log k_{obs}$ profile showed a plateau region between pHs 1.5 and 8.5 The external regions of the plateau, below pH 1 and above pH 9,exhibited pH dependence with a slope of unity. The pH dependence of the rate constant, then, can be represented by the following equation.

$$k_{obs} = k_0 + k_H a_{H^+} + k_{OH} \frac{K_W}{a_{H^+}}, \tag{1}$$

where k_0 represents the rate constant for spontaneous hydrolysis, k_H the rate constant for specific acid catalysis by protons and k_{OH} the rate constant for specific base catalysis by hydroxide ions. In the acid region, $a_{H^+} \gg K_W/a_{H^+}$, therefore, Eq. 1 reduces to

$$k_{obs} = k_0 + k_H a_{H^+}. \tag{2}$$

Similarly, in the alkaline region, $a_H \ll k_W/a_{H^+}$, which gives the relation:

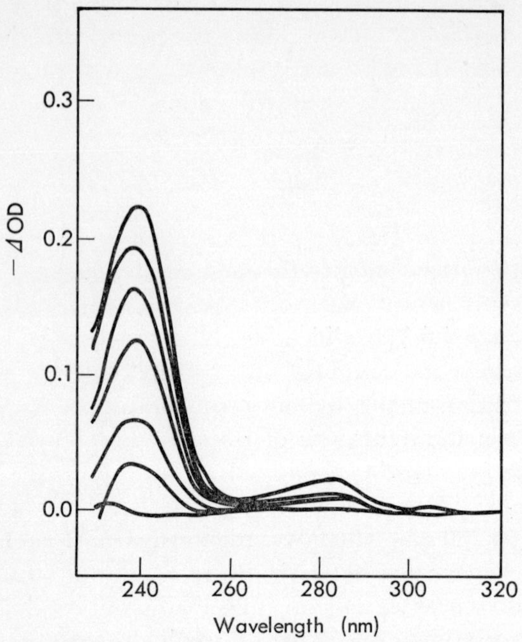

FIG. 1. Difference spectra between BNAG and hydrolysates. Spectra were recorded at reaction times, 0 (lower), 0.5, 1.0, 2.0, 3.0, 4.0 and 5.0 hr (upper), respectively. Hydrolysis was carried out at pH 5.3 and 50°C.

$$k_{obs} = k_0 + k_{OH}\frac{K_W}{a_{H^+}}\,. \tag{3}$$

Plotting of k_{obs} *vs.* a_{H^+} or K_W/a_{H^+} gives the value of k_0 from the intercept on the ordinate and k_H or k_{OH} from the slope of the line shown in Fig. 3. The values of the rate constants thus calculated are listed in Table I, together with those obtained by Pisczkiewicz and Bruice (*9*) for several glycosides. As can be seen in the table, all β-glycosides of N-acetylglucosamine exhibit large values of k_0. It was pointed out that a large k_0 value was due to the participation of the 2-acetamido group as an intramolecular nucleophile in the bond-cleavage of glycoside to form an intermediate with the oxazoline ring in either protonated or neutral form.

It is probably safe to assume analogously that in the neutral region

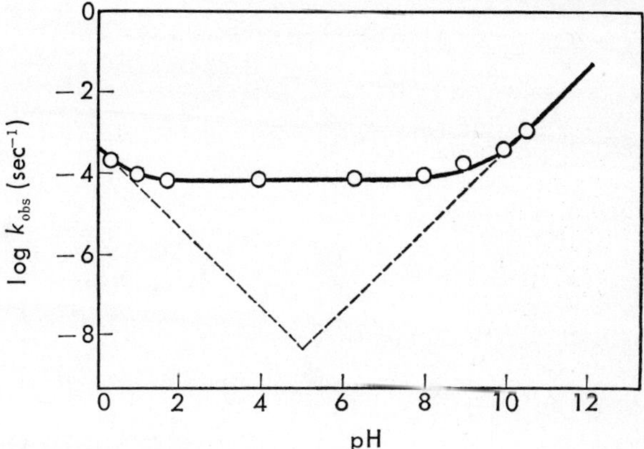

FIG. 2. The pH-log k_{obs} profile of hydrolysis of BNAG at 50°C. Circles show experimental data and solid line represents theoretical value calculated by Eq. 1.

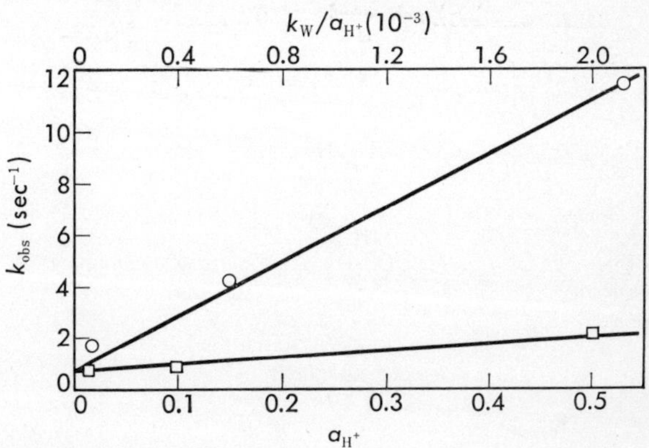

FIG. 3. Linear relationship between k_{obs} and a_{H^+} (□) or K_W/a_{H^+} (○).

TABLE I. Kinetic Parameters in Hydrolysis of Glucosides of 2-Acetamido-2-deoxy-D-glucopyranose

Compound	k_0 (sec^{-1})	k_H (M^{-1} sec^{-1})	k_{OH} (M^{-1} sec^{-1})
1-Benzoyl-β-NAG†	0.70×10^{-4}	2.95×10^{-4}	0.53
p-NP-β-D-NAG	0.171×10^{-4}	2.97×10^{-4}	1.1×10^{-3}
p-NP-α-D-NAG		4.00×10^{-4}	0.9367×10^{-3}
p-NP-β-D-Glucose	0.50×10^{-4}	1.39×10^{-4}	0.033
p-NP-α-D-AG Glucose		6.20×10^{-4}	0.3483

† Abbreviated as BNAG.

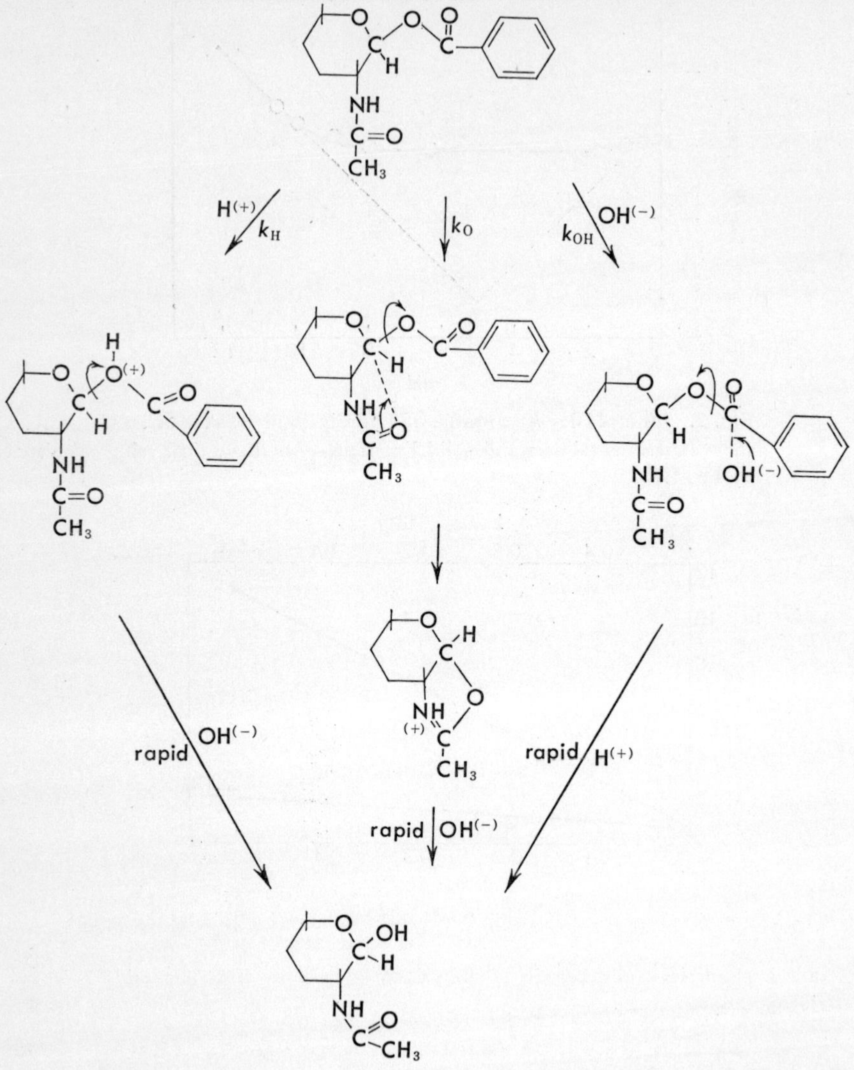

Scheme II

BNAG is hydrolyzed *via* the oxazoline intermediate by the participation of the 2-acetamido group. The reaction pathway corresponding to Eq. 1 may be represented by Scheme II.

In the pH region where lysozyme shows enzymatic activity, BNAG is hydrolyzed mainly by spontaneous reaction. On the other hand, it was observed that protons accelerated the hydrolysis of BNAG. If the glycosyl enzyme formed between Asp-52 and the sugar residue at subsite D is the real intermediate in the lysozyme catalyzed reaction, it may be reasonable to assume that the hydrolyzable bond of glycosyl enzyme should be cleaved by the participation of certain catalytic

TABLE II. Value of k_{obs} in Tris Catalyzed Hydrolysis of BNAG

pH	Conc. of Tris (M)	k_{obs} (10^{-4} sec^{-1})
6.7	0.081	0.804
6.7	0.405	0.784
7.95	0.500	2.280
8.75	0.137	1.735
8.75	0.333	2.770
8.75	0.500	3.000
8.75	0.667	4.660

groups located at the active site of the lysozyme, and these catalytic groups should act as general acids. In order to ascertain the above possibility, the catalytic effects of acetic acid, imidazole and Tris were studied. Neither acetic acid nor imidazole exhibited the catalytic activity for the hydrolysis of BNAG. Tris showed an activity in the pH region above its pK value. At pH 8.75, the k_{obs} was a linear function of the concentration of Tris (Table II), that is, the reaction was of the first order with respect to the catalyzer. This fact and information obtained from catalytic behavior of Tris for general esters (*15*) suggest that at pH 8.75 Tris acted as a nucleophile attacking the carbonyl carbon and forming an amide but not catalyzing the transesterification.

Hydrolysis of 2-Phenyl-4,5-(D-glucopyrano)-Δ^2-oxazoline (PGO)

Many experimental results have suggested the participation of the 2-acetamido group of the substrate as an intramolecular nucleophile in the bond-cleavage of β-1,4-glucosaminide linkage to form an oxazoline intermediate. When the sugar oxazoline is an intermediate, the catalytic groups of lysozyme must exert action in both processes, the formation and breakdown of the oxazoline intermediate. In such a case, the mode of action of catalytic groups, of course, should be consistent with the nature of the intermediate. Elucidation of the hydrolytic behavior of intermediate-analogues may offer some key to an understanding of the actual mode of action of the catalytic group of the enzyme. In connection with the mechanism of lysozyme catalysis, the hydrolytic behavior of oxazolines includes two interesting points; one is the fission type of oxazoline in hydrolysis and the other concerns the rate or ease of nucleophilic displacement at the C_5 atom in the oxazoline ring.

Generally, it was observed that the hydrolysis of oxazolines in acid media formed the corresponding esters (*16*), indicating that the fission took place at the C_2-N_3 bond of the oxazoline ring. Actually, in the lysozyme catalyzed reaction, only an amide compound, N-acetylglucosamine residue at the reducing end, was produced. It is therefore of interest to make clear the mode of action of catalytic groups of lysozyme causing the C_5-O_1 or O_1-C_2 fission in the oxazoline ring. The C_5 atom of the oxazoline ring (C_1 atom of the pyranose ring) has an sp^3 structure in the oxazoline ring, although its original nature is that of carbonyl carbon. It has been recognized that a nucleophilic displacement on an sp^3 carbon usually occurs with difficulty, and this reaction is too difficult accelerate by a general base catalysis alone. Therefore, it is necessary to elucidate the mechanism by which the oxazoline intermediate is catalytically displaced by nucleophiles to produce the N-acetylgucosmine residue or N-acetylglucosaminide residue.

The formation of the oxazoline ring in the sugar residue might depend upon the polarizable nature of carbonyl carbon in the acyl group of 2-substituent. When the formation of a carbonyl anion is assisted by a nucleophilic participation of the catalytic group, the oxazoline ring is formed more easily.

In order to make clear the problems described above, first the behavior of sugar oxazoline in hydrolysis was examined. The sugar oxazoline which might be formed in a lysozyme catalyzed reaction has a structure of 2-methyl-4,5-(D-glucopyrano)-Δ^2-oxazoline. However, this compound could not be synthesized because of its instability. In the present experiment, therefore, 2-phenyl-4,5-(D-glucopyrano)-Δ^2-

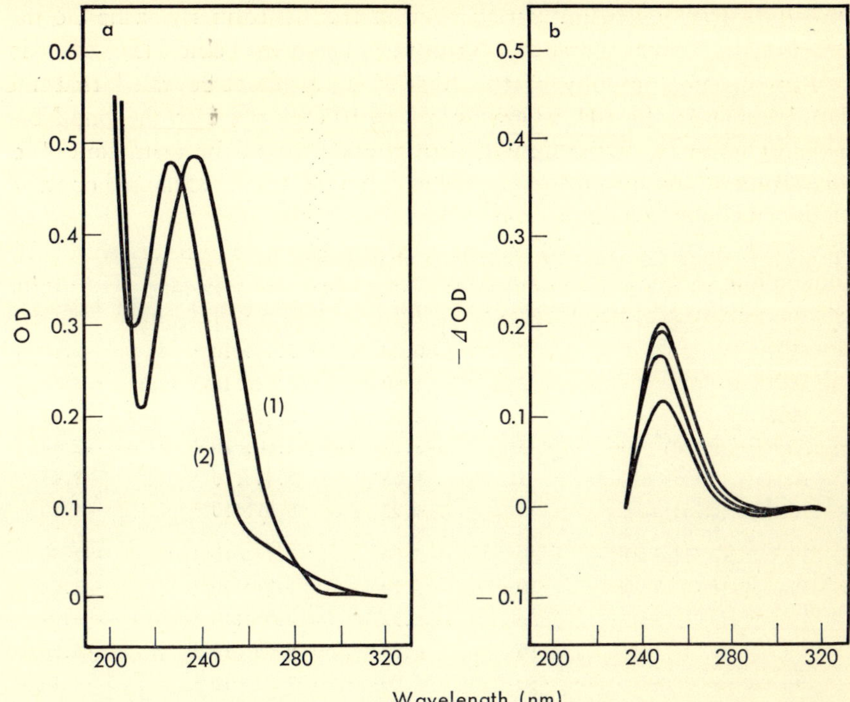

FIG. 4. a: Absorption spectra of PGO (1) and 2-benzamido-2-deoxy-D-glucopyranose (2). b: Difference spectra between PGO and hydrolysates.

oxazoline was synthesized and used as an analogue of the sugar intermediate in the real enzyme hydrolysis.

2-Phenyl-4,5-(D-glucopyrano)-Δ^2-oxazoline (PGO) was synthesized, directly by the reaction of N-acetylglucosamine with acetic anhydride in the presence of zinc chloride according to the methods of Pravdić *et al.* and Micheel and Köchling (*17*, *18*).

The hydrolysis of PGO was carried out at various pHs and 5°C using the difference spectrophotometric techniques. The PGO exhibits a maximum optical density at 239 nm, while the hydrolyzed product shows an absorption spectrum with a peak at around 226 nm. The difference spectrum between PGO and the hydrolyzed product has a negative peak in the region between 246 and 250 nm depending upon the conditions used for the hydrolysis (Fig. 4). The rate constant of hydrolysis k_{obs} was calculated from the optical density at the negative peak of the difference spectrum, assuming that the hydrolysis is a first order reaction. The calculated k_{obs} values are listed in Table III.

Paperchromatography of the hydrolyzed product revealed that the bond-fission in the pH region from 2 to 10 took place at the bond between C_5 and O_1 of the oxazoline ring, resulting in the formation of 2-benzamido-2-deoxy-D-glucopyranose. In contrast, alkaline or acid

TABLE III. Rate Constants of Hydrolysis of PGO

Buffer	Temp. (°C)	pH	k_{obs} (sec^{-1})	logk_{obs}
HCl	5	0.20	7.46×10^{-3}	−2.11
HCl	5	1.80	5.84×10^{-3}	−2.23
HCl	5	2.90	3.52×10^{-3}	−2.45
Acetate (0.1 M)	5	3.22	2.60×10^{-3}	−2.59
Acetate (0.1 M)	5	4.03	3.54×10^{-4}	−3.45
Acetate (0.1 M)	5	4.95	8.01×10^{-5}	−4.10
Phosphate (1/15 M)	50	5.00	2.77×10^{-3}	−2.56
Phosphate (1/15 M)	50	5.20	1.31×10^{-3}	−2.88
Phosphate (1/15 M)	50	6.32	2.95×10^{-4}	−3.53
Phosphate (1/15 M)	50	6.85	2.08×10^{-4}	−3.68
Phosphate (1/15 M)	50	8.00	7.92×10^{-5}	−4.10
Phosphate (1/15 M)	50	8.90	9.89×10^{-4}	−4.00
Carbonate (0.2 M)	50	10.65	1.60×10^{-4}	−3.76

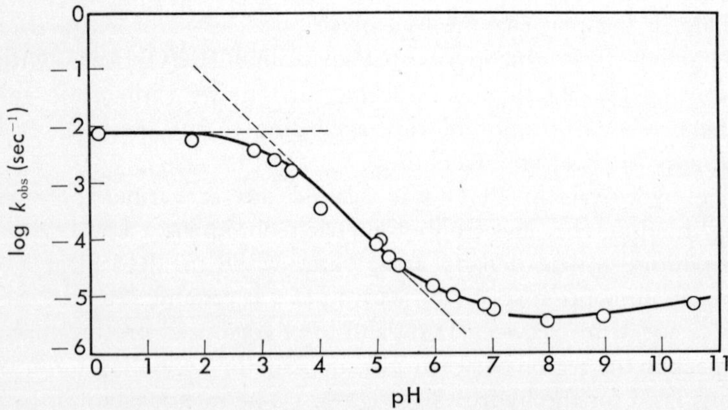

FIG. 5. pH-log k_{obs} profile of hydrolysis of PGO at 5°C.

TABLE IV. Effects of Acetic Acid, Imidazole and Tris on Rate Constant of Hydrolysis of PGO

Catalyst	conc.	Temp. (°C)	pH	k_{obs} (sec^{-1})	log k_{obs}
Acetate	0.01 M	5	3.5	1.29×10^{-3}	−2.89
	0.02			1.62×10^{-3}	−2.78
	0.05			1.54×10^{-3}	−2.81
Imidazole	0.043	50	5.3	0.99×10^{-3}	−3.00
	0.215			1.00×10^{-3}	−3.00
	0.430			1.02×10^{-3}	−2.99
Imidazole	0.125	50	5.9	3.52×10^{-3}	−3.52
	0.375			3.60×10^{-3}	−3.44
	0.500			3.40×10^{-3}	−3.47
Tris	0.038	50	7.0	1.78×10^{-3}	−3.75
	0.380			1.02×10^{-3}	−3.97

hydrolysis of PGO gave unidentified products other than 2-benzamido-2-deoxy-D-glucopyranose.

The pH dependence of logk_{obs} is shown in Fig. 5. Below pH 2 and above pH 7, the values of k_{obs} are nearly independent of the pH value. The shape of the profile is very simple, and it is likely that it is the only factor controlling the rate of hydrolysis.

The catalytic activities of acetic acid, imidazole and Tris for hydrolysis of PGO were not observed as shown in Table IV.

The pH dependence of k_{obs} exhibits two inflection points near pH 7 and pH 3. This fact indicated that the fission type was different in acid and alkaline regions as it was evidenced by paperchromatography. Martine and Parcell (*16*) reported that 2-methyl-Δ^2-oxazoline was hydrolyzed in an acid medium through orthooxazoline to an ester product. Thus, ordinary oxazolines show C_2-N_3 fission in an acid medium, giving rise to esters as hydrolyzed products. In contrast, the hydrolysis of PGO in an acid region formed the amide product. Similar observations were made in the case of the hydrolysis of 2-phenyl-Δ^2-oxazoline 5-one in neutral region (*19*). These indicate the possibility that oxazolines with carbonyl or carbonyl-like carbon at the C_5 position always show the C_5-O_1 fission independently of pH value and substituent at the C_2 position.

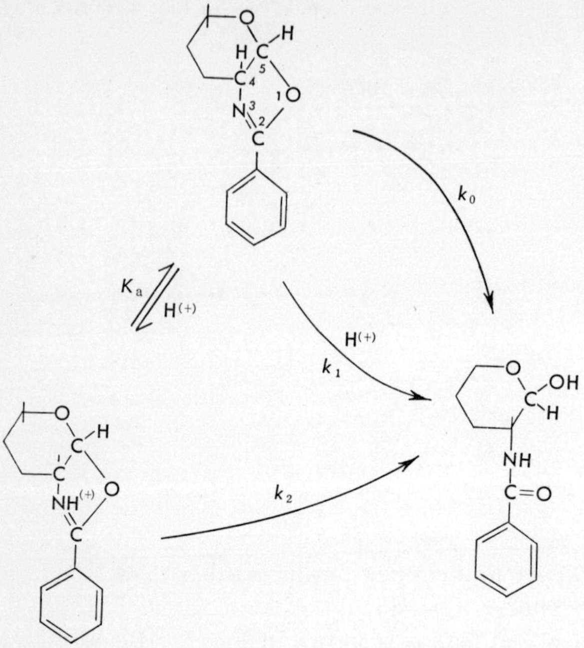

Scheme III

From the pH-logk_{obs} profile, the hydrolytic mechanism of PGO may be represented as above.

The kinetic equation corresponding to the above mechanism may be written as

$$v = (k_0 + k_1 a_{H^+})[S] + k_2[SH], \quad (4)$$

where v is the reaction rate, k_1 and k_2 the rate constants, k_0 that of spontaneous hydrolysis, and S and SH the neutral and protonated oxazoline, respectively. When K_a represents the association constant of the system of

$$S + H^+ \rightleftharpoons SH,$$

the following relations are obtained.

$$\left.\begin{aligned} [\mathrm{S}] &= \frac{K_a}{K_a + a_{\mathrm{H}^+}}[\mathrm{S_T}] \\ [\mathrm{SH}] &= \frac{a_{\mathrm{H}^+}}{K_a + a_{\mathrm{H}^+}}[\mathrm{S_T}] \\ K_a &= \frac{[\mathrm{S}]a_{\mathrm{H}^+}}{[\mathrm{SH}]},\ [\mathrm{S}]+[\mathrm{SH}] = [\mathrm{S_T}] \end{aligned}\right\} \tag{5}$$

The reaction rate v is written as follows by substituting Eq. 5 for Eq. 4.

$$v = \frac{k_0 K_a + (k_1 K_a + k_2)a_{\mathrm{H}^+}}{K_a + a_{\mathrm{H}^+}}[\mathrm{S_T}] \tag{6}$$

Then, k_{obs} is

$$k_{\mathrm{obs}} = \frac{k_0 K_a + (k_1 K_a + k_2)a_{\mathrm{H}^+}}{K_a + a_{\mathrm{H}^+}}\,. \tag{7}$$

Under the conditions of $a_{\mathrm{H}^+} \ll K_a$ (pH$\gg$pK), Eq. 7 is reduced to

$$\left.\begin{aligned} k_{\mathrm{obs}} &= k_0 + \left(k_1 + \frac{k_2}{K_a}\right)a_{\mathrm{H}^+} \\ \log(k_{\mathrm{obs}} - k_0) &= \log\left(k_1 + \frac{k_2}{K_a}\right) - \mathrm{pH}\,. \end{aligned}\right\} \tag{8}$$

On the other hand, if $a_{\mathrm{H}^+} \gg K_a$ (pH$\ll$pK),

$$\left.\begin{aligned} k_{\mathrm{obs}} &= k_1 K_a + k_2 \\ \log k_{\mathrm{obs}} &= \log(k_1 K_a + k_2) = \text{constant}\,. \end{aligned}\right\} \tag{9}$$

The terms $k_1[\mathrm{S}]a_{\mathrm{H}^+}$ and $k_2[\mathrm{SH}]$ cannot be distinguished kinetically. That is, the main pathway of the hydrolysis cannot be decided by kinetics.

TABLE V. Kinetic Parameters in Hydrolysis of PGO

$K_a = 10^{-3}$ M (pK_a=3.0)

$k_0 = 6.0\times10^{-6}$ sec^{-1}

$k_1 + \dfrac{k_2}{K_a} = 7.8$ M^{-1} sec^{-1}

$k_1 = 7.8$ M^{-1} sec^{-1} (assuming that reactive species is neutral form)

$k_2 = 7.8\times10^{-3}$ sec^{-1} (assuming that reactive species is protonated form)

$$k_{\mathrm{obs}} = \frac{6.0\times10^{-9} + 7.8\times10^{-3}\,a_{\mathrm{H}^+}}{10^{-3} + a_{\mathrm{H}^+}}$$

When the neutral oxazoline S is assumed to be reactive, k_2 could be considered to be zero, and k_1 can be calculated from the experimental data. Conversely, when protonated oxazoline SH is assumed to be reactive, k_2 can be calculated. The values of k_1 and k_2 thus calculated are listed in Table V together with that of k_0.

A Possible Mechanism of Lysozyme Catalysis

The pH dependence of the hydrolysis of BNAG showed a plateau in the pH-region between 1.5 and 8.5. This means that in this region BNAG was hydrolyzed spontaneously *via* an oxazoline intermediate which was formed by the intramolecular participation of the 2-acetamido group in the cleavage of the glycoside ester. This conclusion is supported by the fact that various β-glycosides are more rapidly hydrolyzed spontaneously than α-glycosides. In the acid region, BNAG showed a usual fission type in the cleavage of the glycoside ester linkage, *i.e.*, the fission occurred at the bond between C_1 of the pyranose ring and the oxygen atom of the ester group. BNAG was very labile in the alkaline region, and its hydrolysis was catalyzed by an amine such as Tris, resulting in the formation of an amide. This fact indicates that the fission occurred at the bond between the oxygen atom and carbonyl carbon atom of the ester group.

It may be assumed from many experimental results on the hydrolysis of glycosides that in general α-glycosides are more stable in the neutral region than are corresponding β-glycosides. If the glycosyl enzyme intermediate was formed between the sugar residue at subsite D and Asp-52 in a lysozyme catalyzed reaction, the glycosyl bond must take an α-anomeric structure. Because the α-glycosyl enzyme appears to be stable in a weak acid region, the hydrolysis of the intermediate should be catalytically accelerated by some catalytic groups in the active site of the enzyme. This could be done by Glu-35 in the conjugate base as a general base by withdrawing protons from the water molecule. However, in the glycosyl enzyme, it is likely that the carbonyl carbon atom of the ester group is more accessible to the attack of the hydroxide ion than the C_1 atom of the pyranose ring. The transesterification has been widely observed in the reaction of alkoxide ion with esters. In the case of transglycosylation, most acceptors are believed to be activated in the form of oxyanion by the general base catalysis of Glu-35. The proper align-

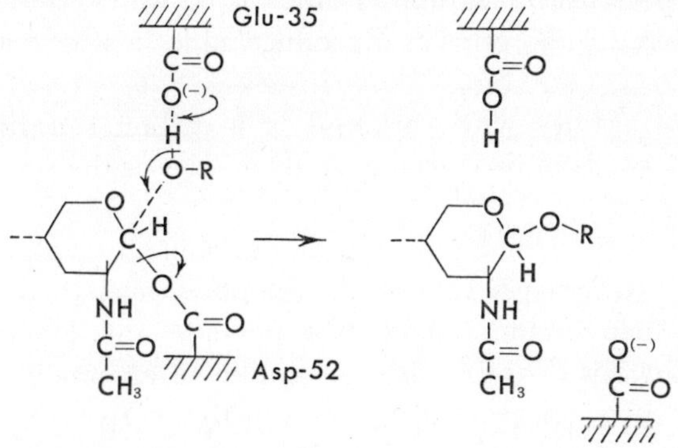

Glu-35

Asp-52

α-Glycosyl enzyme

Scheme IV

ment of Glu-35, the water or acceptor molecule and the C_1 carbon atom of the pyranose ring may, therefore, be necessary for the attack of the nucleophiles only at the C_1 atom of the pyranose ring, resulting in the retention of the β-anomeric structure of the product.

Several authors have pointed out that the formation of the glycosyl enzyme is not expected because the C_1 atom of the sugar residue at subsite D is situated too far from Asp-52 to form a covalent bond between them. The hypothesis of a glycosyl enzyme intermediate also cannot explain the role of the 2-acetamido group. Such findings in the lysozyme catalyzed hydrolysis may not exclude the possibility that other glycosidases (for instance, glucoamylase) form a glycosyl enzyme intermediate in their catalyzed reaction. From the present experiment, it can be concluded that when the glycosyl enzyme is formed as a reaction intermediate, the ester bond between the sugar residue and an enzyme carboxyl group needs to be hydrolyzed catalytically by the participation of the catalytic group in the active site, but not spontaneously.

In the hydrolysis in the pH-region between 2 and 10, the fission of the sugar oxazoline was observed at the bond between the C_5 and O_1 atoms, forming the amide as the final product. It was generally believed that the hydrolysis of oxazolines resulted in the formation of an ester in the acid medium and an amide in the alkaline medium. In the case of sugar oxazoline, the product was only the amide in a wide range of pH values, while in the presence of a limited amount of water, the sugar oxazoline was hydrolyzed to an ester as a product under a strong acidic condition (*20*).

It is likely that oxazolines with carbonyl or a carbonyl-originated carbon atom at the C_5 position produce generally amides as the products of hydrolysis.

The pH-hydrolytic rate constant profile of the sugar oxazoline exhibited a simple curve with the pK value of 3. For obtaining a theoretical curve which fits the experimental data, two possible and kinetically indistinguishable pathways in the hydrolysis could be assumed. One is the pathway *via* a neutral species and the other is that *via* a protonated oxazoline. At the present time, the author does not have any evidence to decide which is the real pathway. It is generally believed, however, that the protonated oxazoline is very unstable and hydrolyzes spontaneously.

Several authors have been emphasizing that in the mechanism of lysozyme catalyzed hydrolysis of a substrate composed of oligo- or polysaccharides of N-acetylglucosamine sugar oxazoline is formed as an intermediate. When the sugar oxazoline is assumed to be the intermediate of the lysozyme catalyzed reaction, the following become easily explainable. 1) Intramolecular participation of a 2-acetamido nucleophile in the cleavage of glycoside linkage. 2) The role of Asp-52 in connection with the formation and hydrolysis of the oxazoline. 3) The stabilization mechanism of the intermediate which seems to be necessary to account for the efficiency of the transglycosylation. 4) Anomeric retention of the product. 5) High efficiency of transglycosylation.*

Piszkiewizz and Bruice (*9, 11*) reported that the withdrawal of protons from the 2-acetamido group by Asp-52 as a general base may not accelerate the oxazoline formation, because the nucleophilic center of negatively charged amido group is always the nitrogen atom, and for the withdrawal of protons by Asp-52 the carbonyl oxygen atom of the amido group should rotate away from the direction of the C_1 atom of the sugar residue at subsite D. One of the possible roles of Asp-52 in the oxazoline formation may be the action as a general base to withdraw the proton from

* Sugar oxazoline has been widely used for the synthesis of oligosaccharides and glycosides.

the water molecule and to permit the attack of the resulting hydroxide ion to form a tetrahedral structure of the carbonyl carbon atom of the amido group, which seems to react easily with the C_1 atom of the sugar residue. The other is the participation of Asp-52 itself as a nucleophile in the oxazoline formation, *i.e.*, an Asp-52 nucleophile may attack the carbonyl carbon atom of the 2-acetamido group to form the ortho-ester, and the resulting oxyanion of carbonyl carbon would attack the C_1 atom of the sugar residue.

PGO was hydrolyzed by specific acid catalysis *via* one of two pathways which are kinetically indistinguishable as shown in Scheme III. However, it is reasonable to assume that the reactive species is protonated oxazoline. If Asp-52 can accelerate the hydrolysis of the oxazoline intermediate, there would be only one way in which Asp-52 may act as a general acid to donate protons, forming the protonated oxazoline intermediate. In this mode of action, Asp-52 in acid form can donate the proton either above or below the plane of the nitrogen atom without rotating the other adjacent atoms. For the working of this mechanism, Asp-52 must act first as a general base in the process of the formation of the oxazoline intermediate.

Because PGO exhibits the fission type by which the amide product is formed, it is clear that lysozyme catalysis does not need to alter the fission type of the oxazoline intermediate. Generally, a nucleophilic displacement of a hydroxide ion on an sp^3 carbon atom has been recognized to be quite difficult. It is, therefore, very likely that the attack of hydroxide ion, which was produced by the action of Glu-35 as a general base from the water molecule, on the C_5 atom of the sugar oxazoline intermediate should be assisted simultaneously by other factors such as the protonation of the oxazoline nitrogen atom.

A tentative mechanism including the oxazoline intermediate may be summarized in Scheme V.

The elucidation of the catalytic mechanisms of enzymes is one of the most difficult tasks in biochemistry, and there are many approaches to this subject. Any one approach cannot attain the goal, but each offers fragmental information on the subject. The use of a model reactions may also offer only part of the answer to the question of the catalytic mechanisms. It is hoped, however, that the physico-chemical studies on the model reactions dealt with in this paper may have shown a new angle for obtaining an insight into the mechanism of enzyme catalyses.

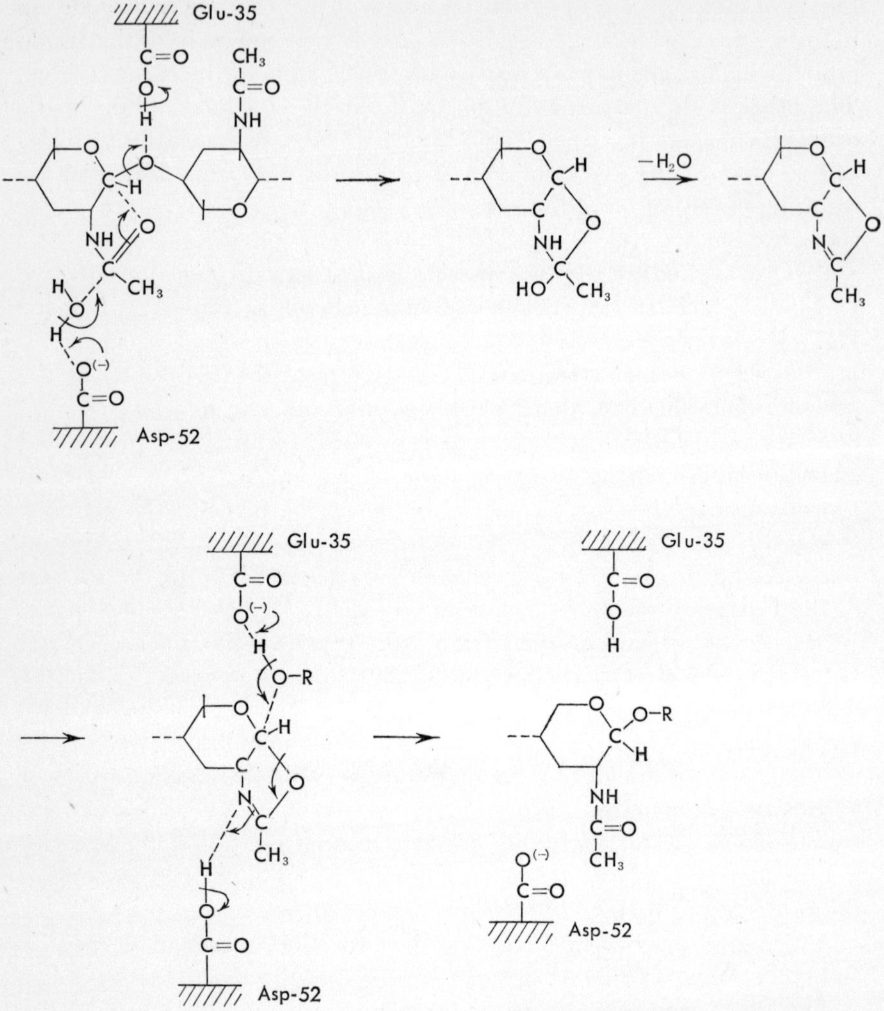
Glu-35
Asp-52
−H2O
Glu-35
Asp-52
Glu-35
Asp-52

Scheme V

SUMMARY

In lysozyme catalyzed hydrolysis and transglycosylation, the sugar residue located in subsite D should be stabilized for a certain interval as a reaction-intermediate. Although many studies have been made on the possible structure of the reaction-intermediate, it is not likely that enough evidence for deciding the real structure of the reaction-intermediate has been accumulated.

In the present experiment, 1-O-benzoyl 2-acetamido-2-deoxy-β-D-glucopyranoside and 2-phenyl-4,5-(D-glucopyrano)-Δ^2-oxazoline were synthesized as model reaction-intermediates and their modes of hydrolytic behavior were investigated. The possibility of glycosyl-lysozyme and the sugar oxazoline being the real intermediates in the lysozyme catalyzed reactions and the mode of action of the catalytic groups of the lysozyme molecule were discussed on the basis of the experimental results.

REFERENCES

1 D. C. Phillips, *Proc. Natl. Acad. Sci. U.S.*, **57**, 484 (1967).
2 C. C. F. Blake, G. A. Mair, A. C. T. North, D. C. Phillips and V. R. Sarma, *Proc. Roy. Soc.*, **B167**, 365 (1967).
3 J. A. Rupley, *Proc. Roy. Soc.*, **B167**, 416 (1967).
4 N. Sharon, *Proc. Roy. Soc.*, **B167**, 402 (1967).
5 M. A. Raftery and T. Rand-Meir, *Biochemistry*, **7**, 3281 (1968).
6 F. W. Dahlquist, C. L. Borders, Jr., G. Jacobson and M. A. Raftery, *Biochemistry*, **8**, 694 (1969).
7 J. A. Rupley, V. Gate and R. Bilbery, *J. Am. Chem. Soc.*, **90**, 5633 (1968).
8 F. W. Dahlquist, T. Rand-Meir and M. A. Raftery, *Biochemistry*, **8**, 4214 (1969).
9 D. Piszkiewicz and T. C. Bruice, *J. Am. Chem. Soc.*, **89**, 6237 (1967).
10 D. Piszkiewicz and T. C. Bruice, *J. Am. Chem. Soc.*, **90**, 2156 (1968).
11 D. Piszkiewicz and T. C. Bruice, *J. Am. Chem. Soc.*, **90**, 5844 (1968).
12 T. C. Bruice and S. J. Benkovic, "Bioorganic Mechanism," Benjamin Inc., New York, Vols. I and II (1966).
13 W. P. Jencks, "Catalysis in Chemistry and Enzymology," McGraw-Hill, New York (1969).
14 H. G. Fletcher, Jr., *in* "Methods in Carbohydrate Chemistry," ed. by R. L. Whistler and M. L. Wolform, Academic Press, New York, Vol. II, p. 231 (1963).
15 T. C. Bruice and J. L. York, *J. Am. Chem. Soc.*, **83**, 1382 (1961).
16 R. B. Martin and A. Parcell, *J. Am. Chem. Soc.*, **83**, 4835 (1961).
17 N. Pravdić, T. D. Inch and H. G. Fletcher, Jr., *J. Org. Chem.*, **32**, 1815 (1967).
18 F. Micheel and H. Köchling, *Ber.*, **93**, 2372 (1960).

19 J. de Jersey, P. Willadsen and B. Zerner, *Biochemistry*, **8**, 1959, 1967, 1975 (1969).
20 F. Micheel, F. P. van de Kamp and H. Peterson, *Ber.*, **90**, 521 (1957).

Received for publication July 4, 1971.

ANALYSIS OF ACTION PATTERNS OF AMYLASES WITH SPECIAL REFERENCE TO THEIR SUBSITE AFFINITIES

Keitaro Hiromi
The Laboratory of Enzyme Chemistry, Department of Food Science and Technology, Faculty of Agriculture, Kyoto University, Kyoto

For the enzymes which catalyze the degradation of homopolymer substrates, it is often observed that the rate of reaction is largely dependent on the degree of polymerization (DP) of substrates of linear chain molecules especially in the range of lower DP's (*1–7*). Since the interactions between substrate molecules and the active sites of enzymes become stronger as DP increases, it would naturally be expected that the Michaelis constant K_m will decrease with DP, which would lead to enhanced rates especially at substrate concentrations lower than the K_m values. In some cases, however, it has been noticed that the maximum velocity V also increases with DP in parallel with the decrease in K_m. Two typical examples obtained for hydrolyses of linear substrates catalyzed by an exo- and endo-amylase, glucoamylase from *Rhizopus delemar* (*8*) and Taka-amylase A (*2*), respectively, are shown in Figs. 1 and 2. In fact, tenfold and ten thousandfold differences in V are seen between maltose (DP=2) and maltoheptaose (DP=7) for glucoamylase and Taka-amylase A, respectively.

The chemical nature of the glucosidic linkages to be hydrolyzed can-

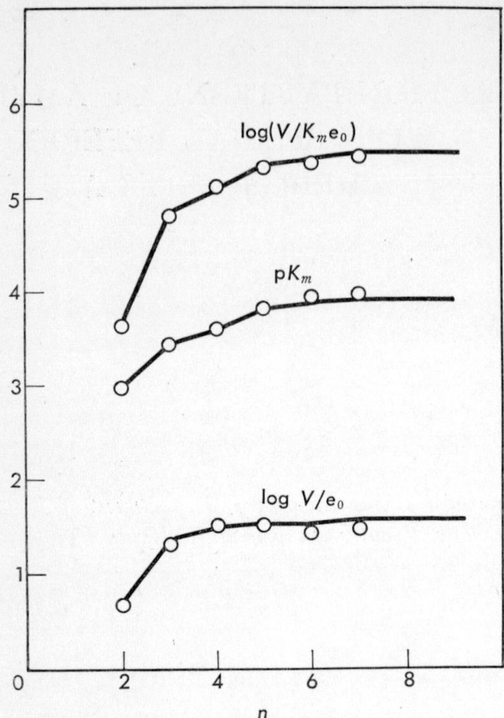

FIG. 1. Dependence of rate parameters on the degree of polymerization n for glucoamylase-catalyzed hydrolysis of linear malto-oligosaccharides (*7*). The open circles are the experimental points obtained at 25°C and pH 4.5 (*8*) (K_m in M, V/e_0 in sec^{-1}). The solid lines are theoretical curves calculated from Eqs. 15 through 17 by using the values of subsite affinities and k_{int} listed in Fig. 6 (see text).

not be considered to be very much dependent on DP, as is evidenced by the similarity in the rate constants of acid-catalyzed hydrolyses of maltose and amylose (0.37×10^{-3} min^{-1} and 0.14×10^{-3} min^{-1}, respectively, in 8% sulfuric acid at 55°C (*9*)). Therefore, it seems *a priori* more reasonable to assume that the intrinsic rate of hydrolysis of substrate linkage in the enzyme-substrate complex is essentially independent of DP of linear substrates (*7*). However, this assumption is apparently incompatible with the experimental results, in so far as the maximum velocity is regarded simply as the breakdown rate of the enzyme-substrate complex into products per unit enzyme concentration as the conventional

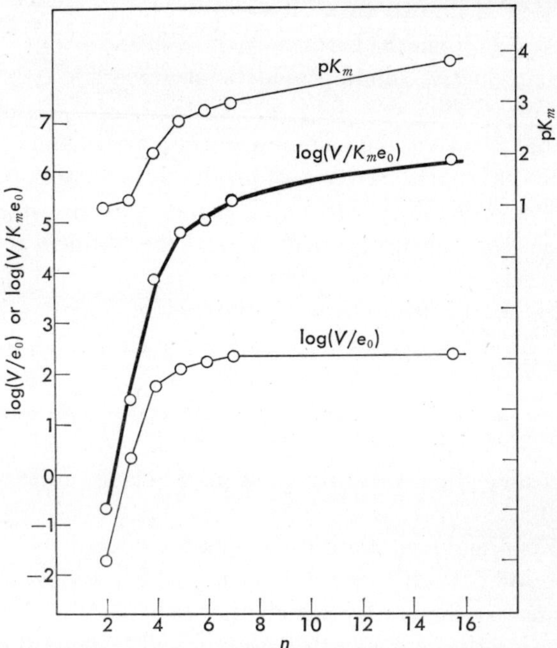

FIG. 2. Dependence of rate parameters on the degree of polymerization n for Taka-amylase A-catalyzed hydrolysis of linear malto-oligosaccharides (*2*). The open circles are the experimental points obtained at 25°C and pH 5.3 (*2*) (K_m in M, V/e_0 in $\sec^{-1}$). The thick solid line for $\log(V/K_m e_0)$ is the theoretical curve calculated from Eq. 17 by using the values of subsite affinities and k_{int} listed in Table II (see text).

Michaelis-Menten mechanism would predict. If we accept the above assumption, the remarkable DP dependence of V must be attributed to some other reasons.

The most likely possibility seems to be the multiplicity in binding modes of the substrate molecule to the active site of the enzyme (*10–12*). The recent success in X-ray analysis of lysozyme and its complex with tri-*N*-acetyl glucosamine (tri-NAG) revealed that the tri-saccharide, which acts as a competitive inhibitor to natural substrates, is bound in the active center cleft a few residues away from the catalytic site at which the substrate linkage is cleaved (*13*). Such an inactive complex is called a " nonproductive " complex and is distinguished from a " productive " complex from which products can be formed (*5, 10*). On the other hand, tri-NAG itself is known to be hydrolyzed by lysozyme, though much more slowly than longer substrates such as hexa-NAG (*5*). This indicates that the productive binding mode is also possible even for tri-NAG, though its probability may be low. In this stage, we may hope to account for the apparent dependency of rate parameters on DP, based on the probabilities of occurrence of productive and nonproductive complexes. Such a probability will be determined by the strength of interaction between the substrate molecule and the enzyme active site for a particular mode of binding. The strength of interaction may be ascribed to the contributions from individual interactions between substrate residues (monomer units constituting the substrate molecules) and local regions of enzyme active sites specifically interacting with the residues, which are called " subsites." Therefore, the arrangement of the subsites, each of which may have its specific affinity to a substrate residue, will determine the probability of a particular binding mode of a substrate.

The arrangement of subsite affinities is thus considered to be closely related to the " action pattern " of the enzyme, which involves the DP dependence of rate parameters by which the distribution of products of various DP's during the course of the reaction will be determined, and also the mode of cleavage of substrates (at which linkage a substrate molecule is hydrolyzed). In this article, a quantitative approach to the problem will be made for amylase-catalyzed hydrolyses of linear substrates based on simple models and assumptions. Most amylases, unlike lysozyme, have no appreciable transfer action which may complicate the analysis of rate data. Moreover, amylase is one of the few enzymes whose natural substrates consist of identical monomer units. For these reasons, amylases are considered to be especially well suited for studies of this kind.

Basic Concepts and Theory

In this section, the basic models and assumptions employed for the theoretical treatment including the definition of subsite affinities will be described first. Then, a theory will be developed which relates the subsite affinities with the measurable rate parameters and the mode of cleavage of substrates (*2*, *7*).

1. *Subsites and subsite affinities*

As suggested from the results of X-ray analysis of lysozyme and its complex with oligosaccharides (*13*), the active site of amylase may be assumed to be a cleft in which several glucose residues of linear substrates can be accommodated. Each glucose residue may be specifically

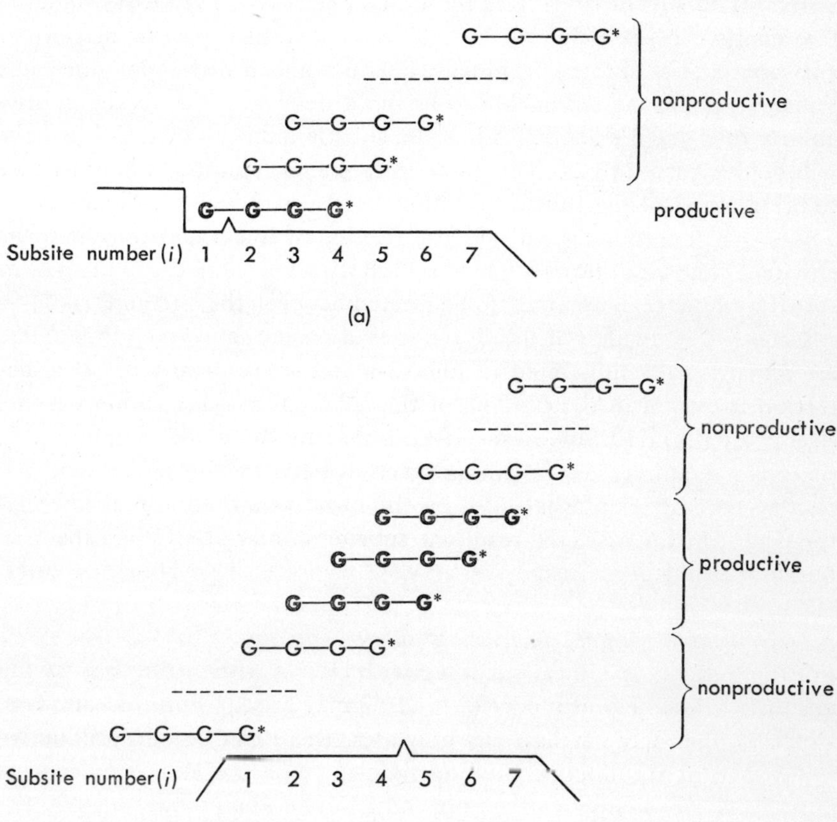

FIG. 3. Schematic models showing the active sites of an exo-enzyme (glucoamylase) and an endo-enzyme (Taka-amylase A) and the productive and nonproductive binding modes. The number refers to the subsite number. The wedge represents the catalytic site at which the substrate linkage is cleaved. G represents the glucose residue of a substrate (maltotetraose (n=4) in this case). Productive complexes are distinguished by bold letters. The reducing end is marked with an asterisk. a) Glucoamylase, b) Taka-amylase A.

bound to a corresponding " subsite," a local region complementary to each substrate residue. The number of subsites in the active center of an amylase will be denoted by m. A " subsite affinity " is defined for each individual subsite as the strength of the interaction between the subsite and a substrate residue which is expressed in free energy units. The subsite affinity is the unitary part (*14*, *15*) to the standard free energy decrease due to the interaction. Since linear substrates of amylases consist of identical glucose residues linked with α-1,4 glucosidic linkages, it may be reasonable to assign a definite value to the subsite affinity of a given subsite. Thus the subsite affinity of the i-th subsite will be designated by A_i (kcal/mole in free energy units), where i refers to the number of the subsite counting from a particular terminal subsite where the nonreducing end glucose residue of linear substrate is to be situated. The catalytic site, which is directly involved in the hydrolysis of substrate linkage, is assumed to be located between the r-th and $(r+1)$-th subsites. The models of the active sites of an exo- and an endo-amylase are schematically illustrated in Figs. 3-a and -b, respectively. If a distortion is caused in the binding of the substrate residue with a subsite, as suggested in lysozyme-hexa-NAG interaction by model building (*13*), the positive affinity due to the interaction between the subsite and the residue may be counterbalanced by the negative affinity due to the distortion. In this case, the resultant subsite affinity A_i of the subsite in question could be negative.*

2. *Association constant in terms of subsite affinities*

For the binding of a linear substrate with DP$=n$ with an amylase having m subsites, a variety of modes of binding may arise. For glucoamylase, which is one of the typical exo-amylases, there can be one productive and $(m-1)$ nonproductive binding modes. For the endo-amylases, at most $(n-1)$ productive and m nonproductive binding modes may be expected (see Fig. 3.)

Consider a binding equilibrium between an n-mer substrate S_n and

* In an earlier paper (7) dealing with exo-enzymes, the distortion-free energy, D, was assumed to arise at a terminal subsite only in a productive complex, which was defined separately from the net interaction affinity expressed as A_i, the resultant subsite affinity being (A_i-D). However, it is more convenient to define the subsite affinity A_i to include the possible distortion-free energy, D, as in a subsequent paper (2) and also in this article. Therefore, D does not appear explicitly here.

an enzyme E in a particular binding mode j, where j is the number specifying the mode of binding, either productive or nonproductive. (The number j is conveniently taken to equal the number of subsites at which the nonreducing or reducing end of the linear substrate is situated in that binding mode.) Thus,

$$E+S_n \rightleftharpoons ES_{n,j}\,, \tag{1}$$

where $ES_{n,j}$ is the enzyme-substrate complex of the j-th binding mode.

Let $K_{n,j}$ be the association constant defined by

$$K_{n,j}=[ES_{n,j}]/[E][S_n]\,. \tag{2}$$

The standard free energy change $\Delta G_{n,j}$ for the binding equilibrium then becomes

$$\Delta G_{n,j}=-RT\ln K_{n,j}\,, \tag{3}$$

where R and T are the gas constant and the absolute temperature, respectively. The standard free energy change $\Delta G_{n,j}$, or the standard affinity $-\Delta G_{n,j}$, can be divided into " unitary " and " cratic " parts (*14*, *15*). The unitary part of the standard affinity will be termed the " molecular binding affinity " and is designated by $B_{n,j}$, which represents the net specific interaction between the substrate molecule and the enzyme active site in that particular binding mode. The cratic part of the standard affinity arises merely from the nonspecific random mixing of solute species with solvent, which is common for solution equilibria accompanied by a change in number of species. This corresponds to the difference between the free energy of mixing of two species, $E+S_n$, with water and that of a single species, $ES_{n,j}$, and is denoted by $-\Delta G_{mix}$. It amounts to -2.4 kcal/mole at 25°C in water as solvent. Thus we have

$$-\Delta G_{n,j}=B_{n,j}-\Delta G_{mix}=B_{n,j}-2.4\text{ kcal/mole}\,. \tag{4}$$

Since $B_{n,j}$ is the net affinity due to the interaction between S_n and E in the j-th binding mode, we may assume, to the first approximation, that it is simply expressed by the sum of the subsite affinities of the subsites occupied by the substrate molecule in that binding mode.

$$B_{n,j}=\left(\sum_i^{\text{cov.}} A_i\right)_{n,j}, \tag{5}$$

where $\sum\limits_i^{\text{cov.}}$ indicates that the sum is to be taken for all the subsites covered by the binding. This is one of the fundamental assumptions on which the theory is based.

By combining Eqs. 3 through 5, the association constant can be expressed in terms of the subsite affinities as follows:

$$RT \ln K_{n,j} = \left(\sum_{i}^{\text{cov.}} A_i\right)_{n,j} - \Delta G_{mix}$$
$$= \left(\sum_{i}^{\text{cov.}} A_i\right)_{n,j} - 2.4 \text{ kcal/mole} \tag{6}$$

or

$$K_{n,j} = (0.018) \exp (B_{n,j}/RT) = (0.018) \exp (\sum_{i}^{\text{cov.}} A_i/RT)_{n,j} \,. \tag{7}$$

Now Eq. 7 allows us to calculate the association constant $K_{n,j}$ for any linear substrate in any binding mode, if the subsite affinities are known.

3. Rate parameters in terms of substite affinities

As the second fundamental assumption, we assume, as was mentioned earlier, that the intrinsic rate constant for the hydrolytic cleavage of substrate linkage in a productive complex, k_{int}, is constant, irrespective of either DP of the substrate or the binding mode. Then, the reaction scheme for the hydrolysis of an n-mer substrate S_n, involving multiple binding modes for both productive and nonproductive complexes, may be written as follows:

$$E + S_n \begin{cases} \overset{K_{n,p}}{\rightleftharpoons} ES_{n,p} \xrightarrow{k_{int}} E + P \\ \underset{K_{n,q}}{\rightleftharpoons} ES_{n,q} \end{cases} \tag{8}$$

where $ES_{n,p}$ and $ES_{n,q}$ represent the productive and nonproductive complexes of the n-mer substrate, respectively. The subscripts p and q refer to numbers used to discriminate the binding modes in both types of complexes, respectively.* $K_{n,p}$ and $K_{n,q}$ are the association constants for productive and nonproductive complexes, respectively, as defined in Eq. 2.

By assuming rapid equilibria between E and S_n, the initial rate v can

* The sum of the set consisting of p and that of q is equal to the set of j, which is used to specify a binding mode in general.

be expressed in the familiar form of the Michaelis-Menten type as follows:

$$v = \frac{V[\mathrm{S}_n]}{K_m + [\mathrm{S}_n]}, \tag{9}$$

where the Michaelis constant K_m and the maximum velocity V are given by the following expressions (*11*, *12*).

$$1/K_m = \sum_j K_{n,j} = \sum_p K_{n,p} + \sum_q K_{n,q} \tag{10}$$

$$V/e_0 = k_{int} \sum_p K_{n,p} / \sum_j K_{n,j} = k_{int} \sum_p K_{n,p} / (\sum_p K_{n,p} + \sum_q K_{n,q}) \tag{11}$$

and hence

$$V/K_m e_0 = k_{int} \sum_p K_{n,p}, \tag{12}$$

where e_0 is the enzyme concentration, and $\sum_p$ and $\sum_q$ mean that the sum is taken for all the productive and nonproductive complexes, respectively.*

Thus both K_m and V involve the terms arising not only from the productive complexes but from the nonproductive ones; the participation of the nonproductive terms tends to decrease both K_m and V in a similar manner. The maximum velocity is proportional to the fraction of productive complexes out of all the complexes. Since the probability of occurrence of a complex is dependent on the DP of substrate n, this fraction may also be dependent on n. Therefore, even if the intrinsic rate constant is independent of n as was assumed, V as well as K_m could be strongly n-dependent. Equation 11 shows that if the nonproductive term $\sum_q K_{n,q}$ is negligible compared with the productive term $\sum_p K_{n,p}$, V/e_0 becomes equal to k_{int}. This situation is realized for endo-amylases when n exceeds the number of subsites m, as will be discussed below in the case of Taka-amylase A.

It should be noted that, in contrast to K_m and V, the ratio V/K_m

* For simplicity, the possibility that more than one substrate molecule are bound at the active site has been ignored in the present treatment. The theory is apparently oversimplified in this sense. More rigorous treatment involving the binding of two substrate molecules has been made (*16*), which will be published elsewhere. This effect, however, is not appreciable except for the case of substrates with $n=2$ and 3.

involves the terms for productive complexes only. As will be seen shortly, this is a useful relationship for evaluating subsite affinities.

Since the association constants, $K_{n,j}$ have already been expressed in terms of subsite affinities, the combination of Eqs. 10 through 12 with Eq. 7 gives the expression of the rate parameters in terms of subsite affinities.

$$1/K_m = (0.018) \sum_j \exp(B_{n,j}/RT)$$

$$= (0.018) \sum_j \exp\Big(\sum_i^{\text{cov.}} A_i/RT\Big)_{n,j} \tag{13}$$

$$V/e_0 = k_{int} \frac{\sum_p \exp\Big(\sum_i^{\text{cov.}} A_i/RT\Big)_{n,p}}{\sum_j \exp\Big(\sum_i^{\text{cov.}} A_i/RT\Big)_{n,j}} \tag{14}$$

$$V/K_m e_0 = (0.018) k_{int} \sum_p \exp\Big(\sum_i^{\text{cov.}} A_i/RT\Big)_{n,p} \tag{15}$$

According to these equations, one can calculate all the rate parameters for a given n-mer substrate, if all of the subsite affinities A_i are known. Conversely we may be able to determine the subsite affinities from the rate parameters of various substrates.

4. *Evaluation of subsite affinities*

For exo-amylases, there can exist only one productive complex for a given substrate. Thus, Eq. 15 reduces for an n-mer substrate to

$$(V/K_m e_0)_n = (0.018) k_{int} \exp(B_{n,p}/RT)$$

$$= (0.018) k_{int} \exp\Big(\sum_i^{\text{cov.}} A_i/RT\Big)_{n,p} . \tag{16}$$

When two substrates, n-mer and $(n+1)$-mer, are bound in productive forms in such a manner as is shown in Fig. 4-a, where the only difference is that one particular subsite is occupied by $(n+1)$-mer but not by n-mer, it is apparent that the difference between the molecular binding affinities for the two complexes gives the subsite affinity of the subsite in question (A_{n+1} in this case). Thus we have

$$\ln (V/K_m e_0)_{n+1} - \ln (V/K_m e_0)_n = (B_{n+1,p} - B_{n,p})/RT$$

$$= \Big(\sum_i^{\text{cov.}} A_i/RT\Big)_{n+1,p} - \Big(\sum_i^{\text{cov.}} A_i/RT\Big)_{n,p} = A_{n+1}/RT . \tag{17}$$

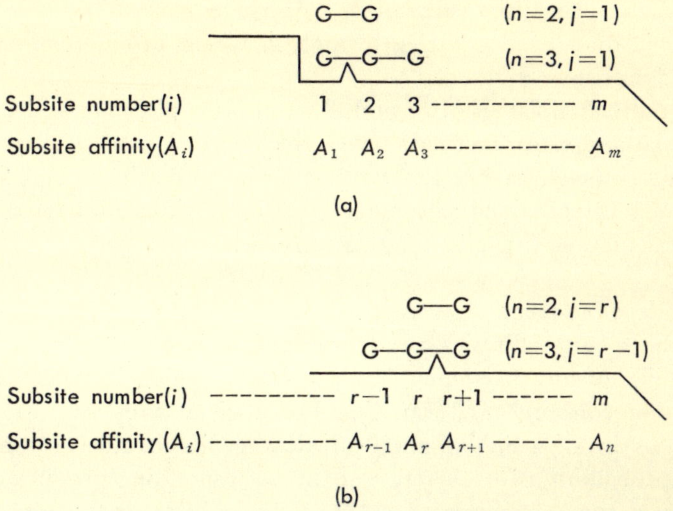

G—G (n=2, j=1)

G—G—G (n=3, j=1)

Subsite number(i) 1 2 3 ------------ m

Subsite affinity(A_i) A_1 A_2 A_3 ------------ A_m

(a)

G—G (n=2, j=r)

G—G—G (n=3, j=r−1)

Subsite number(i) ---------- r−1 r r+1 ------ m

Subsite affinity (A_i) ---------- A_{r-1} A_r A_{r+1} ------ A_n

(b)

FIG. 4. Examples for productive binding modes of n-mer and $(n+1)$-mer substrates for an exo-amylase (glucoamylase) (a) and for an endo-amylase (b); n is taken to be 2. The binding mode j is specified by the number of the subsite at which the nonreducing end glucose residue (on the left side) of the substrates is situated.

In this way, by using a series of substrates with n ranging from 2 to m, it is possible to determine the subsite affinities, A_3, A_4, ... , A_m, other than those of the two subsites on both sides of the catalytic site.

The same procedure can be applied to endo-amylases (Fig. 4-b), provided that only one productive complex is predominant for each of the n-mer and $(n+1)$-mer substrates, and that their binding modes can be inferred from the product analysis. An example will be shown later for Taka-amylase A.

The two subsite affinities of the subsites adjacent to the catalytic site, A_r and A_{r+1}, cannot be evaluated in this way, since they are common for all productive complexes. However, the sum (A_r+A_{r+1}) can be obtained if the value of k_{int} is known. Since there is only one productive complex for maltose ($n=2$) either for exo- or endo-amylase, where only the r-th and the $(r+1)$-th subsites are occupied, $(V/K_m e_0)$ for maltose is given by

$$\ln (V/K_m e_0)_2 = (A_r+A_{r+1})/RT + \ln k_{int} + \ln (0.018)\,. \tag{18}$$

For endo-amylases, the value of k_{int} may be estimated from the saturation value of V/e_0 at sufficiently high n, where productive complexes are predominant.

The individual values of A_r and A_{r+1}, in principle, may be determined by a trial-and-error method, using the expression of K_m or V/e_0 in terms of A_i (such as Eq. 13 or 14), in order to reach the best fit to the experimentally obtained rate parameters for various substrates. Computer calculation is useful for this purpose.

5. *Modes of cleavage*

Once the subsite affinities are determined, it is possible to predict the probability of any particular binding mode, which is reflected in the association constant $K_{n,j}$ (Eq. 7). For a given substrate, the rate of hydrolysis in a particular mode of cleavage is obviously proportional to the probability of occurrence of the corresponding productive mode of binding, since the intrinsic rate constant for hydrolysis in a productive complex is assumed not to be influenced by the binding mode. The relative probability of the modes of cleavage of a substrate, therefore, is directly related to the difference in the molecular binding affinities for the binding modes concerned.

For example, when the two binding modes of maltotetraose with an endo-amylase are as shown in Fig. 5, the relative rate of the following two modes of cleavage

$$\text{G–G–G–G*} \begin{cases} \nearrow \text{G–G + G–G*} & \text{(I)} \\ \searrow \text{G + G–G–G*}. & \text{(II)} \end{cases}$$

v_I/v_{II} is determined by the difference in the subsite affinities A_3 and A_7 which are not common in the two binding modes. Thus

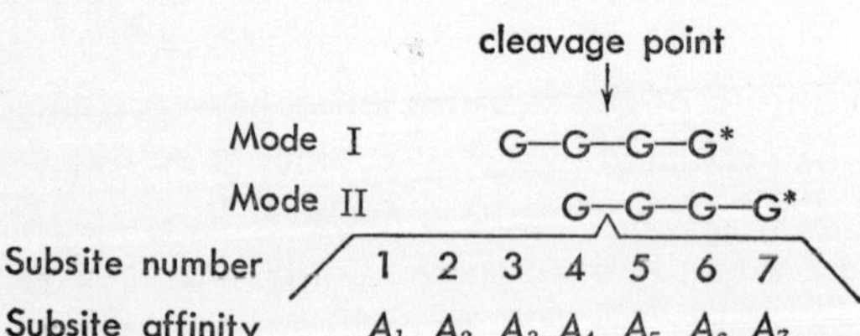

FIG. 5. Two productive binding modes leading to two different modes of cleavage of maltotetraose for an endo-amylase (Taka-amylase A). The reducing end glucose residue is labeled by an asterisk.

$$v_{I}/v_{II} = \exp(B_{I}-B_{II})/RT = \exp(A_3-A_7)/RT, \quad (19)$$

where B_I and B_{II} are the molecular binding affinities corresponding to the two modes I and II, respectively, the contribution from the common subsite affinities, A_4 to A_6 being cancelled by taking the difference. If the difference (A_3-A_7) amounts to 2 kcal/mole, for example, v_I/v_{II} becomes 30. Thus the mode of cleavage I will be dominant.

Conversely, the difference in the two subsite affinities may be evaluated, if the ratio v_I/v_{II} is experimentally obtainable. It is possible to determine the relative rates of various modes of cleavage by the use of terminally labeled substrates (*17*, *18*). In this way, Thoma *et al.* determined the relative values of subsite affinities of liquefying α-amylase from *Bacillus subtilis*, except for the two subsites adjacent to the catalytic site (*18*).

Application to Exo- and Endo-Amylases

The theory developed in the foregoing section has been applied to the hydrolyses of linear substrates catalyzed by three kinds of amylases; glucoamylase from *Rh. delemar* (exo-amylase) (*7*, *8*), Taka-amylase A from *Aspergillus oryzae* (*2*) and liquefying α-amylase from *B. subtilis* (*3*, *4*) (endo-amylases). Of these, the application to the former two amylases will be described briefly.

1. Glucoamylase

Glucoamylase is one of the typical exo-amylases, which splits off glucose from the nonreducing ends of starch (*19*). The values of rate param-

TABLE I. Rate Parameters for Glucoamylase-Catalyzed Hydrolyses of Malto-oligosaccharides (*8*)

DP (n)	K_m (M)	V/e_0 (sec^{-1})	$V/K_m e_0$ (M^{-1} sec^{-1})
2	1.1×10^{-3}	4.6	4.2×10^{3}
3	3.6×10^{-4}	23	6.4×10^{4}
4	2.5×10^{-4}	33	1.3×10^{5}
5	1.6×10^{-4}	32	2.0×10^{5}
6	1.2×10^{-4}	28	2.3×10^{5}
7	1.1×10^{-4}	31	2.8×10^{5}

At 25°C and pH 4.5.

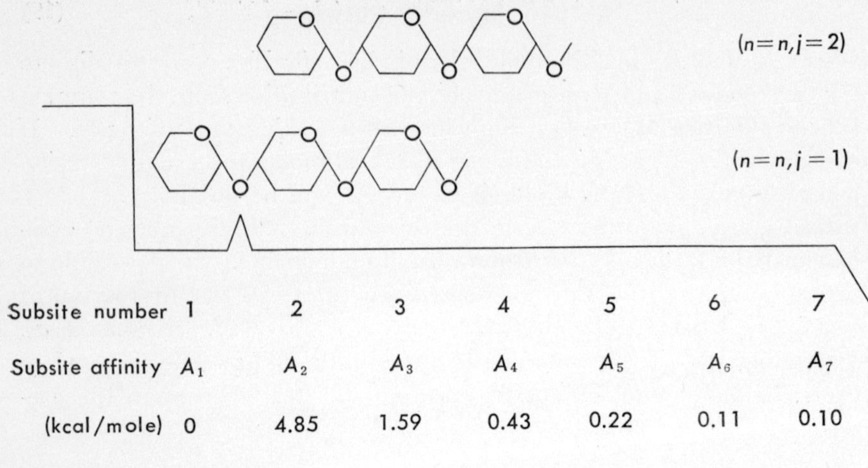

FIG. 6. Arrangement of subsite affinities of glucoamylase (*7*, *8*). The two dominant binding modes of linear substrates are also shown. The numerical values show the subsite affinities evaluated (see text). The intrinsic rate constant of hydrolysis of substrate linkage in a productive complex was determined to be $k_{int}=77\ \text{sec}^{-1}$.

eters, K_m, V/e_0 and V/K_me_0, obtained for a series of linear substrates (oligosaccharides with $n=2$–7) at pH 4.5 and 25°C are summarized in Table I (*8*). Figure 6 shows a schematic model for the active site of glucoamylase, where the subsites are numbered as indicated. The number specifying the binding mode j of a linear substrate is taken equal to the number of the subsite at which the nonreducing terminal glucose residue of the substrate is situated.

First, the subsite affinities A_3, A_4, . . . A_7 are determined according to the procedure stated in the former section (Eq. 17) by using the log (V/K_me_0) values of maltose through maltoheptaose. The values thus obtained are listed in Fig. 6. The largest of them, A_3, amounts only to 1.59 kcal/mole, and is appreciably lower than the " apparent " molecular binding affinity of maltose $B_{2,app}$, which is calculated to be

$$B_{2,app} = RT \ln (1/K_m) + \Delta G_{mix} = 4.01 + 2.4 = 6.4 \text{ (kcal/mole)}.$$

This implies that either A_1 or A_2, or at least their sum, should be appreciably greater than A_3; otherwise the large $B_{2,app}$ value could not be accounted for. Therefore we may conclude that for all substrates, the major binding modes are $j=1$ (productive) and $j=2$ (nonproductive); the binding modes ($j \geq 3$) which do not involve A_1 and A_2 will have minor contribution. Thus we have

$$1/K_m = \sum_j K_{n,j} = K_{n,1}+K_{n,2}\,, \tag{20}$$

where the nonproductive terms $K_{n,3}$, $K_{n,4}$, . . . are neglected. As readily seen from Fig. 6,

$$B_{n,2}-B_{n,1} = A_{n+1}-A_1\,. \tag{21}$$

Then substituting Eqs. 7, 20 and 21 into Eq. 11, we have

$$[k_{int}/(V/e_0)_n-1]\exp(A_1/RT) = \exp(A_{n+1}/RT)\,. \tag{22}$$

Since the values of A_{n+1} ($n \geq 2$) have been determined, the plot of $(e_0/V)_n$ against $\exp(A_{n+1}/RT)$ for various n allows us to evaluate k_{int} and A_1 (*8*). These values were determined to be $k_{int}=77\ \text{sec}^{-1}$ and $A_1=0$ kcal/mole. The apparent zero value for A_1 may suggest that the interaction between the subsite and glucose residue is counterbalanced by distortion brought about in the vicinity of the catalytic site.

Finally, the remaining subsite affinity A_2 can be evaluated from the K_m value and other subsite affinities determined above. For example, with maltose ($n=2$) the following equation

$$\begin{aligned} 1/K_m &= K_{2,1}+K_{2,2} \\ &= (0.018)[\exp(A_1+A_2)/RT+\exp(A_2+A_3)/RT] \end{aligned} \tag{23}$$

may be solved for A_2. It is also possible to similarly evaluate A_2 with higher substrates. A satisfactorily constant value of A_2 was obtained for $n=2$–6: $A_2=4.85$ (± 0.10) kcal/mole (*7, 8*).

Thus all the subsite affinities and the intrinsic rate constant k_{int} for the hydrolysis of substrate linkage have been determined (*7*). These values are included in Fig. 6. The decreasing tendency of A_i (except for A_1) with increasing distance from the catalytic site should be noticed.

In this stage, it is interesting to see how closely these values of A_i and k_{int} can reproduce the rate parameters to be compared with experimental values, when they are substituted into Eqs. 13 through 15. The theoretical values of rate parameters as functions of n are shown by solid lines in Fig. 1. The good agreement between theoretical and experimental values seems to support the validity of the theory and the basic assumptions.

In terms of the A_i values thus obtained, the large discrepancy between the rates of hydrolyses of phenyl α-maltosides and of phenyl α-glucosides catalyzed by glucoamylase (the ratio of V/e_0 for the two substrates amounts to 500–1000) has also reasonably been accounted for (*20*).

2. *Taka-amylase A*

Taka-amylase A is one of the typical endo-amylases which hydrolyzes α-1,4 glucosidic linkages of amylose nearly (but not strictly) in a random manner. Maltose had generally been believed not to be hydrolyzed by α-amylases. However, it has been shown that Taka-amylase A actually hydrolyzes maltose (*21*), though at an extremely lower rate compared with higher oligosaccharides. As seen in Fig. 7, the dependence of rate parameters on the substrate chain length is much more remarkable than in glucoamylase. The reciprocal Michaelis constant increases sharply with DP up to $n=7$, then less rapidly increases with decreasing increment.* The maximum velocity, on the other hand, increases crucially with n, and then levels off above $n=7$.

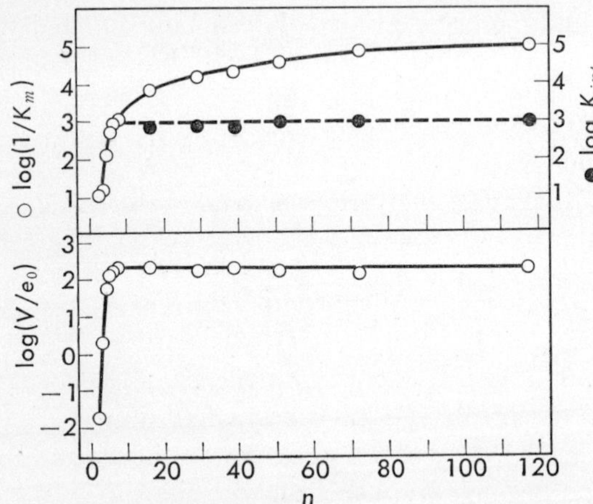

FIG. 7. Plots of K_m, V/e_0 and K_{int} *vs.* DP (n) for Taka-amylase A catalyzed hydrolyses of linear substrates (*2*). K_{int} was calculated from Eq. 24 assuming $m=7$. ○ log $(1/K_m)$ or log V/e_0; ● log K_{int}.

* The Michaelis constant K_m is expressed in units of moles of molecules (not linkages) per liter throughout this paper.

The most important difference in the binding modes of endo-amylases from that of exo-amylase resides in the multiplicity of productive complexes in the former. Since linear substrates tend to occupy as many subsites as possible to acquire the highest molecular binding affinity, all the m subsites will be occupied by the substrates with the n value larger than m. Moreover, these productive complexes could occur in $(n-m+1)$ ways. (This factor, which increases linearly with increasing n, may be termed " degeneracy.") In contrast, nonproductive complexes, in which a reduced number of subsites are occupied, may occur only in m ways, where m is a fixed number. Therefore, when n exceeds m, productive complexes will become dominant over nonproductive ones. Hence Eqs. 10 and 11 are simplified to

$$1/K_m = \sum_p K_p = (n-m+1)K_{int} \tag{24}$$

$$V/e_0 = k_{int}\,, \tag{25}$$

where K_{int} denotes the association constant of a productive complex in which all the subsites are covered by the substrate residues. Equation 24 suggests that $K_m(n-m+1)$ should be constant when n exceeds m. This property may be used to estimate the number of subsites m, together with the value of n above which V/e_0 becomes constant ($=k_{int}$). In this way, we estimated the number of subsites m of Taka-amylase A to be 7(*2*). The value of $1/K_{int}$ ($=K_m(n-m+1)$) calculated with $m=7$ for various n are plotted by closed circles in Fig. 7. The constancy of K_{int} shows the validity of the estimation.

Next the subsite affinities A_i were evaluated with the help of the knowledge about the modes of splitting of oligosaccharides determined by Okada *et al.* with terminally labeled substrates (*22*). A single main mode of splitting has been observed for each of the oligosaccharides with $n=4$–6, which is shown by the arrows in Fig. 8. These results strongly suggest that the catalytic site, at which the substrate linkage is cleaved, is situated between the fourth and the fifth subsites counted from the nonreducing terminal (Fig. 8). The cleavage point for maltotriose, for which the data were not available, was estimated by analogy with that of phenyl α-maltoside which liberates phenol but not phenyl α-glucoside. Maltoheptaose has been found to be split mainly at two positions giving rise to reducing end-labeled maltotriose and maltopentaose (*22*).

The sum of the subsite affinities (A_4+A_5) of the subsites adjacent to

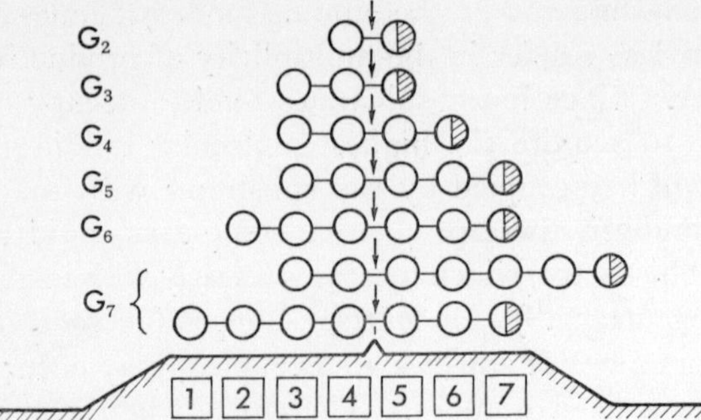

FIG. 8. Main modes of productive binding of malto-oligosaccharides with Taka-amylase A (*2*). The arrow shows the point of cleavage as suggested from the product analysis of terminally labeled substrates (*22*) (except for maltose and maltotriose). The subsites are numbered as indicated. G_n represents a linear substrate with DP=n. ○ and ◍ refer to the glucose residue and the reducing end glucose residue, respectively.

the catalytic site can readily be obtained using Eq. 18, from the value of $V/K_m e_0$ for maltose (=0.22 M^{-1} sec^{-1}) and k_{int} obtained from the saturation value of V/e_0 (=2.0×10^2 sec^{-1}). Thus we have $(A_4+A_5)=-1.7$ kcal/mole.* The negative value may again be indicative of some strain (distortion) brought about in the productive complex near the catalytic site.

Other subsite affinities were evaluated according to the procedure

* Otherwise, the sum (A_4+A_5) can be evaluated by curve-fitting procedure to obtain the best fit to the experimentally obtained $V/K_m e_0$ values of various DP with Eq. 15, after determining other A_i values. A value of -1.9 kcal/mole, rather than -1.7 kcal/mole, for (A_4+A_5) was found to give better fit to the experimental points (*23*) (see the thick line for $\log(V/K_m e_0)$ in Fig. 2). ** The determination of individual values of A_4 and A_5 will not be attempted here. The equations for the rate parameters (Eqs. 13 and 14) which should be used for this purpose have been derived by neglecting the binding of more than one substrate molecule to the enzyme active site. It leads in effect to overestimation of the nonproductive terms especially for lower n. Therefore, appreciable deviation in K_m and V/e_0 may be expected for maltose and maltotriose for which the above effect may not be ignored. With more elaborate rate equations considering the binding of two substrate molecules (*16*), the individual values of A_4 and A_5 were determined to fit all the rate parameters for all the substrates (*23*). The result showed that a reasonable agreement between the calculated and experimental values of rate parameters was obtained for a certain set of A_4 and A_5. *** Only maltoheptaose is exceptional; the reason why G_2+G_5* is more frequent than G_3+G_4* is not easily accounted for.

described in the foregoing section (Eq. 17(. For example, the subsite affinity A_3 can be obtained (Figs. 4-b and 8)

$$A_3 = RT[\ln (V/K_m e_0)_{\mathrm{maltotriose}} - \ln (V/K_m e_0)_{\mathrm{maltose}}]$$

and so on. A_1 was determined with the value for maltoheptaose and other A_i values assuming the two predominant productive binding modes (*2*).

The values of subsite affinities and k_{int} thus determined are summarized in Table II. The theoretical curve for $V/K_m e_0$ drawn with these values of A_i and of k_{int} and Eq. 15 (*23*) are in good agreement with the experimental points as shown in Fig. 2.**

It is seen that the modes of cleavage of oligosaccharides actually observed for maltotetraose to maltohexaose are essentially consistent with those predicted from the evaluated subsite affinities.*** For example, a possible mode of cleavage of maltopentaose:

$$\text{G-G-G-G-G*} \longrightarrow \text{G-G-G} + \text{G-G*} \qquad \text{(I)}$$

would be appreciably less frequent than the predominant mode:

$$\text{G-G-G-G-G*} \longrightarrow \text{G-G} + \text{G-G-G*}, \qquad \text{(II)}$$

since the molecular binding affinity corresponding to Mode I is less than that for Mode II by

$$A_7 - A_2 = 1.3 - 0.3 = 1.0 \text{ (kcal/mole)},$$

which leads to the cleavage rate ratio $v_{\mathrm{I}}/v_{\mathrm{II}}$ of about 1/5.4, as seen from Eq. 19 (see Fig. 8 and Table II).

TABLE II. Subsite Affinities and k_{int} for Taka-amylase A (*2*)

Subsite number (i)	1	2	3	4 5	6	7
Subsite affinity (A_i)	A_1	A_2	A_3	$(A_4+A_5)_8$	A_6	A_7
(kcal/mole)	0.3	0.3	3.0	−1.7[a] −1.9[b]	3.2	1.3
k_{int} (sec^{-1})				200		

At 25°C and pH 5.3.
The catalytic site is situated between the fourth and the fifth subsites counted from the nonreducing end side. [a] Evaluated from $(V/K_m e_0)$ value of maltose. [b] Evaluated from the curve fitting for $(V/K_m e_0)$ values of n=2–15.5 (*23*). (This value was used to draw the theoretical curve for $(V/K_m e_0)$ in Fig. 2.)

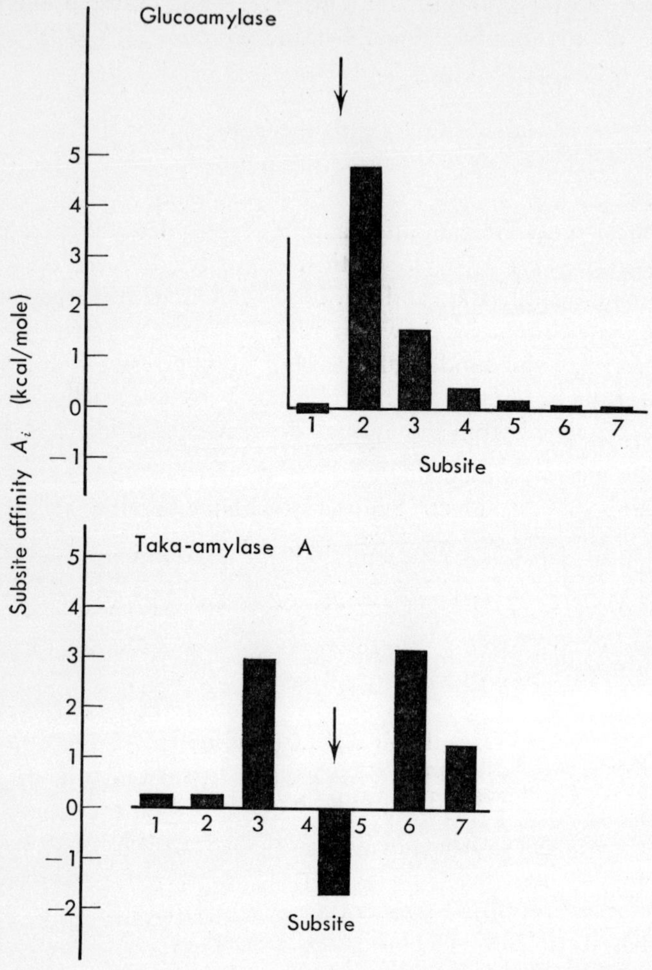

FIG. 9. Comparison between the arrangements of subsite affinities of glucoamylase and Taka-amylase A. The arrow shows the position of the catalytic site. For the 4th and 5th subsites of Taka-amylase A the sum A_4 and A_5 was shown.

The remarkable difference between the arrangements of subsite affinities for the two amylases, glucoamylase and Taka-amylase A, is clearly visualized in Fig. 9. It is interesting to note that on both sides of the catalytic site of Taka-amylase A are located two subsites with higher affinities which are sufficient to compensate for the negative affinities near the catalytic site.

Concluding Remarks

The application of the treatment to the two kinds of amylase has shown that the dependence of rate parameters on DP of linear substrates and the mode of cleavage of substrates can reasonably be interpreted on the basis of the present theory. It should be emphasized that the observed large increase in the maximum velocity with the increase in DP in lower DP range could be accounted for solely in terms of the probability of occurrence of productive complexes, even when the intrinsic rate of hydrolysis of substrate linkage in a productive complex is assumed to be entirely independent of DP.

The magnitudes of subsite affinities evaluated for the two amylases seem to be reasonable in view of the interaction involved in the binding (presumably hydrogen bonding) and of the distortion free energy which is estimated to be approximately 3–10 kcal/mole. Thus the results may be considered to support the validity of the theory and the basic assumptions employed.

Various extensions of the present treatment may be considered. For example, a change in action pattern of an enzyme may be brought about, if the subsite affinity of a subsite is altered by a chemical modification. Calculation has shown the possibility that the introduction of a bulky group into a certain subsite not far from the catalytic site could lead to apparent enhancement of enzyme activity (increase in V/e_0) by a factor of a few hundred for small substrates, but to virtual loss of activity for longer substrates (*24*). Alteration of this kind may possibly be of some practical use.

Subsites may be regarded as structural units constituting the active site of an enzyme. Their magnitudes and arrangement govern its function as reflected upon the action pattern toward various substrates. In this sense, therefore, the present treatment may be said to provide a new approach towards elucidating the structure-function relationship of enzymes at the level of subsites as structural units.

Summary

An attempt has been made to interprete quantitatively the dependency of rate parameters on the degree of polymerization (DP) of linear homopolymer substrates catalyzed by polymer-degrading enzymes based on a simple theory, which is derived from the following basic assumptions.

1. The active site of the enzyme consists of a certain number of subsites, each of which has its own subsite affinity A_i (in free energy units) for a substrate residue.
2. The probability of formation of an enzyme-substrate complex in a particular mode of binding j (either productive or nonproductive) is proportional to exp $(B_{n,j}/RT)$, where the molecular binding affinity $B_{n,j}$ is simply expressed by the sum of A_i of the subsites which are covered by the substrate in that mode of binding.
3. The intrinsic rate constant k_{int} for the cleavage of substrate linkage in a productive complex is constant, irrespective of the binding mode and DP of substrate.

Considering the multiple binding modes of a substrate, the rate parameters are expressed in terms of the subsite affinities, the arrangement of which determines the characteristic action pattern of the enzyme, including the DP-dependence of rate parameters and the modes of cleavage of substrates. A procedure for evaluating subsite affinities from the DP-dependence of the rate parameters has been described.

The theory was applied to the hydrolyses of linear substrates catalyzed by glucoamylase and Taka amylase A, and their subsite affinities were evaluated. The observed strong dependence of the maximum velocity on DP of substrates, which is difficult to reconcile with the simple Michaelis-Menten mechanism, was shown to be reasonably accounted for by the theory. Possible extension of the treatment has been discussed.

References

1 S. Ono, K. Hiromi and M. Zinbo, *J. Biochem.*, **55**, 315 (1964).
2 Y. Nitta, M. Mizushima, K. Hiromi and S. Ono, *J. Biochem.*, **69**, 567 (1971).
3 S. Iwasa, K. Hiromi and H. Hatano, *Seikagaku*, **42**, 755 (1970) (in Japanese).

4 H. Aoshima, K. Hiromi and H. Hatano, *Seikagaku*, **42**, 543 (1970) (in Japanese).
5 J. A. Rupley and V. Gates, *Proc. Natl. Acad. Sci. U. S.*, **57**, 496 (1967); J. A. Rupley, *Proc. Roy. Soc., Ser. B*, **167**, 416 (1967).
6 C. J. Yang and M. F. Singer, *J. Biol. Chem.*, **245**, 995 and 1005 (1970).
7 K. Hiromi, *Biochem. Biophys. Res. Commun.*, **40**, 1 (1970).
8 K. Hiromi, Y. Nitta, C. Numata and S. Ono, unpublished.
9 J. N. BeMiller, *in* " Starch: Chemistry and Technology," ed. by R. L. Whistler and E. Paschall, Academic Press, New York, Vol. I, p. 495 (1965).
10 J. A. Thoma and D. E. Koshland, Jr., *J. Am. Chem. Soc.*, **82**, 3329 (1960).
11 K. R. Hanson, *Biochemistry*, **1**, 723 (1962).
12 G. E. Hein and C. Niemann, *J. Am. Chem. Soc.*, **84**, 4495 (1962).
13 C. C. F. Blake, L. N. Johnson, G. A. Mair, A. C. T. North, D. C. Phillips and V. R. Sarma, *Proc. Roy. Soc., Ser. B*, **167**, 378 (1967); D. C. Phillips, *Proc. Natl. Acad. Sci. U. S.*, **57**, 484 (1967).
14 R. W. Gurney, " Ionic Processes in Solution," McGraw-Hill, New York, pp. 90 (1953).
15 W. Kauzmann, *Advan. Protein Chem.*, **14**, 1 (1959).
16 S. Shibata, M. Sakoda, K. Hiromi and H. Hatano, Abstracts of the 24th Annual Meeting of Chemical Society of Japan, Vol. III, p. 1853 (1971) (in Japanese).
17 J. F. Robyt and D. French, *J. Biol. Chem.*, **245**, 3917 (1970).
18 J. A. Thoma, C. Brothers and J. Spradlin, *Biochemistry*, **9**, 1768 (1970).
19 Y. Tsujisaka, J. Fukumoto and T. Yamamoto, *Nature*, **131**, 770 (1958).
20 N. Suetsugu, E. Hirooka, H. Yasui, K. Hiromi and S. Ono, unpublished.
21 Y. Nitta, K. Hiromi and S. Ono, *J. Biochem.*, **63**, 632 (1968).
22 S. Okada, S. Kitahata, M. Higashihara and J. Fukumoto, *Agr. Biol. Chem.*, **33**, 900 (1969).
23 S. Shibata. Thesis for Masters Degree, Faculty of Science, Kyoto University (1972).
24 K. Hiromi, S. Shibata and M. Ohnishi, to be published.

Received for publication August 11, 1971.

MODE OF ACTION OF ATP CITRATE LYASE

Fujio Suzuki
Department of Biochemistry, Osaka University Dental School, Osaka

ATP citrate lyase (EC 4.1.3.8, also known as citrate cleavage enzyme) catalyzes the following reaction (*1*, *2*).

$$\text{citrate} + \text{ATP} + \text{CoA} \xrightarrow{Mg^{2+}} \text{acetyl-CoA} + \text{oxaloacetate} + \text{ADP} + P_i$$

This enzyme has been found in the soluble fractions of a variety of animal tissues (*3*, *4*) and has been shown to play an important role in lipogenesis (*5*, *6*) and gluconeogenesis (*7*) as a supplier of acetyl-CoA and oxaloacetate, respectively, in the extramitochondrial compartment of the cell. In earlier studies in this laboratory (*8*), ATP citrate lyase was purified from rat liver as a single homogeneous protein in sedimentation, in moving boundary electrophoresis, and in immunochemical analysis.

On the mechanism of this enzyme reaction, Eggerer and Remberger (*9*) reported in 1963 that synthetic citryl-CoA is cleaved to acetyl CoA and oxaloacetate by a partially purified enzyme preparation from chicken liver in the absence of added ATP and Mg^{2+}. In 1964, Srere and Bhaduri (*10*) confirmed the enzymatic splitting of synthetic citryl-CoA to acetyl-CoA and oxaloacetate, but they failed to show the formation of citrylhydroxamate. Accordingly, they proposed that citryl-CoA is an intermediate which is probably enzyme-bound in the cleavage reaction. These results strongly suggested that the ATP citrate lyase reaction involves activation of citrate by ATP to form citryl-CoA, but the experimental data available until 1966 were still insufficient for elucidation of its reaction mechanism. From that time, this laboratory has attempted to obtain a penetrating insight into the mode of action of ATP citrate lyase with a purified preparation from rat liver.

Reaction Mechanism

The first experiment was to examine the participation of ATP. It has been shown that ATP is degraded to ADP and P_i during the reaction. This type of ATP splitting has been reported to occur in several reactions, including those of glutathione synthetase (*11*), succinyl-CoA synthetase (*12*), and pyruvate carboxylase (*13*). In these reactions, it

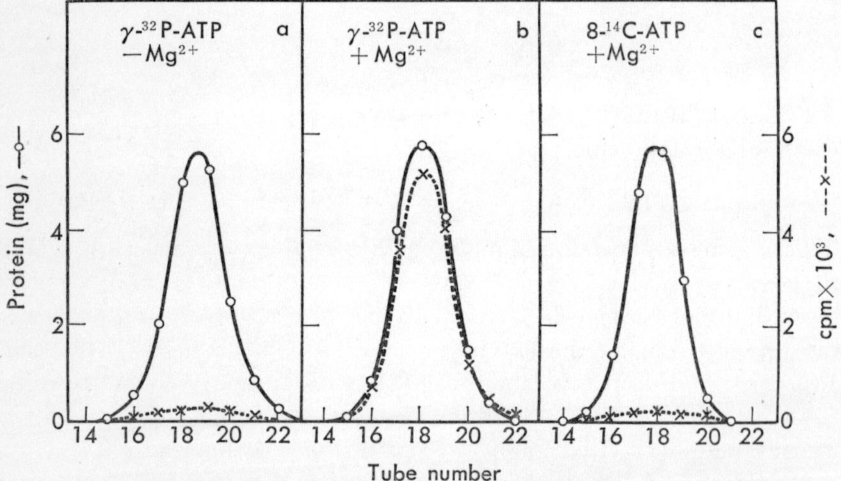

FIG. 1. Formation of phosphorylated enzyme (*14*). The reaction mixture contained (in μ moles): Tris-HCl buffer (pH 8.0) 50, 2-mercaptoethanol 10, $MgCl_2$ 1, ATP 1 (8-^{14}C-ATP, 235,000 cpm or γ-^{32}P-ATP, 215,000 cpm) and enzyme (18.5 mg protein) in a total volume of 1 ml. Mg^{2+} was omitted as indicated. After incubation for 5 min at 37°C, the reaction mixture was chilled and then passed through a column of Sephadex G-50 (1.5 × 30 cm).

has been shown that there are interactions between the enzyme and ATP in the absence of added substrate. Thus, 8-^{14}C-ATP and γ-^{32}P-ATP were separately incubated with the enzyme in the presence of Mg^{2+}, but in the absence of citrate. After incubation for 5 min at 37°C, the reaction mixture was immediately filtered through a column of Sephadex G-50. As shown in Fig. 1, the radioactivity associated with the protein was obtained only with γ-^{32}P-ATP and none was found with 8-^{14}C-ATP. This finding indicated that the enzyme protein itself was phosphorylated by ATP to form an activated, *i.e.*, phosphorylated (E~P) form of the enzyme, and that this process required Mg^{2+}. To confirm this, an ATP-ADP exchange experiment was carried out. When 8-^{14}C-ADP or $^{32}P_i$ was incubated with the enzyme in the presence of unlabeled ATP but in the absence of citrate, the enzyme catalyzed a rapid exchange of the isotope between 8-^{14}C-ADP and ATP, whereas it was incapable of catalyzing an exchange between ATP and $^{32}P_i$. These results provided additional evidence for the formation of E~P.

The next question was whether this E~P was a true intermediate of the reaction. Accordingly, E~^{32}P was prepared and then incubated with 1,5-^{14}C-citrate in the presence or absence of CoA and the formation of 4-^{14}C-oxaloacetate was tested. It was found that the formation of oxaloacetate was quantitative, judging from the phosphate bound to the enzyme, the ratio being close to unity. In this reaction, oxaloacetate was formed from citrate only in the presence of added CoA. Neither ATP nor Mg^{2+} was necessary in this conversion. Moreover, the bound phosphate was completely released from the enzyme on incubation with excess citrate in the absence of added CoA.

The next experiment was to determine whether the E~S complex was formed as a reactive intermediate during the reaction. The enzyme was incubated with 1,5-^{14}C-citrate in the presence or absence of ATP and Mg^{2+}, and after incubation the protein portion was isolated by filtration on Sephadex G-50 columns. As can be seen from Fig. 2, no significant labeling was associated with the protein in the absence of either ATP or Mg^{2+}. On the other hand, a substantial quantity of radioactivity emerged from the column with the enzyme in the presence of both ATP and Mg^{2+}. This coincidence between ^{14}C-labeling and the protein in the Sephadex elution peaks suggested the formation of citrylated enzyme (E~citrate). In the next experiment, E~citrate-^{14}C was isolated and then incubated with and without CoA. It was found

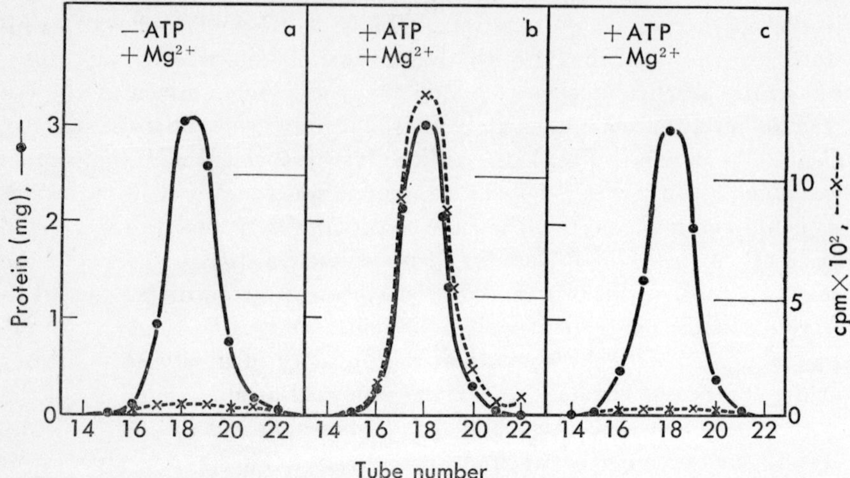

FIG. 2. Formation of citrylated enzyme (*14*). The reaction mixture contained (in μ moles): Tris-HCl buffer (pH 8.0) 100, 2-mercaptoethanol 10, $MgCl_2$ 10, ATP 5, 1,5-^{14}C-citrate 2 (2,460,000 cpm) and enzyme (9 mg protein) in a total volume of 1 ml. ATP or Mg^{2+} was omitted as indicated. After incubation for 10 min at 37°C, the reaction mixture was chilled and then passed through a column of Sephadex G-50 (1.5 × 30 cm).

that acetyl-CoA was formed from E∼citrate only in the presence of CoA, and that neither ATP nor Mg^{2+} was essential in this reaction.

To elucidate the final step of ATP citrate lyase reaction, analysis was made of the backward reaction of this step, since studies on the forward reaction were complicated by the instability of the substrate, E∼citrate, together with the rapid cleavage of bound citrate in the presence of CoA. Thus, the enzyme was incubated either with 1-^{14}C-acetyl-CoA or with 1-^{14}C-acetate, and after incubation the protein portion was isolated by filtration on a Sephadex G-50 column. The radioactivity associated with the protein was obtained only with ^{14}C-acetyl-CoA and none was found with ^{14}C-acetate. To verify the formation of E∼acetate, and further, to characterize the type of binding of acetate to the enzyme, an acetyl-CoA—CoA exchange reaction was carried out. CoA-^{3}H and 1-^{14}C-acetate were separately incubated with the enzyme in the presence of unlabeled acetyl-CoA, and after incubation the incorporation of radioactivity into acetyl-CoA was determined.

The enzyme was found to catalyze a rapid exchange of the isotope between CoA-^{3}H and acetyl-CoA, whereas it was incapable of catalyzing an exchange between 1-^{14}C-acetate and acetyl-CoA. These results indicated that the acetyl moiety of acetyl-CoA was transferred to the enzyme, with the formation of a high energy linkage, and also that this acetylation step was reversible. The next question was whether this E$\sim$acetate was a true intermediate of the reaction. The E$\sim$acetate-1-^{14}C was isolated by gel filtration and incubated with and without added oxaloacetate. After incubation, the ^{14}C-citrate formed was isolated and counted. This analysis showed that citrate was really formed from E$\sim$acetate only in the presence of oxaloacetate. In the next experiment, the enzyme was first incubated with oxaloacetate and after incubation the protein portion was isolated through a column of Sephadex G-50 in exactly the same manner as that applied to the E$\sim$acetate complex. The protein portion thus obtained was then incubated with 1-^{14}C-acetyl-CoA and the formation of ^{14}C-citrate was determined. However, no radioactive citrate could be formed in this case. Based on these data, it was concluded that E$\sim$acetate, but not E$\sim$oxaloacetate, was formed as a reactive intermediate, and that this E$\sim$acetate complex reacted with oxaloacetate to form citrate (more exactly, E$\sim$citrate) in the backward reaction.

As mentioned above, when acetyl-CoA was incubated with the enzyme, its acetyl moiety was transferred to the enzyme to form E$\sim$acetate with a high energy linkage. However, the fate of the remaining part of the acetyl-CoA, CoA, still remains to be determined. When the enzyme was incubated with acetyl-CoA-^{3}H and after incubation the protein portion was isolated by gel filtration through a column of Sephadex G-50, a substantial quantity of radioactivity emerged from the column with the enzyme protein. The bound radioactivity was almost completely liberated from the protein on treatment with perchloric acid at a concentration of 0.2 N. Analysis of the radioactivity released by this treatment showed that over 90% was recovered as free CoA and the remainder as acetyl-CoA. The combination of CoA with the enzyme was more directly demonstrated by using CoA-^{3}H. When free CoA-^{3}H was incubated with the enzyme and after incubation the protein was isolated by gel filtration in a similar manner, a significant amount of radioactivity was again associated with the protein. This combination process did not require added ATP. These data indicated that CoA, even though it was released from acetyl-CoA during the formation of

E∼acetate, could easily combine with the enzyme. Therefore, the true product formed through the interaction between the enzyme and acetyl-CoA should be written as $E^{\sim acetate}_{-CoA}$ instead of E∼acetate.

To characterize the type of binding of CoA to the enzyme, the enzyme was first incubated with CoA-^{3}H for 5 min at 37°C, and incubation was continued with or without unlabeled CoA for an additional 2 min. After the second incubation, the protein portion was isolated by gel filtration and the CoA associated with the protein was determined. The results showed that although there was no significant change in the total amount of CoA bound per unit of enzyme, the bound radioactivity was greatly decreased on the addition of unlabeled CoA in the second incubation, the decreased value simply reflecting dilution by the added unlabeled CoA. Conversely, when the enzyme was first incubated with unlabeled CoA and then CoA-^{3}H was added in the second incubation, a substantial quantity of labeling was found in the bound CoA. Together with the fact that the enzyme was capable of catalyzing the acetyl-CoA—CoA exchange but not the acetyl-CoA—acetate exchange, these results indicate that, unlike the linkage between the enzyme and acetate, the binding of CoA to the enzyme was not energy-rich, and that the bound CoA was freely exchangeable with exogenous CoA in the reaction system.

These findings clearly support the view that the enzyme is actually

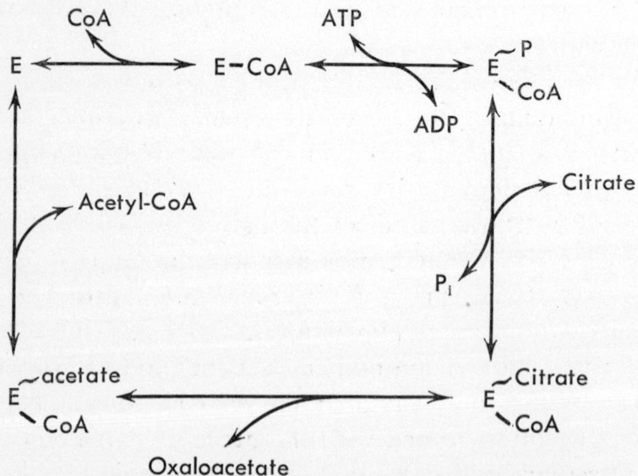

FIG. 3. Reaction mechanism of ATP citrate lyase (*15*).

associated with CoA in the overall reaction. Accordingly, the reaction of ATP citrate lyase can be depicted as in Fig. 3. Thus, during the reaction the high bond energy derived from ATP is retained and moves in the order: $E^{\sim P}_{-CoA}$, $E^{\sim citrate}_{-CoA}$, and $E^{\sim acetate}_{-CoA}$, finally being transferred to give rise to acetyl-CoA.

Characterization of the Phosphorylated Enzyme

In order to characterize the first intermediate, E∼P, the following experiments were carried out. On adding citrate to E∼P, bound phosphate was released from the enzyme. However, this type of phosphate release was not specific to citrate. When various organic acids were added in large excess to E∼P, all tricarboxylic acids tested completely released the phosphate from the enzyme. On the other hand, mono- and dicarboxylic acids such as acetate, succinate, fumarate or L-malate, had little effect at the same concentration. The effectiveness of various tricarboxylic acids in releasing the bound phosphate was compared at a lower concentration. At a molar ratio of tricarboxylic acids to the complex of 2, the liberation of phosphate from the enzyme was most rapid with citrate, followed in order by DL-isocitrate, tricarballylate and *cis*- and *trans*-aconitates. From these results, it is evident that the release of phosphate from the E∼P complex is specifically caused by tricarboxylic acids and that citrate is most effective among the tricarboxylic acids tested. Hydroxylamine also caused complete liberation of phosphate from the enzyme.

When E∼^{32}P was incubated at 37°C for 7 min in 0.02 M Tris-HCl buffer (pH 8.0) containing 0.01 M 2-mercaptoethanol, essentially no release of $^{32}P_i$ was detected. In 1 N KOH, only 7% of the bound phosphate was liberated on incubation at 37°C for 7 min, whereas in 1 N HCl, over 80% was released under the same conditions. This indicates that the linkage between the enzyme and phosphate is more labile in acid than in alkali. However, treatment at 100°C for 7 min resulted in almost complete destruction of the complex either in 1 N KOH or in 1 N HCl. These characteristics make it improbable that the binding site of phosphate is a serine or histidine residue of the enzyme protein because it has been shown that phosphoserine is very stable in acid at 100°C (*16*), and also that phosphohistidine is resistant to hydrolysis in 3 N NaOH at 100°C over a period of 2 to 3 hr, though it is labile in acid (*17*). To

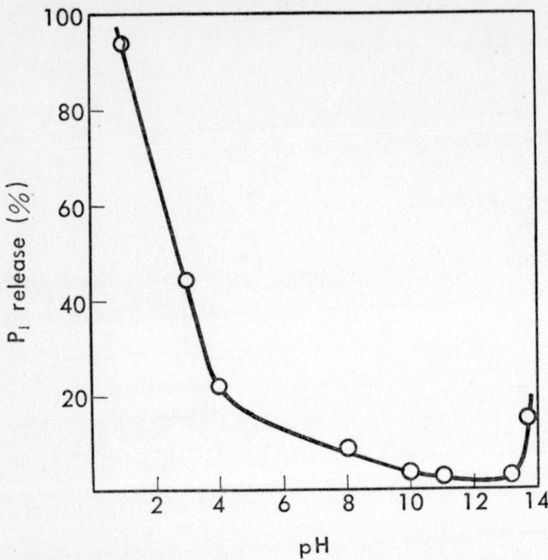

FIG. 4. Effect of pH on the rate of hydrolysis of E∼P (*22*). E∼^{32}P (9,640 cpm) was incubated for 2 hr at 37°C with 80 mM buffers of pH 1 to 11. The following buffer systems were employed. pH 1 to 2, HCl-KCl; pH 4, acetate; pH 7 to 8, Tris–HCl; pH 10 to 11, bicarbonate-carbonate. At pH 13.2 and 13.6, 0.16 N KOH and 0.64 N KOH were used, respectively. Controls in water were kept at 0°C during incubation. After incubation, the mixture was immediately chilled in an ice bath and neutralized, if necessary, with either 2 N $HClO_4$ or 2 N KOH. Then the protein was removed by addition of perchloric acid to a final concentration of 0.2 N in the cold. By this treatment, less than 5% of the bound ^{32}P was liberated as P_i. Therefore, to test the stability of the complex, it was precipitated after incubation under various conditions by adding perchloric acid and the amount of $^{32}P_i$ released in the supernatant was determined.

confirm this, the effect of pH on the rate of hydrolysis of the bound phosphate was studied and the percentage of $^{32}P_i$ released from E∼^{32}P on incubation for 2 hr at 37°C was plotted against pH.

The results are given graphically in Fig. 4. Hydrolysis of the complex is little affected by changes in pH near neutrality, but it is greatly accelerated on decreasing the pH value. The pH-stability curve of E∼P is very similar to that reported by Black and Wright (*18*) for aspartylphosphate which shows sharp increases in the hydrolysis rate at pH's 3 and 13, and a broad flat minimum in the region near neutrality. It is also

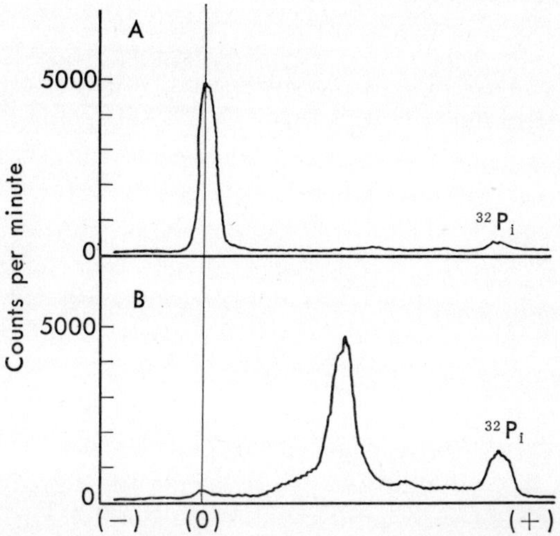

FIG. 5. High voltage paper electrophoresis of E∼^{32}P and its pronase digest (*22*). a) E∼^{32}P; b) pronase digest of E∼^{32}P.

similar to that of citrylphosphate (*19*) but differs from that of butyl thiophosphate (*20*). Therefore, the binding site of phosphate in ATP citrate lyase is probably a carboxyl group of an aspartic acid or glutamic acid residue.

To further characterize the type of linkage between the enzyme and phosphate, an attempt was made to isolate a phosphorylated peptide by pronase digestion. E∼^{32}P was denatured by the addition of cold ethanol and then incubated with pronase-P for 20 hr at 37°C. By this treatment, most of the radioactivity associated with the protein was made soluble. The resulting digest was lyophilized and then subjected to high voltage paper electrophoresis at pH 6.4. The electrophoretic pattern (Fig. 5) showed that the digestion of E∼^{32}P by pronase liberated two major radioactive fragments which migrated towards the anode at pH 6.4. The faster moving peak was identified as P_i. This was presumably released by degradation of E∼P or phosphopeptide during digestion or electrophoresis. The slower moving peak was ninhydrin-positive and represented a phosphopeptide(s). Undigested E∼^{32}P remained at the origin on electrophoresis.

To determine the phosphate-bound amino acid residue in E∼^{32}P, the Lossen rearrangement method was applied after conversion of the phosphopeptides to the corresponding peptidylhydroxamates. According to Gallop *et al.* (*21*), after the Lossen rearrangement, each residue of β-aspartylhydroxamate yields one residue of α, β-diaminopropionic acid, while that of γ-glutamylhydroxamate gives one residue of α,γ-diaminobutyric acid. Thus, E$\sim$^{32}P was prepared on a large scale from 900 mg of purified enzyme. After incubation, the E$\sim$^{32}P formed was precipitated by addition of ethanol and digested with pronase-P. The digest thus obtained was then subjected to paper electrophoresis at pH 6.4. The ^{32}P-phosphopeptide fraction was located on the chromatogram with an Actigraph, eluted with distilled water while cold, and lyophilized. The samples of ^{32}P-phosphopeptide obtained from 9 runs were combined and stored at -20°C. The combined phosphopeptide was further purified by a second paper electrophoresis at pH 6.4 The final phosphopeptide fraction obtained was divided into two parts. One part was treated with hydroxylamine yielding peptidylhydroxamate. After treatment, the mixture was lyophilized and then subjected to paper electrophoresis at pH 6.4. On the electrophoregram, the area corresponding to peptidylhydroxamate was detected by spraying with $FeCl_3$ reagent, cut out and eluted with distilled water while cold. After lyophilization, the material was dissolved in 1 ml of 0.1 M sodium bicarbonate and mixed with 1 ml of 1% ethanolic fluorodinitrobenzene to yield peptidyl dinitrophenylhydroxamate.

During the reaction, the mixture was maintained at pH 8.0 by continuous titration with 5% sodium bicarbonate. After 5 min at room temperature, the mixture was extracted three times with 1 ml volumes of ether to eliminate the residual fluorodinitrobenzene as completely as possible. The peptidyl dinitrophenylhydroxamate thus obtained was then subjected to the Lossen rearrangement by heating for 10 min at 100°C with NaOH at a final concentration of 0.1 N. The remaining part of the phosphopeptide fraction was boiled for 15 min to liberate the bound phosphate completely. The dephosphorylated peptide thus obtained was subjected to treatment with hydroxylamine, to paper electrophoresis at pH 6.4 and finally to the Lossen rearrangement in exactly the same manner as the phosphorylated peptide. After the Lossen rearrangement, the product was hydrolyzed in 6 N HCl for 20 hr at 100°C, and the hydrolysate was evaporated to dryness under reduced pressure. Basic amino acids in the hydrolysate were isolated by paper electrophoresis at pH 6.4 and analyzed by an amino acid analyzer. The

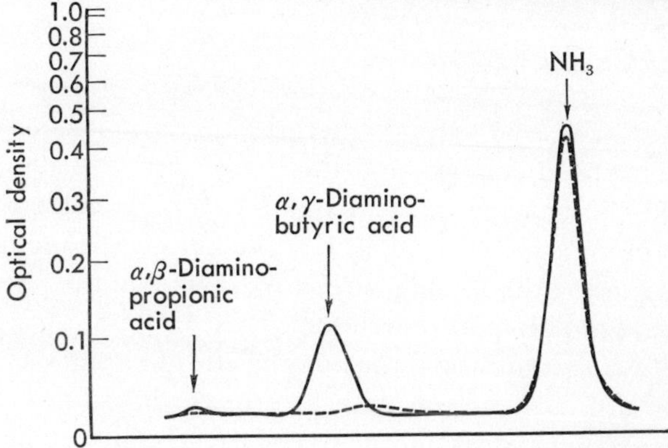

FIG. 6. Identification of α, γ-diaminobutyric acid (*22*). The samples were applied on an 100 cm column of Dowex 50, and elution was carried out with 0.2 M sodium citrate buffer (pH 8.70) at 57°C. ——— phosphopeptide; - - - - - dephosphopeptide.

results shown in Fig. 6 indicate clearly that α,γ-diaminobutyric acid was actually formed from the hydroxylamine-treated phosphopeptide fraction after the Lossen rearrangement and acid hydrolysis, whereas it was not formed from the dephosphorylated peptide fraction after the same treatments. No detectable amount of α,β-diaminopropionic acid was formed from either the phosphorylated or dephosphorylated peptide fraction.

The recovery of the product α,γ-diaminobutyric acid, was 15 to 20% on the basis of the amount of bound ^{32}P in the isolated phosphopeptide-^{32}P. However, this recovery could be corrected to 30 to 40% because the yield during the Lossen rearrangement step, including the dinitrophenylation, was found to be 40 to 50% in the model experiment using synthetic N carbobenzoxy glutamyl-γ-hydroxamic acid as the substrate. The yield on digestion of $E{\sim}^{32}P$ with pronase-P was about 30%; this may be due to loss of bound phosphate from $E{\sim}P$ during the preparation procedures, because the phosphate linkage is unstable. During electrophoresis, some of the bound phosphate (less than 20%) was also liberated as P_i. Thus, the γ-carboxyl group of the glutamic acid residue was present as hydroxamic acid and was, therefore, originally involved in the hydroxylamine-sensitive bond of the peptide. Since α,β-diaminopropionic acid was not detectable, participation of aspartic acid residues in the hydroxylamine-sensitive linkages can be excluded.

Nature of the Citrylated Enzyme

The citrylated enzyme, the second intermediate, is much more labile than the phosphorylated enzyme, and the bound citrate is almost completely lost from the enzyme by acid treatment while cold or during incubation with pronase for 20 hr at 30°C. Therefore, the method used for the determination of the binding site of phosphate in the phosphorylated enzyme could not be applied to the citrylated enzyme. Accordingly, another method was used in which the binding site of citrate was trapped directly from the citrylated enzyme without pronase digestion. In 1968, a method for converting hydroxamic acids to corresponding amines under mild conditions was developed by Hoare *et al.* (*23*). Using this method, treatment of the hydroxamic acid with a water-soluble carbodiimide at pH 5 was found to result in quantitative conversion to amine by a Lossen rearrangement at room temperature. First, the applicability of Hoare's method was tested using γ-hydroxamyl-N-carbobenzoxy-glutamyl-glycinamide as a model substrate. Treatment of this hydroxamic acid derivative with the water-soluble carbodiimide, 1-ethyl-3-dimethylaminopropylcarbodiimide, at pH 5 was found to result in good conversion to α,γ-diaminobutyric acid by the Lossen rearrangement. This experiment with a model substrate suggested that the method of Hoare *et al.* is suitable for the present purpose. Thus, this method was used to determine the amino acid residue binding citrate in the citrylated enzyme, namely, to test whether the γ-carboxyl group of glutamic acid residue is responsible for the binding site of citrate as in the phosphorylated enzyme.

A purified sample of ATP citrate lyase (960 mg) was incubated with ATP in the presence of Mg^{2+} to obtain phosphorylated enzyme. Each run was carried out with about 100 mg of the enzyme. After reaction with ATP for 10 min, a large excess of potassium citrate was added to the mixture and incubation was continued for an additional 2 min. This treatment converted the phosphorylated enzyme almost completely to the citrylated enzyme with concomitant release of P_i (*14*). The mixture was further incubated with hydroxylamine, and the protein portion was isolated by a column of Sephadex G-50. Free hydroxylamine, which interferes with the separation of α,γ-diaminobutyric acid in subsequent steps, was removed by this procedure. Then the mixture was

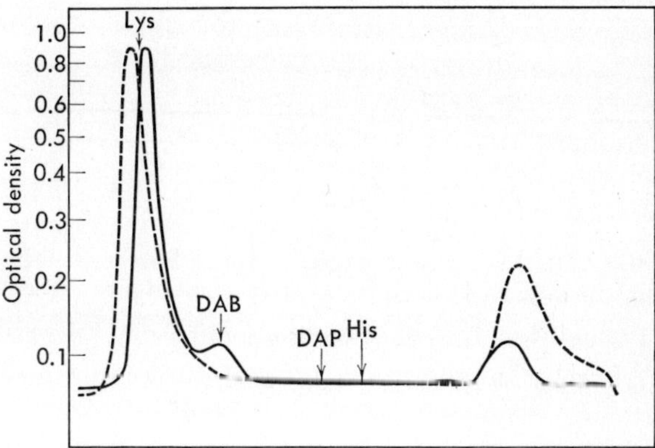

FIG. 7. Identification of α, γ-diaminobutyric acid (*24*). The samples were applied on a 45-cm column of Chromosorb No. 3105, and elution was effected with 0.35 M sodium citrate buffer (pH 5.28) at 50°C. This analysis was carried out by Dr. M. Ebata of the Institute of the Shionogi Pharmaceutical Co., Ltd. ——— sample obtained from the citrylated enzyme; - - - - control sample.

lyophilized and hydrolyzed by heating with 6 N HCl in a sealed, evacuated tube for 20 hr at 110°C. The acid hydrolysate was evaporated to dryness under reduced pressure, and the residue was dissolved in water. The aqueous solution was decolored by treatment with charcoal and filtered. The filtrate was chromatographed on a column of Dowex-50 (H^+ form). The fractions eluted with 2 N aqueous ammonia were combined and concentrated to dryness *in vacuo*. The dried material was then subjected to preparative paper electrophoresis to isolate basic amino acids. The area corresponding to the positions of α,γ-diamino butyric acid and α,β-diaminopropionic acid was cut out and eluted with water.

After concentration, the eluate was subjected to preparative paper chromatography, using a solvent system of *n*-butanol: pyridine: water (1: 2: 2, v/v). The area corresponding to these amino acids on the paper chromatogram was cut out and eluted with water. As a control, a mixture of enzyme and citrate was treated in exactly the same manner as the citrylated enzyme. The final eluates, from the citrylated enzyme and the control, were subjected to amino acid analysis. As seen in Fig.

7, α,γ-diaminobutyric acid was formed from the hydroxylamine-treated citrylated enzyme after treatment with 1-ethyl-3-dimethylaminopropyl-carbodiimide and acid hydrolysis, whereas it was not formed from the mixture of enzyme and citrate after the same treatments. No detectable amount of α,β-diaminopropionic acid was formed from either the citrylated enzyme or the control. The recovery of the product was about 7% judging from the amount of enzyme used under the conditions (amount of the enzyme) used in the present experiment. However, the low recovery was probably due to the lability of the citrylated enzyme intermediate, some of which was degraded during incubation. In accordance with this, it was found that the molar ratio of citrate to the enzyme in the complex was low, ranging from 0.1 to 0.24, after isolation through a column of Sephadex G-50 (*14*).

From these results, it was concluded that the binding site of citrate in the citrylated intermediate of ATP citrate lyase is the γ-carboxyl group of glutamic acid residue as in the phosphorylated intermediate. It is conceivable that the formation of an anhydride linkage between citrate and the γ-carboxyl group of the glutamyl residue of the enzyme results in electron withdrawal into the enzyme and in electron deficiency in the bond between C-2 and C-3 of the bound citrate molecule. This would enhance the breaking of the C-C bond between C-2 and C-3 of the citrate and facilitate the splitting off of the oxaloacetate moiety by the action of CoA.

Summary

The mode of action of ATP citrate lyase was investigated using a single homogeneous preparation from rat liver. It was shown that the initial step of the reaction is the activation of the enzyme by ATP to form the phosphorylated enzyme which has a high energy linkage, followed by conversion of the latter to citrylated enzyme in the presence of citrate with concomitant liberation of P_i. The high bond energy between enzyme and citrate is then retained and transferred to give rise to acetyl-CoA. The enzyme easily combines with free CoA or CoA that is released from acetyl-CoA during the formation of the acetylated enzyme. This binding of CoA to the enzyme does not depend on the presence of added ATP, and the bound CoA is freely exchangeable with exogenous CoA. These data indicate that the linkage between the enzyme and

CoA is not energy-rich. Therefore, it is concluded that $E^{\sim P}_{-CoA}$, $E^{\sim citrate}_{-CoA}$ and $E^{\sim acetate}_{-CoA}$ are involved as intermediates in the reaction.

The chemical properties of the activated intermediates, the phosphorylated and the citrylated enzymes, were then investigated. It was shown that the binding site of phosphate in the phosphorylated enzyme is the γ-carboxyl group of the glutamic acid residue and that this carboxyl group is also responsible for the binding site of citrate in the second intermediate.

References

1 P. A. Srere and F. Lipmann, *J. Am. Chem. Soc.*, **75**, 4874 (1953).
2 P. A. Srere, *J. Biol. Chem.*, **236**, 50 (1961).
3 M. Wolleman, *Acta Physiol. Acad. Sci. Hung.*, **10**, 171 (1956).
4 P. A. Srere, *J. Biol. Chem.*, **234**, 2544 (1959).
5 A. Spencer, L. Corman and J. M. Lowenstein, *Biochem. J.*, **93**, 378 (1964).
6 Y. Takeda, K. Adachi, H. Inoue and H. Tanioka, *Proc. Symp. Clin. Physiol. Pathol.*, **4**, 59 (1964).
7 A. F. D'Adamo, Jr. and D. E. Haft, *J. Biol. Chem.*, **240**, 613 (1965).
8 H. Inoue, F. Suzuki, H. Tanioka and Y. Takeda, *J. Biochem.*, **60**, 543 (1966).
9 H. Eggerer and U. Remberger, *Biochem. Z.*, **339**, 62 (1963).
10 P. A. Srere and A. Bhaduri, *J. Biol. Chem.*, **239**, 714 (1964).
11 J. E. Snoke and K. Bloch, *J. Biol. Chem.*, **231**, 825 (1955).
12 G. Kreil and P. D. Boyer, *Biochem. Biophys. Res. Commun.*, **16**, 551 (1964).
13 M. C. Scrutton and M. F. Utter, *J. Biol. Chem.*, **240**, 3714 (1965).
14 H. Inoue, F. Suzuki, H. Tanioka and Y. Takeda, *J. Biochem.*, **63**, 89 (1968).
15 H. Inoue, T. Tsunemi, F. Suzuki and Y. Takeda, *J. Biochem.*, **65**, 889 (1969).
16 L. J. Kleinsmith, V. G. Allfrey and A. E. Mirsky, *Proc. Natl. Acad. Sci. U. S.*, **55**, 1182 (1966).
17 P. D. Boyer, M. DeLuca, K. E. Ebner, D. E. Hultquist and J. B. Peter, *J. Biol. Chem.*, **237**, PC 3306 (1962).
18 S. Black and N. G. Wright, *J. Biol. Chem.*, **213**, 27 (1955).
19 C. T. Walsh, Jr. and L. B. Spector, *J. Biol. Chem.*, **244**, 4366 (1969).
20 E. B. Herr, Jr. and D. E. Koshland, Jr., *Biochim. Biophys. Acta*, **25**, 219 (1957).

21 P. M. Gallop, S. Seifter, M. Lukin and E. Meilman, *J. Biol. Chem.*, **235**, 2619 (1960).
22 F. Suzuki, K. Fukunishi and Y. Takeda, *J. Biochem.*, **66**, 767 (1969).
23 D. G. Hoare, A. Olson and D. E. Koshland, Jr., *J. Am. Chem. Soc.*, **90**, 1638 (1968).
24 F. Suzuki, *Biochemistry*, **10**, 2707 (1971).

Received for publication August 3, 1971.

STRUCTURAL CHANGE IN MYOSIN INDUCED BY ATP

Fumi Morita

Department of Chemistry, Faculty of Science, Hokkaido University, Sapporo

Myosin is a principal structural protein of muscle having ATPase activity. It has been established that the interaction between myosin and actin with ATP is a fundamental event in muscle contraction. Accordingly ATP is expected to induce a structural change in the myosin molecule.

There are some reports indirectly suggesting more or less a structural change in myosin induced by the substrate. Some workers report a protective action of substrate or its analogs against denaturation of myosin and of heavy meromyosin (*1–4*), or a different reactivity of reagents for chemical modification in the presence of a substrate (*5–11*). Using the chemically modified proteins, a structural change is also suggested from the changes in the electronic (*12*) or electron paramagnetic spectrum (*13*, *14*) of the attached reporter groups. However, there was no decisive evidence directly showing structural change although there have been efforts made by using certain physico-chemical techniques (*3*, *11*, *15–17*) to establish such evidence.

In 1966 we observed that a difference absorption spectrum of heavy meromyosin, an active fragment of myosin, was induced by the addition of ATP, ADP, and PP_i (*18*). It was ascribed to a red shift of the absorption bands due to some tyrosyl and tryptophanyl chromophores in heavy meromyosin. The phenomenon indicates that a structural change in heavy meromyosin is caused by substrate so that the chromophores are buried into the interior of the protein moiety. The same phenomenon was also observed with purified myosin (*19*). This article will describe some of our work performed with particular emphasis on elucidating the mechanism and role of the structural change of myosin reflected in the difference spectrum.

Changes in Electronic Spectra of Myosin Induced by ATP

An active fragment with molecular weight of 3.65×10^5, heavy meromyosin, is obtained from myosin by tryptic digestion. Because of a turbidity lower than that of myosin and of high solubility even at low ionic strengths, heavy meromyosin is a suitable protein for study of minor changes in its electronic spectra. Moreover, the characteristics of myosin ATPase are well preserved in heavy meromyosin.

The ultraviolet absorption difference spectrum induced by ATP is most easily observed with heavy meromyosin in the presence of $MgCl_2$ (*20, 21*) since the ATPase activity of heavy meromyosin as well as myosin is considerably lower in the presence of $MgCl_2$. The trace of the difference spectrum obtained in the presence of $MgCl_2$ by the double cell method is shown in Fig. 1 (*20*). Immediately after the addition of ATP the difference spectrum showed two maxima at 289 and 281 mμ with a shoulder in the vicinity of 293 mμ, and a minimum at around 250 mμ (traces a and b in Fig. 1). After the hydrolysis of ATP the two maxima shifted to 288 and 280 mμ, respectively, accompanying the disappearance of the shoulder near 293 mμ (traces c and d in Fig. 1). The latter difference spectrum was the same in size and shape as the one induced by the addition of ADP. The values of the difference molar absorbancy, $\Delta\varepsilon$, vary somewhat with the preparations and are 5 to $6 \times 10^3\ \text{M}^{-1}\ \text{cm}^{-1}$ at 289 mμ for that induced by ATP, and 2.5 to $3.5 \times 10^3\ \text{M}^{-1}\ \text{cm}^{-1}$ at 288 mμ for that by ADP.

With a carefully purified preparation of myosin, the same difference spectra as above, both in shape and in difference molar absorbancy, were observed (*19*). Moreover, the same shape in difference spectra

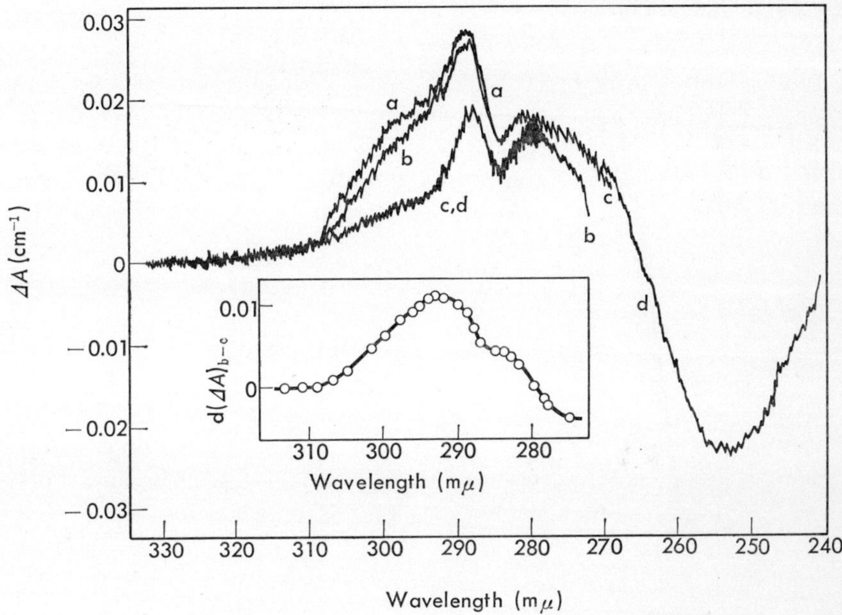

FIG. 1. Difference spectrum of heavy meromyosin induced by ATP. Experimental conditions were 2.17 mg heavy meromyosin per ml, 10.6 mM $MgCl_2$, 0.08 M KCl, 0.07 M Tris-HCl (pH 8), and 71 μM ATP at 25°C. The periods after adding ATP to heavy meromyosin at which the wavelength of 289 mμ was reached were 11, 115, 220, and 370 sec for traces a, b, c, and d, respectively. The inset is a plot against the wavelength of the difference in absorbancy obtained by subtracting the ΔA value of trace c from that of trace b.

as those of myosin or heavy meromyosin was also observed with subfragment-1 with molecular weight of 1.3×10^5 which is an active fragment of heavy meromyosin (*22*). The difference molar absorbancy of subfragment-1 was, however, about one-half that of heavy meromyosin or of myosin (*22*). This means that the necessary part of the protein which induces the difference spectrum is well preserved in subfragment-1 since two moles of subfragment-1 are yielded per mole of heavy meromyosin (*23*, *24*). That is, the structural change of myosin reflected in the difference spectrum occurs in a very restricted region near the active site and the change ought to be within the portion of subfragment-1. It should be noted that the part of myosin to combine specifically with actin is also localized in this subfragment-1 region.

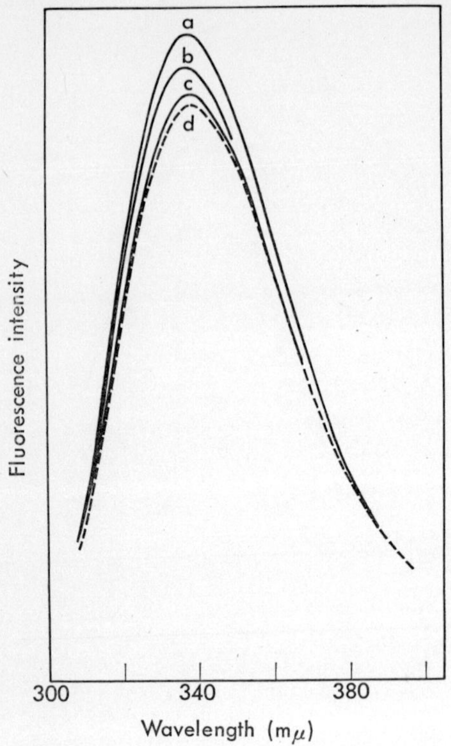

FIG. 2. Fluorescence emission spectrum of heavy meromyosin after the addition of ATP. Experimental conditions were 0.1 mg heavy meromyosin per ml, 1 mM $MgCl_2$, 0.6 M KCl, 20 mM phosphate buffer (pH 6.6), and 2.22 μM ATP at 20°C. The periods after adding ATP to heavy meromyosin at which the maximum was reached were 50, 170, and 247 sec for traces a, b, and c, respectively. Trace d is the emission spectrum of heavy meromyosin without adding ATP. Excitation was made at 293 mμ.

The structural change in heavy meromyosin induced by ATP was also shown by the change in the fluorescence spectrum. As shown in Fig. 2, the fluorescence emission spectrum of heavy meromyosin excited at 293 mμ increased the intensity by about 15% with a tendency of blue shift on the addition of ATP. After the hydrolysis of ATP, the spectrum recovered a form similar to the original one.*

The red shift of the absorption spectum, and the blue shift as well as the increase in intensity of the fluorescence emission spectrum, indicate mainly that some tyrosyl and tryptophanyl chromophores in the protein molecule have become buried in the nonpolar protein interior after the addition of ATP. It was suggested from the study using the solvent perturbation method that 4 tyrosyl and 1 tryptophanyl residues were buried out of 77 tyrosyl and 19 tryptophanyl residues per 3.65×10^5 g of heavy meromyosin (*26*, *27*).

It was attempted, then, to detect the structural change by other means. However, neither change in the reflactive index increment nor in the viscosity of heavy meromyosin was detected on the addition of ATP (F. Morita, M. Yazawa and H. Yoshino, unpublished experiments). It was also reported that changes in the helix content of heavy meromyosin (*11*) and in the sedimentation coefficient of myosin (*11*, *28*) were not detected on the addition of ATP. These results appear to indicate that there is no gross conformational change in the polypeptide backbone. The change may mainly be a displacement of the side chain groups.

Transient Kinetics of the Structural Change

For an understanding of the role of the structural change of the protein molecule reflected in the electronic spectra, it seems to be important to know what molecular species is associated with the structural change during the process of ATPase reaction. Formation of the difference spectrum after mixing ATP with heavy meromyosin was followed by the stopped flow method at a fixed wavelength, 293 mμ (*29*). As shown in Fig. 3, ΔA_{293} increased rapidly after the addition of ATP and reached a steady value. As shown in Fig. 4, the initial velocity of the increase of the absorbancy was proportional to the initial concentration of ATP, indicating that the difference spectrum of heavy meromyosin was formed as a second order reaction. This result indicates that the structural change occurs at the same time as the formation of the enzyme-ATP complex. The second order rate constant obtained from the slope such

* Work on the absorption and fluorescence spectrum of myosin performed with a similar intention was reported independently by Burshtein (*25*). His results, however, differ from ours in various respects. The myosin preparation which he used might have been inadequate because the turbidity change of crude myosin easily exceeded the real changes in the spectra.

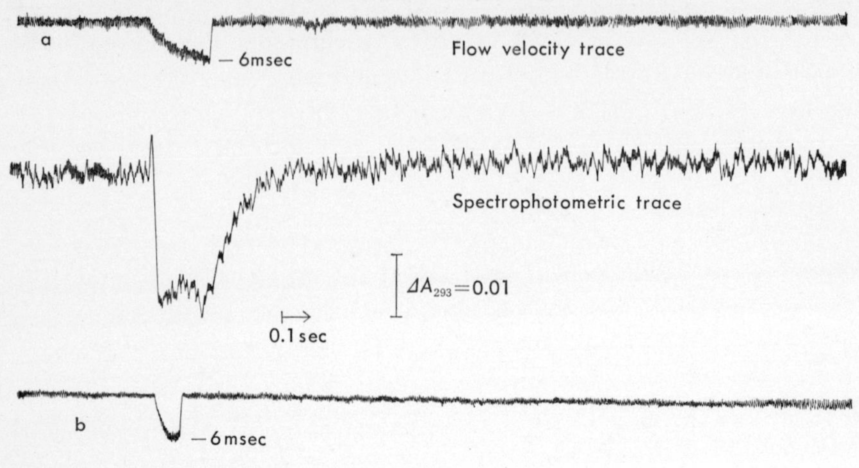

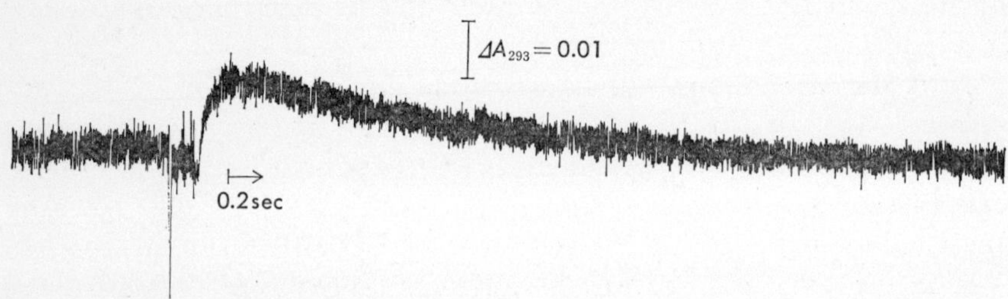

FIG. 3. Traces of the stopped flow measurement of the change in absorbancy at 293 mμ of heavy meromyosin after mixing with ATP. Experimental conditions were 2.05 mg heavy meromyosin per ml, 1 mM $MgCl_2$, 0.1 M KCl, and 20 mM Tris-HCl (pH 8) at 6°C for a, and were 1.85 mg heavy meromyosin per ml, 2 mM $CaCl_2$, 0.25 M KCl, and 20 mM Tris-HCl (pH 8) at 10°C for b. Concentration of ATP was 30.2 μM.

as shown in Fig. 4 was, for example, 5.7×10^5 M^{-1} sec^{-1} at 8°C and 3.6×10^6 M^{-1} sec^{-1} at 25°C in the presence of 0.1 M KCl, 1 mM $MgCl_2$, and 20 mM Tris-HCl (pH 8). The value was fairly independent of the kind of divalent cation present (*29*, *30*). The order of these rate constants is in the range of generally reported values of enzyme-substrate complex formation (*31*, *32*). Since the helix-coil transition in polypeptides occurs in periods as short as 10^{-7} sec (*31*), the rate of the change

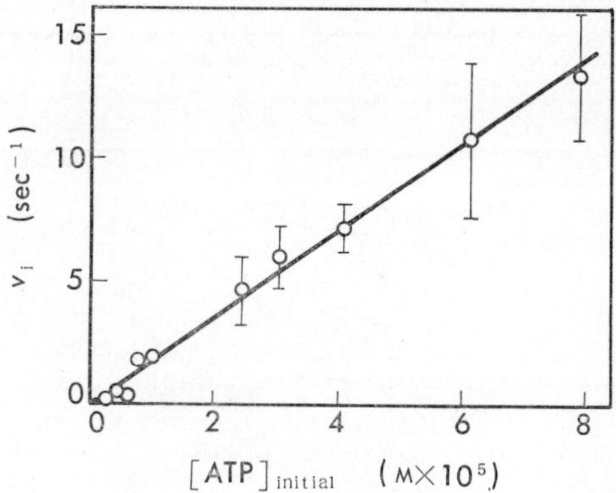

FIG. 4. Dependence of the initial velocity of the formation of the difference spectrum on the initial concentration of ATP. Experimental conditions were 1.79 mg heavy meromyosin per ml, 0.2 mM $MgCl_2$, 0.25 M KCl, and 20 mM Tris-HCl (pH 8) at 10°C.

of protein structure in the present study may be determined by the rate of binding of ATP to the protein.

After the steady state, the decay of the difference spectrum was observed. As shown in trace b of Fig. 3, immediate decay was observed by the stopped flow measurement in the presence of $CaCl_2$. On the other hand, it was slow enough to be followed by a usual recording spectrophotometer in the presence of $MgCl_2$ (as shown in Fig. 1). Taking the time required for the difference spectrum to fall from the maximum value to one-half the maximum, τ, the value of the equation,

$$s_0/\int_0^\infty p\,dt \sim s_0/p_m\tau\,,$$

which has been derived by Chance (*33*), was calculated. In the equation s_0 is the initial concentration of ATP, p the concentration of the species whose spectrum is changed, and p_m the maximum value of p. In the presence of $MgCl_2$, the calculated value agreed well under various conditions with the value of V_m/e which was obtained from the initial velocity of ATPase reaction at the steady state (*19*, *20*, *22*, *29*). The result indicates that the lifetime of the difference spectrum induced by ATP is the same as that of the Michaelis-complex of ATPase reaction which follows a simple Michaelis-Menten kinetics.

The results obtained from these kinetic measurements indicate that the structural change reflected in the difference spectrum is induced at the same time as formation of the Michaelis-complex and persists with the intermediate. Tonomura and his collaborators have reported that one mole of initial burst of P_i liberation occurs per mole of myosin or heavy meromyosin ATPase (*34*, *35*). We have shown that the difference spectrum of heavy meromyosin induced by ADP gives the maximum ΔA at one mole of ADP binding out of two moles of maximum binding of ADP (*36*). It appears, therefore, that the ATPase reaction accompanying the structural change is operated by one mole of ATP binding out of two moles. The heterogeneity in the binding of the substrate will be discussed later.

Displacement of a Tryptophanyl Residue

As shown in Fig. 1, the difference absorption spectrum induced by ATP decayed with time especially near the 293 mμ region. The inset of Fig. 1 shows the difference between the traces of b and c of the figure plotted against the wavelength. It shows two maxima at 293 and 284 mμ and is similar in shape to the tryptophan difference spectrum which is known to be induced by a change in the reflactive index of the environment (*37*). This result appears to suggest that a tryptophanyl residue which is buried by ATP recovers after the hydrolysis of ATP to ADP.

The mechanism of this movement of the tryptophanyl residue was further investigated. The difference spectrum of heavy meromyosin after the addition of various ATP analogs was measured (H. Yoshino, F. Morita and K. Yagi, in preparation). The maximum value of the difference spectrum induced by nucleoside triphosphate was at 289 mμ, while that induced by nucleoside diphosphate, ribose triphosphate, TP_i, and PP_i was at 288 mμ. Then as a measure of the contribution of the tryptophanyl residue to the difference spectrum, the ratio of the $\Delta\varepsilon$ value at 293 mμ, $\Delta\varepsilon_{293}$, to that at 289 mμ, $\Delta\varepsilon_{289}$, was taken for that induced by nucleoside triphosphate, and the ratio of $\Delta\varepsilon_{293}$ to $\Delta\varepsilon_{288}$ was taken for that induced by nucleoside diphosphate, ribose triphosphate, TP_i, and PP_i. As shown in Table I, the difference spectra induced by the nucleoside triphosphates gave higher ratios than those for the corresponding diphosphate. Since the ribose triphosphate moiety is the common structure of all nucleoside triphosphates, the simplest explanation seemed to be that the ribose triphosphate moiety plays a role in inducing

TABLE I. Characteristics of Difference Absorption Spectra of Heavy Meromyosin Induced by ATP Analogs

Triphosphate	$\Delta\varepsilon_{289}$ (M^{-1} cm^{-1}) $\times 10^{-3}$	$\Delta\varepsilon_{293}/\Delta\varepsilon_{289}$ $\times 10^2$	Diphosphate	$\Delta\varepsilon_{288}$ (M^{-1} cm^{-1}) $\times 10^{-3}$	$\Delta\varepsilon_{293}/\Delta\varepsilon_{288}$ $\times 10^2$
ATP	5.8 (12)	79 (8)	ADP	3.7 (20)	49 (19)
CTP	6.9 (7)	80 (6)	CDP	2.2 (12)	53 (12)
TTP	6.0 (3)	89 (3)	TDP	2.0 (14)	56 (14)
UTP	4.5 (1)	68 (1)	UDP	1.6 (10)	15 (10)
ITP[a]	3.1 (4)	58 (1)	IDP	—	—
GTP[a]	1.7 (3)	58 (1)	GDP	—	—
Ribose triphosphate	2.4[c] (2)	30[d] (4)			
TP_i	1.8[c] (8)	26[d] (6)	PP_i[b]	1.6 (21)	28 (21)

Experimental conditions were 0.71 to 2.0 mg heavy meromyosin per ml, 10 mM $MgCl_2$, 0.25 M KCl, and 20 mM Tris-HCl (pH 8) at 25°C. Concentrations of nucleoside tri- and diphosphate varied from 32 to 150 μM, ribose triphosphate from 0.14 to 0.54 mM, TP_i from 50 to 200 μM, and PP_i from 6 μM to mM. Numbers in parentheses indicate the number of determinations.

[a] The difference spectrum was obtained by the stopped flow method at 8°C. [b] The experimental conditions were 1.5 to 1.8 mg heavy meromyosin per ml, 0.75 mM $MgCl_2$, and 15 mM Tris-HCl (pH 8) at 10°C. [c] $\Delta\varepsilon_{288}$. [d] $\Delta\varepsilon_{293}/\Delta\varepsilon_{288}$.

such displacement of the tryptophanyl residue. However, both values of $\Delta\varepsilon_{293}/\Delta\varepsilon_{288}$ of the difference spectra induced by TP_i and ribose triphosphate were very small as shown in Table I. These results indicate that the base moiety rather than ribose triphosphate makes a large contribution in inducing the displacement of the tryptophanyl residue. Furthermore, it seems that the base moiety must be connected to some part of ribose triphosphate because the addition of adenosine plus TP_i to heavy meromyosin did not change the size and shape of the difference spectrum observed on the addition of TP_i only (*20*).

An interaction between the tryptophanyl residue and the base moiety of the nucleoside triphosphate molecule might occur at the active site, vanishing after hydrolysis to diphosphate. The broad specificity concerning the structure of the base suggests a rather nonspecific interaction such as charge transfer. Tryptophan may be considered to be an electron donor. When N-acetyltryptophan amide was used as a model compound, a new absorption band suggesting a charge transfer interaction between

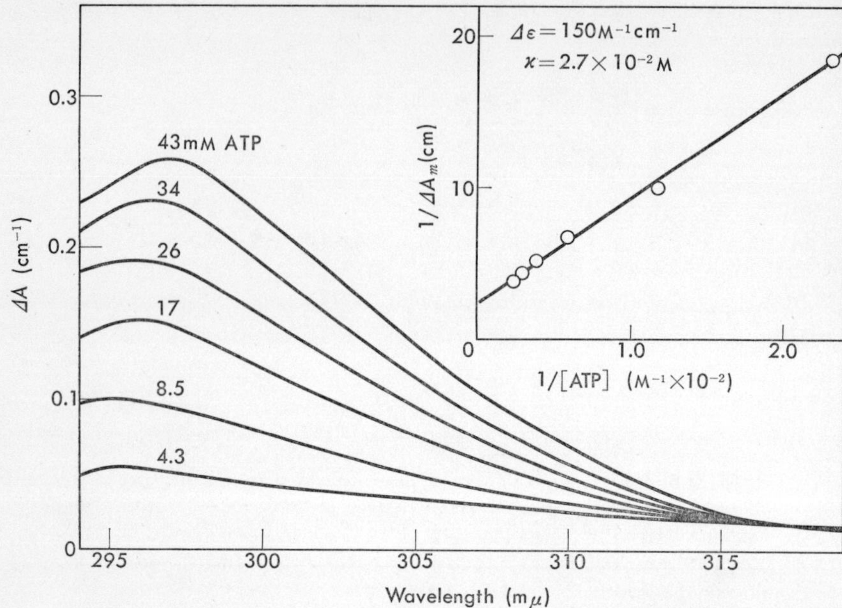

FIG. 5. Absorption spectra indicating interaction between N-acetyltryptophan amide and ATP. Experimental conditions were 50 mM Tris-HCl (pH 8) at 20°C. 2.73 mM N-acetyltryptophan amide was mixed with ATP of concentrations indicated. Four matched 5 mm cells were used in a double cell way. The inset shows the double reciprocal plot of maximum ΔA, ΔA_m, and the concentration of ATP. $\Delta\varepsilon$ and the dissociation constant, K, shown in the inset were those calculated from the slope and intercept of the plot assuming $\Delta A_m = \Delta\varepsilon(t)/(1+K/ATP)$, where (t) represents the concentration of N-acetyltryptophan amide.

the indole group and the adenine moiety of ATP was observed by using the double cell method (Fig. 5). Recently, Montenary-Garestier and Hèléne reported a charge transfer interaction between tryptophan and various nucleosides in a frozen mixture observed by the change in luminescence (*38*).

It has also been shown by us that the displacement of the tryptophanyl residue requires the presence of divalent cation such as Mg^{2+} or Ca^{2+} (*30*). Furthermore, the divalent cation may be bound to the active site through coordination from two imidazole groups of histidine residues (*30*). Taking these results into account together with the knowledge of the structure of the ATP molecule, a tentative mechanism of the displace-

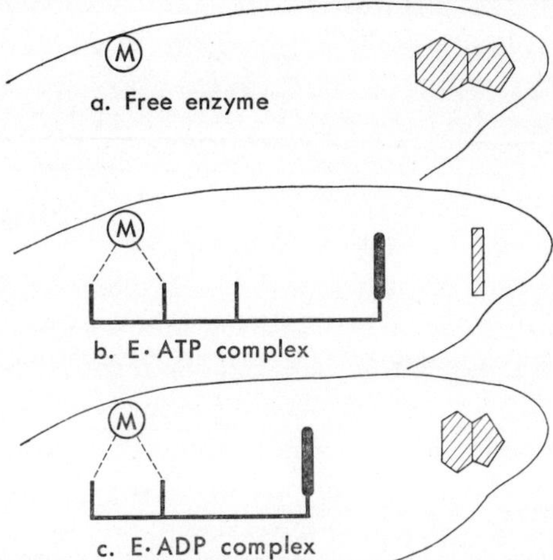

FIG. 6. Tentative mechanism showing the displacement of the tryptophanyl residue. a : Active site of free enzyme. M represents a divalent cation such as Mg^{2+} or Ca^{2+} which is bound to the protein by two imidazole groups. The hatched plane is the accessible indole group of the tryptophanyl residue discussed. b : Michaelis-complex. The blackened figure represents ATP molecule, the three small vertical bars the triphosphate moiety, and the bold part the base moiety. The hatched bar indicates that the indole group is buried after interaction with the base moiety of ATP. c : E·ADP complex. The base moiety is moved to the left side and the indole group partially returns to the original position.

ment of the tryptophanyl residue is illustrated in Fig. 6. Cohn and Hughse Jr. have shown from the NMR study that the γ and β phosphate groups in ATP coordinate with divalent cations such as Mg^{2+} and Ca^{2+}, while in ADP β and α phosphate groups coordinate with those cations (*39*). Accordingly, the difference in the degree of burying of the tryptophanyl chromophore due to ATP and ADP in heavy meromyosin may be attributed to the difference in the modes of the coordination between the two nucleosides with the divalent cations. As illustrated in Fig. 6, ATP might be bound to the protein through coordination of γ and β phosphate groups with the metal cation which is bound by two imidazole groups. At the same time, the tryptophanyl residue at the active site which is accessible before binding of ATP is stacked with the base moiety of ATP and buried after interaction between them. After hydrolysis of ATP, β and α phosphate groups of ADP make the coordination with the divalent cation in substitution for γ and β phosphate groups of ATP. As a result, the base moiety becomes more remote from the tryptophanyl indole group than it was in ATP. Consequently the interaction between the base and the indole group is removed, and the buried indole returns close to its original position.

As shown in the inset of Fig. 5, the $\Delta\varepsilon$ value of the charge transfer band observed with the model compounds was about 150 $\text{M}^{-1}\text{cm}^{-1}$. This is less than one-tenth of $\Delta\varepsilon$ in the difference spectrum of heavy meromyosin or of myosin induced by ATP at the same wavelength region. Accordingly, if the interaction between the tryptophanyl chromophore and the base moiety of ATP occurs as discussed above, the contribution of $\Delta\varepsilon$ due to the interaction may be minor. It is interesting that the values of $\Delta\varepsilon_{289}$ induced by various nucleoside triphosphates (Table I) are in the order of energy of the lowest empty molecular orbital of bases reported by Pullman (*40*) except for the case of thymidine triphosphate. This correlation suggests that the charge transfer interaction between the tryptophanyl indole group and the base moiety of substrate may participate in enlarging the structural change, though it may not be a prerequisite for inducing the change.

As shown in Fig. 1, the tyrosyl residues remain buried either in the presence of ATP or of ADP. The difference spectra of heavy meromyosin induced by ribose triphosphate, TP_i, and PP_i are nearly the same and resemble the tyrosine difference spectrum reported (*37*). The four tyrosyl residues which are buried may relate mainly to the phosphate moiety of ATP.

Role of the Structural Change

As already evident, the V_m value of ATPase reaction catalyzed by heavy meromyosin, as well as by myosin, changes with conditions such as concentration of KCl and temperature. As described above, the structural change of heavy meromyosin in the presence of $MgCl_2$ is associated with the Michaelis-complex under the conditions to give various V_m values. However, the value of $\Delta\varepsilon$ induced by ATP is almost independent of those conditions. We have been unable to find any definite correspondence between the values of V_m and of $\Delta\varepsilon$. There is no reason, therefore, to expect at present that the structural change of protein reflected in the electronic spectrum plays a role in the ATPase reaction only.

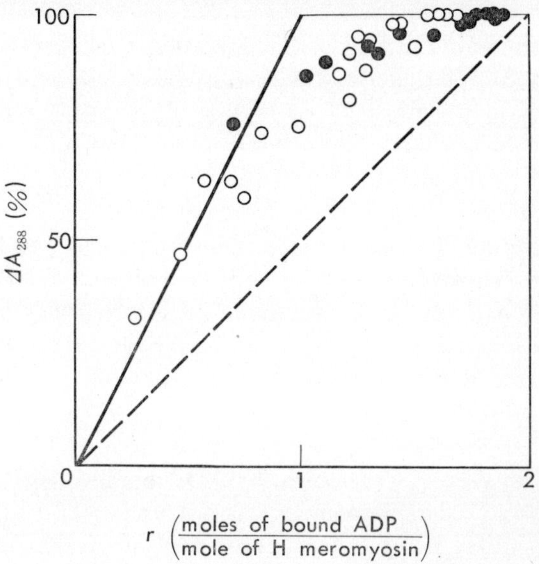

FIG. 7. Comparison between ΔA_{288} of heavy meromyosin induced by ADP and the number of moles of bound ADP per mole of heavy meromyosin (r). r was determined by Sephadex G 25 gel filtration in the presence of 10 mM $MgCl_2$, and 20 mM Tris-HCl (pH 8) at 25°C. Concentration of KCl was 0.08 M (●) and 0.7 M (○).

The primary event of muscle contraction is the sliding of myosin filament along the actin filament (*41*). A structural change in the region of subfragment-1, which forms a cross bridge between both filaments, is expected to occur at the sliding. As described above, the structural change discussed in this paper is restricted to a portion of subfragment-1. A possible contribution of the structural change may be related to the change of the cross bridge, directly giving rise to the sliding of filaments.

Another possibility is that it plays a role in the regulatory mechanism of muscle contraction. It seems to be almost established that the maximum number of the binding of substrate is 2 moles per mole of myosin or heavy meromyosin (*24*, *36*, *42*, *43*). We have attempted to find out the relation between the number of moles of bound substrate and the structural change of protein induced by the substrate. The number of moles of bound ADP per mole of heavy meromyosin (r) was directly measured by using the gel filtration method and was compared with the ΔA_{288} value induced by ADP (*36*). By measuring *r* in as wide a range as possible it was found that the binding of ADP was not a simple first order reaction in the presence of 0.08 M KCl and 10 mM $MgCl_2$ at 25°C, while it seemingly was a first order reaction when KCl concentration was increased to 0.7 M. In Fig. 7 the value of ΔA_{288} induced by ADP is plotted against *r*. In spite of a distinct difference between the modes of ADP binding in the presence of 0.08 M and 0.7 M KCl, the plots of ΔA_{288} *vs*. *r* are indistinguishable from each other. Furthermore, about 90% of the maximum change in ΔA_{288} is observed at *r* of 1. This result indicates heterogeneity between the 2 moles of ADP binding at least with respect to inducing the difference spectrum. Only 1 mole of ADP binding out of the 2 moles is responsible for inducing the structural change of heavy meromyosin. There is a possibility that an interaction exists between the two sites for ADP binding, although the mechanism for this peculiar phenomenon is not understood at present. Whether the two subfragment-1 parts consisting of myosin molecules are identical or not should be a point for further discussion.

It is already known that the maximum value of *r* of PP_i binding decreases from 2 to 1 in the presence of actin (*24*, *42*). Furthermore, the myosin or heavy meromyosin ATPase is activated by the addition of actin (*44–49*). The binding of substrate to myosin, therefore, should be closely related to the binding of actin to myosin. The structural change induced by the substrate reported in this paper, then, may play a role in a regulatory mechanism of the interaction between myosin and actin. Further study is, of course, necessary to obtain a clear answer to this problem.

Summary

The structural change of myosin or heavy meromyosin induced by ATP is detected by means of the spectrophotometric method of absorption and of fluorescence emission of proteins. The change appears to be restricted to the subfragment-1 region of the protein molecules. The structural change is associated with the Michaelis-complex of the ATPase reaction. It may be the displacement of side chains of four tyrosyl residues and one tryptophanyl residue in a way that the chromophores are buried in the interior of the protein moiety. An interaction, probably of charge transfer type, between the tryptophanyl residue and the base moiety of the ATP molecule, is assumed to occur, and it may participate in enlarging the structural change. Only 1 mole of bound substrate out of the 2 moles is responsible for inducing the structural change which may play a role either in the sliding between myosin and actin filaments or in the regulatory function through the binding between myosin and actin.

Acknowledgments
I wish to thank Miss Takiko Shimizu, Mr. Michio Yazawa and Mr. Hidenori Yoshino who have collaborated in various parts of this work. I also wish to thank Dr. Takao Nakamura (Osaka University) and Professor Yasuyuki Ogura for their kind guidance and valuable suggestions on the flow experiments. Thanks are also due to Professor Yuji Tonomura (Osaka University) for his encouragement, and to Professor Koichi Yagi (Hokkaido University) for his continued interest and extensive support in this work.

REFERENCES

1 T. Yasui, Y. Hashimoto and Y. Tonomura, *Arch. Biochem. Biophys.*, **87**, 55 (1960).
2 A. Stracher, *J. Biol. Chem.*, **236**, 2467 (1961).
3 K. Sekiya, S. Mii, K. Takeuchi and Y. Tonomura, *J. Biochem.*, **59**, 584 (1966).
4 H. Takashina and M. Kumagai, *J. Biochem.*, **61**, 768 (1967).
5 T. Sekine and M. Yamaguchi, *J. Biochem.*, **54**, 196 (1963).
6 H. M. Levy, P. D. Leber and E. M. Ryan, *J. Biol. Chem.*, **238**, 3654 (1963).
7 Y. Yamaguchi and T. Sekine, *J. Biochem.*, **59**, 24 (1966).
8 F. Fàbiàn and A. Mühlrad, *Biochim. Biophys. Acta*, **162**, 596 (1968).
9 G. Bailin and M. Bàràny, *Biochim. Biophys. Acta*, **168**, 282 (1968).
10 M. Bàràny, C. Bailin and K. Bàràny, *J. Biol. Chem.*, **244**, 648 (1969).
11 W. B. Gratzer and S. Lowey, *J. Biol. Chem.*, **244**, 22 (1969).
12 H. C. Cheung, *Biochim. Biophys. Acta*, **194**, 478 (1969).
13 J. C. Seidel, M. Chopek and J. Gergely, *Biochemistry*, **9**, 3265 (1970).
14 D. B. Stone, *Arch. Biochem. Biophys.*, **141**, 378 (1970).
15 Y. Tonomura, K. Sekiya, K. Imamura and T. Tokiwa, *Biochim. Biophys Acta*, **69**, 305 (1963).

16 C. M. Kay, W. A. Green and K. Oikawa, *Arch. Biochem. Biophys.*, **108**, 89 (1964).
17 M. R. Iyengar, S. C. Glauser and R. E. Davies, *Biochem. Biophys. Res. Commun.*, **16**, 379 (1964).
18 F. Morita and K. Yagi, *Biochem. Biophys. Res. Commun.*, **22**, 297 (1966).
19 H. Yoshino, F. Morita and K. Yagi, *J. Biochem.*, **71**, 351 (1972).
20 F. Morita, *J. Biol. Chem.*, **242**, 4501 (1967).
21 K. Sekiya and Y. Tonomura, *J. Biochem.*, **61**, 787 (1967).
22 F. Morita and T. Shimizu, *Biochim. Biophys. Acta*, **80**, 545 (1969).
23 H. Mueller, *J. Biol. Chem.*, **240**, 3816 (1965).
24 K. M. Nauss, S. Kitagawa and J. Gergely, *J. Biol. Chem.*, **244**, 755 (1969).
25 E. A. Burshtein, *J. Appl. Spectroscopy*, **6**, 81 (1967) (in Russian).
26 T. Shimizu, F. Morita and K. Yagi, *J. Biochem.*, **69**, 453 (1971).
27 T. Shimizu, F. Morita and K. Yagi, *J. Biochem.*, **69**, 447 (1971).
28 J. E. Godfrey and W. F. Harrington, *Biochemistry*, **9**, 886 (1970).
29 F. Morita, *Biochim. Biophys. Acta*, **172**, 319 (1969).
30 M. Yazawa, F. Morita and K. Yagi, *J. Biochem.*, **71**, 301 (1972).
31 M. Eigen and G. G. Hammes, *Advan. Enzymol.*, **25**, 1 (1963).
32 G. G. Hammes and P. R. Schimmel, *in*" The Enzymes," ed. by P. D. Boyer, Academic Press, New York and London, Vol. II, p. 67 (1970).
33 B. Chance, *J. Biol. Chem.*, **151**, 553 (1943).
34 T. Kanazawa and Y. Tonomura, *J. Biochem.*, **57**, 604 (1965).
35 K. Imamura, M. Tada and Y. Tonomura, *J. Biochem.*, **59**, 280 (1966).
36 F. Morita, *J. Biochem.*, **69**, 517 (1971).
37 S. Yanari and F. A. Bovey, *J. Biol. Chem.*, **235**, 2818 (1960).
38 T. Montenay-Garestier and C. Hèlène, *Biochemistry*, **10**, 300 (1971).
39 M. Cohn and T. R. Hughes, Jr., *J. Biol. Chem.*, **237**, 176 (1962).
40 B. Pullman, *J. Chem. Phys.*, **43**, S233 (1965).
41 H. E. Huxley and W. Brown, *J. Mol. Biol.*, **30**, 383 (1967).
42 Y. Tonomura and F. Morita, *J. Biochem.*, **46**, 1367 (1959).
43 L. H. Schliselfeld and M. Bàràny, *Biochemistry*, **7**, 3206 (1968).
44 L. Leadbeater and S. V. Perry, *Biochem. J.*, **87**, 233 (1963).
45 K. Yagi, T. Nakata and I. Sakakibara, *J. Biochem.*, **58**, 236 (1965).
46 K. Sekiya, K. Takeuchi and Y. Tonomura, *J. Biochem.*, **61**, 567 (1967).
47 E. Eisenberg and C. Moos, *Biochemistry*, **7**, 1486 (1968).
48 E. M. Szentkiralyi and A. Oplatka, *J. Mol. Biol.*, **43**, 551 (1969).
49 E. Eisenberg and C. Moos, *J. Biol. Chem.*, **245**, 2451 (1970).

Received for publication October 15, 1971.

STRUCTURE AND MODE OF ACTION OF Na^+, K^+-ATPase

Makoto Nakao and Kei Nagano*

Department of Biochemistry, School of Medicine, Tokyo Medical and Dental University, Tokyo

Gain of potassium and loss of sodium against the electrochemical gradient is one of the most remarkable features of animal cells. Since Skou (*1*) discovered Na^+, K^+-sensitive ATPase in 1957, a large number of papers on the subject have been published (for reviews, see (*1*) and (*2*)). Today the enzyme is believed to play an important role not only in the active exchange of Na^+ and K^+ but also in the carrier-bound transport of many other substances across cell membranes. In the latter cases, the specific carriers are assumed to be driven by the transcellular gradient of Na^+ concentration which, in turn, is maintained by Na^+, K^+-ATPase (*3*, *4*).

The enzyme Na^+, K^+-ATPase is essentially bound to cell membranes. There are, however, various other ATPases in different cells and subcellular structures other than cell membranes, *e.g.*, actomyosin, sarcoplasmic reticulum, or mitochondrial ATPases. These ATPases differ from each other in their roles in cellular activity, but they share several charac-

* Present address: Laboratory of Biology, Jichi-Medical School. Minamikō-chi-machi, Kōchi-gun, Tochigi.

teristics of the molecular mechanism. For example, their ATP-splitting function is regulated by changes in their allosteric sites. These are the action site(s) of myosin, Ca^{2+} site(s) of the relaxing factor, and Na^+ and K^+ sites of Na^+, K^+-ATPase. The subtle biological control of the function of these enzyme is thus related to allosteric sites rather than to active sites where ATP is hydrolyzed.

Another common property of these ATPases is that they are mechanochemical enzymes, translocating certain other molecules along or across the enzyme molecules themselves. The molecule that is translocated may be of low or high molecular weight. In the case of myosin, it is the companion protein, actin. In the case of Na^+, K^+-ATPase, coupled Na^+ and K^+ translocation occurs. The molecules that are translocated in these cases are, at the same time, allosteric effectors of the enzymes, as noted above.

Thus, while the study of the reaction mechanism of these ATPases aims at elucidating the ATP hydrolysis mechanism, it must also give due attention to the behavior of allosteric sites. Studies along the latter line may be said to be just begun.

In this report the results obtained in this laboratory as well as other basic information on Na^+, K^+-ATPase are summarized. Some of the conclusions must be regarded as tentative and in need of reexamination after sufficient purification of the enzyme, which will be achieved in the near future.

Location of Na^+, K^+-ATPase

The first point to be discussed is the distribution among living organisms and cellular location of the enzyme.

1. *Phylogenetic distribution*

The enzyme has been found in all animal cells after extensive investigations (*2*). Mammalian cells have been most thoroughly investigated, but a considerable amount of data has been accumulated for other classes of animals. The existence of the enzyme in plants has been occasionally reported (*5, 6*) but the data thus far obtained are not very conclusive with descriptions not precise enough for identification of the enzyme.

Evidence has been reported for *Escherichia coli* (*7*). In this case, Na^+, K^+-sensitive activity comprises only about 10% of the total ATPase activity, so that it is not possible to reach a final conclusion at present. Thus, existence of the enzyme in bacteria in general has not yet been confirmed.

2. *Cellular location*

Most investigators agree with the view that the enzyme exists only in cell membranes. However, absence of the enzyme, its proenzyme or modified forms of the enzyme in membranous structures other than cell membranes has not been critically examined.

An interesting unequal distribution on the cell membrane has been observed in the case of asymmetrically situated cells. By " asymmetrical situation " we mean the case in which the cells on one side face outward, and on the other side bordering the " milieu interieur " (Fig. 1).

Epithelial cells of intestinal mucosa are one example. In these cells,

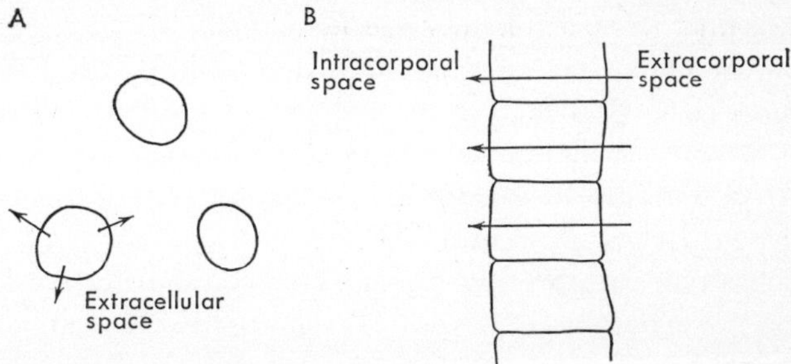

FIG. 1. Schematic representation of (A) red blood cells and (B) epithelial cells as models of symmetrically and asymmetrically situated cells.

TABLE I. Distribution of Sucrase and Na^+, K^+-ATPase in Homogenate of Rat Intestinal Epithelial Cells

Fraction	Protein (mg)	Sucrase (μmoles substrates/hr)	Ouabain-sensitive ATPase (μmoles substrates/hr)	*A*/*S*†
Starting homogenate	100	520 (100%)	580 (100%)	1.1
Combined supernatant	85	187 (36%)	580 (100%)	3.1
Final pellet	15	320 (62%)	25 (4%)	0.08
Recovery	100	507 (98%)	605 (104%)	

" Final pellet " corresponds to partially purified brush border fraction. See Ref. (*8*) for further details. † *A*/*S*: Ratio of the activity of sucrase to that of ouabain-sensitive ATPase.

" outer side " cell membranes facing the intestinal canal have numerous microvilli, while " inner side " cell membranes abutting on connective tissue, neighboring cells, or body fluid do not. Careful preparation followed by sequential sucrose gradient centrifugation separates the microvilli fraction from the inner side cell membranes (*8*).

Examination of each fraction indicates that the fraction containing sucrase, a supposed marker enzyme of microvilli (*9*), is different from the Na^+, K^+-ATPase-rich fraction (Table I). Thus it is concluded that Na^+, K^+-ATPase is present only in the inner side cell membrane.

Animal cells in symmetrical situations (Fig. 1), typically represented by red blood cells but actually including most of the tissue-composing cells, probably have an equal distribution of Na^+, K^+-ATPase all around the cell membrane, although conclusive evidence seems to be difficult to obtain.

Another example of unequal distribution that is expected and tentatively proved is that of toad bladder epithelial cells. The premise and the outline of the experimental design are described in the following.

The short circuit current across the bladder of the toad, *Rana catesbiana*, is satisfactorily explained as being due to the activity of the Na^+, K^+-ATPase in this tissue (*10*), since a close parallel relation is observed between the short circuit current and the activity of Na^+, K^+-ATPase when both are partially inhibited by ouabain (K_i for short circuit current, 4 and 8×10^{-7} M ouabain at 22°C and 14°C; K_i for Na^+, K^+-ATPase,

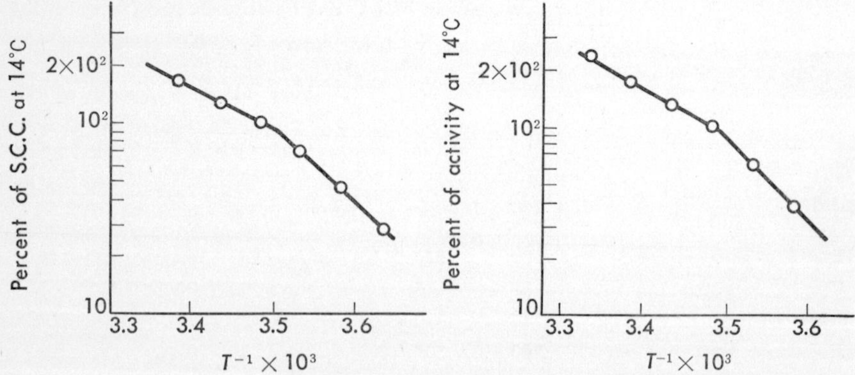

FIG. 2. Arrhenius plot for short circuit current (S.C.C.) and for Na^+, K^+-ATPase activity (*Rana catesbiana*) (*10*). S.C.C. and the enzyme activity are expressed as percent of the values at 14°C.

3 and 6×10^{-7} M ouabain at 22°C and 14°C, respectively). Moreover, both the Na^+, K^+-ATPase activity and the short circuit current show identical temperature dependency (Fig. 2). The Q_{10} values are 2.1 (over 13°C) and 3.7 (under 13°C) for the enzyme activity, and 2.0 (over 13°C) and 3.5 (under 13°C) for the short circuit current. In the determination of Na^+, K^+-ATPase, toad kidney was used instead of bladder as the material tissue for technical reasons, but the above conclusion seems to be valid as discussed in our original paper (*10*).

One of our preliminary results is that the short circuit current, and, consequently, the activity of Na^+, K^+-ATPase, of the toad bladder, was not inhibited by a large molecule inhibitor of Na^+, K^+-ATPase (aminoethyldextran (T-70)-*p*-chloromercuribenzoate, 70,000 daltons, *cf.* (*14*)) if it was applied from the mucosal side of the bladder stretched on a short circuit apparatus. The inhibitor, being a large molecule, is supposedly unable to penetrate the cell membrane, so the above result is again consistent with the view that Na^+, K^+-ATPase is not present on the mucosal side (" outer side ") of the bladder epithelial cells.

Polar Orientation of the Enzyme in Cell Membranes

The polar, or vectorial (*11*), orientation of the functional sites of Na^+, K^+-ATPase on the cell membrane is well established. The enzyme activity of red blood cells is accelerated *in situ* by K^+ outside and Na^+ inside the cells, as shown by the experiments using reversibly hemolyzed red cells (*12*, *13*). Using the same technique, it was confirmed that ATP as substrate, ADP and P_i as products, and the indispensable divalent cation, Mg^{2+}, must interact from inside the cell membrane. Thus some of the essential sites seem to be simultaneously attainable from outside and inside the cell.

A specific inhibitor, ouabain, affects the enzyme when it is added outside the cell, as is the case for K^+. However, it is still uncertain whether the molecules of the drug attack the enzyme really from the outside of the cell membrane since they are of relatively small molecular size. The possibility of interaction within the membrane or penetration of ouabain or K^+ into the cell across the membrane does not yet seem to be excluded. To check this crucial point, a high molecular mercury compound was newly synthesized in our laboratory from aminoethyldextran T-2500 and *p*-chloromercury benzoate. The high

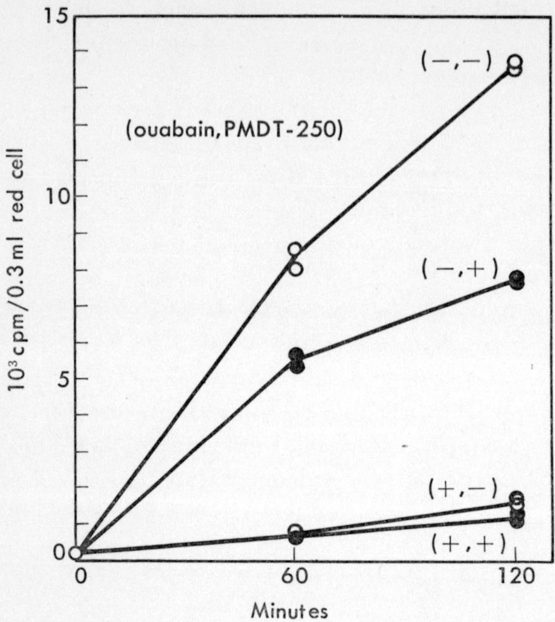

FIG. 3. Effect of PMDT-250 on the cation uptake of human red cells (*14*). K^+-depleted red cells were preincubated with or without 55 μM PMDT-250 and then suspended in media containing 1 mM ^{86}RbCl with or without 2 $\times 10^{-5}$ M ouabain.

molecular SH-inhibitor (PMDT) has an average molecular weight of 250,000.

The inhibitor was applied to intact erythrocytes to examine its effect on the membrane ATPases and the active transport of $^{86}Rb^+$ as a substitute for K^+ ions (*14*). Figure 3 indicates clear inhibition of the ouabain-sensitive fraction of the total $^{86}Rb^+$ influx.

When the reagent was applied to the suspending medium of intact red blood cells, the activity of Na^+, K^+-ATPase measured as the ouabain-senstivie fraction of the total ATPase activity of the cell membrane was strongly inhibited, whereas the ouabain-insensitive ATPase remained intact (Table II-A).

On the contrary, when the reagent was added to the fragmented red cell membrane, the ouabain-insensitive ATPase was also inhibited, so that the (IS/S) ratio was slightly affected in contrast to the case observed

TABLE II. Effects of PMDT-250 on the ATPase Activities of the Red Cell Membrane (*14*)

A. Experiments in intact red cells.

	PMDT-250	*IS*	*S*	(*IS*/*S*)	Relative (*IS*/*S*)
Exp. 1	−	0.54	0.47	1.14	1.00
	+	0.61	0.14	4.5	3.94
Exp. 2	−	0.77	0.33	2.32	1.00
	+	1.00	0.08	13.3	5.73
Exp. 3	−	1.15	0.75	1.54	1.00
	+	0.91	0.27	3.4	2.21

B. Experiments in fragmented ghosts.

	PMDT-250	*IS*	*S*	(*IS*/*S*)	Relative (*IS*/*S*)
Exp. 1	−	0.61	0.87	0.69	1.00
	+	0.28	0.47	0.60	0.87
Exp. 2	−	1.07	1.35	0.79	1.00
	+	0.50	1.02	0.49	0.62

ATPase activities are expressed as μmoles P_i liberated/ml red cells/hr. IS: Activity of ouabain-insensitive ATPase. S: Activity of ouabain-sesitive ATPase.

in the intact cell experiment (Table II-A *vs.* B). This indicates that the site vulnerable to the inhibitory reagent was attainable from outside the cell membrane in the case of ouabain-sensitive ATPase (Na^+, K^+-ATPase), but not in the case of ouabain-insensitive ATPase (Mg^{2+}-ATPase).

From the above results it seems that the molecule of Na^+, K^+-ATPase stretches the whole width from the inside to the outside of the intact cell membrane and probably has a rather bulky molecular dimension.

Molecular Size of Na^+, K^+-ATPase

The molecular size of the enzyme has not yet been precisely determined since purification of the enzyme has not been achieved. Estimate of molecular size has been reported from several laboratories (*15–18*), but the values differ widely between 250,000 and 1,000,000. Combination of several means of determination enabled us to estimate a plausible molecular weight (*15*).

Three methods were used in our experiments. The first was the β-ray inactivation method which depends on the "target theory" (*15*, *17*). The log (dose)-inactivation plot gave a straight line indicating that the molecular weight was about 500,000 daltons (*19*). Using the same technique, formation of phosphorylated enzyme (EP) from ATP (see below) was found to require a higher dosage for inactivation, and the plot gave another straight line giving a molecular size of about 300,000. This result allowed us to surmise the possible subunit structure of the enzyme (*19*, *20*).

The second method of estimation was gel chromatography and re-chromatography of the active enzyme which had been solubilized in the presence of a mild detergent, deoxycholate (DOC) (*15*). The elution yielded a broad plateau in the curve of Na^+, K^+-ATPase activity (Fig. 4), suggesting a multiple-size mixture of the active enzyme. The sharp drop in the activity at the right edge of the plateau corresponded to the smallest particle size of the active Na^+, K^+-ATPase entity and again indicated a value of about 500,000 daltons.

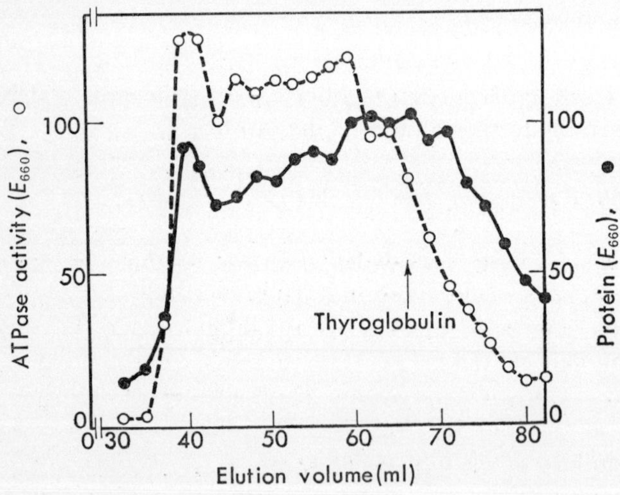

FIG. 4. Gel filtration of "solubilized" Na^+, K^+-ATPase. Sepharose 4B column in 0.1% DOC (*15*).
● protein concentration; ○ Na^+, K^+–ATPase activity.

The third approach was determination of the molecular size from the sedimentation coefficient and partial specific volume (*15*). The sedimentation coefficient was obtained from sucrose density gradient centrifugation of the DOC-solubilized enzyme in a solution containing DOC. A gel chromatography fraction equivalent to a molecular size of 500,000 was analyzed by the above technique.

In order to calculate the apparent molecular weight from this S-value, the partial specific volume was measured in the following way. The sample was mixed with various concentrations of sucrose solution. After centrifugation for several hours, enzyme activities in the upper, middle, and lower layers of the centrifuged solution were compared with each other. If the enzyme was less dense than the sucrose solution, it would float, and if otherwise, it would sink. If the density of the enzyme was equal to that of the sucrose solution, no difference would be observed among the activities in the three layers. A partial specific volume of 0.85 was obtained, since the activity moved upwards in sucrose solutions of $d=1.16$ and 1.17 and moved downwards in a solution of $d=1.18$. This value and the sedimentation coefficient of 14 S gave an apparent particle size of nearly 400,000.

The values obtained by three independent methods coincided quite well. The final conclusion still awaits the purification of the enzyme. However, it may be safe to say that the enzyme has a fairly large size. It also requires three different cations, Mg^{2+}, Na^+, and K^+, probably at least at three different sites. These facts suggest quite a complex structure and mode of reaction of the enzyme.

Phosphorylated Intermediate

The Na^+, K^+-ATPase was labeled with the terminal phosphate of (γ-^{32}P)ATP (*21*, *22*). The phosphorylated enzyme (EP) was obtained when the reaction was stopped by either perchloric acid (PCA) or sodium dodecyl sulphate (SDS). The incorporated phosphate was released immediately as inorganic phosphate (P_i) at alkaline pH but was relatively stable at weakly acidic pH (Fig. 5) (*21*). The hydrolysis at a relatively stable pH followed the first order kinetics. The result indicates that the phosphorylated site on the protein seems to be chemically identical. The half-hydrolysis time under these conditions was about 41 min and 45 min when EP was obtained by SDS or PCA termination of the labeling

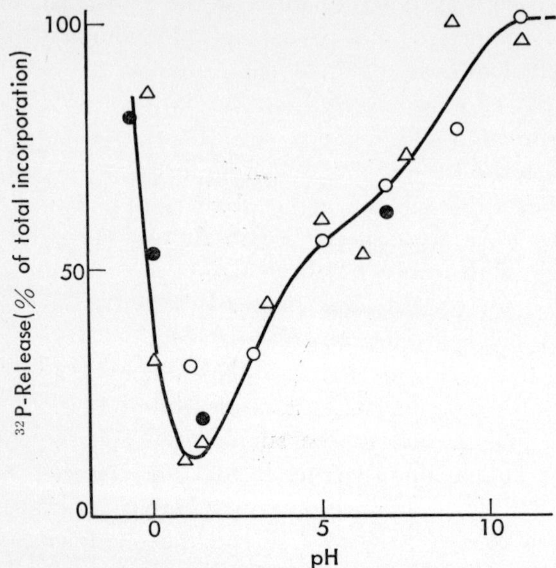

FIG. 5. Effect of pH upon the rate of phosphate liberation from phosphorylated Na^+, K^+-ATPase (EP) (*22*). Different symbols represent three different lots of Na^+, K^+-ATPase preparations.

reaction, respectively. Thus, EP's obtained under the two different conditions seem to be identical.

EP was also homogeneous as a macromolecular entity. When the dissolved EP in SDS was run through a Sephadex G-200 or Sepharose 6B column, a symmetrical labeled peak was obtained at a position different from the bulk protein peak (Fig. 6, (*23*)). The labeled fraction appeared at the same position on rechromatography. From the elution curve the apparent molecular size of the solubilized EP was estimated to be about 300,000 daltons. This value corresponds well with the estimate made based on the β-ray inactivation of the phosphorylation activity described above.

At neutral pH EP was liable to hydroxylaminolysis (*22*). This result, considered together with the pH-stability relationship described above, suggests the acylphosphate nature of the incorporated phosphate.

The amount of EP was maximum when the labeling was performed in the presence of Na^+ alone as a monovalent cation. When K^+ was added to the reaction mixture, the level of phosphorylation rapidly

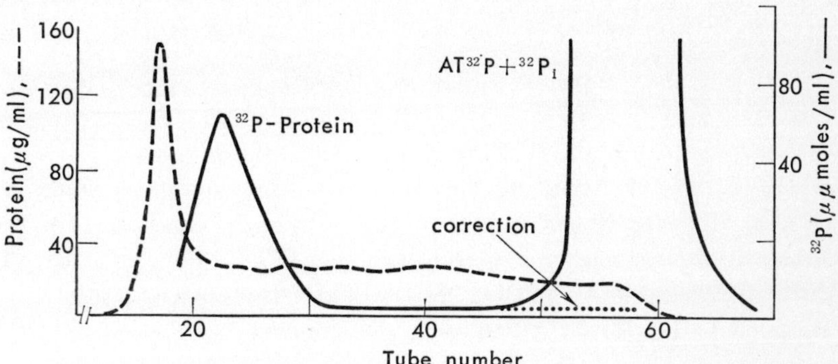

FIG. 6. Separation of the phosphorylated protein as a single radioactive peak by gel filtration through Sephadex G-200 (*23*). —— protein concentration; ---- ^{32}P (cpm)

lowered. However, the amount of EP obtained in the presence of both Na^+ and K^+ was identical with that obtained in the presence of Na^+ alone.

These results are consistent with the following reaction scheme where EP is the intermediate, and Na^+ and K^+ are required for its formation (phosphorylation) and decomposition (dephosphorylation), respectively,

$$E + ATP \xrightarrow{Na^+} EP \xrightarrow{K^+} E + P_i . \tag{1}$$

K^+-p-Nitrophenylphosphatase Activity

The Na^+, K^+-ATPase preparation showed K^+-dependent *p*-nitrophenylphosphatase (K^+-pNPPase) activity (*24–26*). The K^+-pNPPase was inhibited by ouabain and was supposed to be intimately related to the Na^+, K^+-ATPase activity. The simplest assumption is that K^+-pNPPase represents the latter half of the Na^+, K^+-ATPase reaction (Reaction 1), *i.e.*, a K^+-dependent dephosphorylation step.

The fact that ATP enhances the K^+-pNPPase activity under certain conditions (*24*, *25*, *27*), however, makes this interpretation unlikely, since ATP should always inhibit it if ATP and pNPP are hydrolyzed *via* the same EP. Moreover, Na^+ inhibits the K^+-pNPPase activity in sharp contast to the ATPase activity which requires the simultaneous presence of the two cations (Fig. 7-B). Revised reaction schemes may be proposed (*28*) where ATP and pNPP form different EP's. The simplest form may be

$$ATP+E_1 \xrightarrow{Na^+} E_1P \longrightarrow E_2P \xrightarrow{K^+} E_1+P_i \quad (2)$$

$$pNPP+E_1 \longrightarrow E_2P \xrightarrow{K^+} E_1+P_i, \quad (3)$$

where E_1 and E_2 indicate two different conformational states of the enzyme. A mechanism similar to Reaction 3 is also assumed for another substrate of this enzyme, acetylphosphate (*29*). The presence of two different states of EP in Na^+, K^+-ATPase reaction (Reaction 2) is also supported (*30*, *31*) by other lines of evidence.

Na^+, K^+-ITPase Activity

ITP and CTP can substitute for ATP as substrates of the enzyme. The

TABLE III. Comparison of K_m's and K_i's for Ions in Na^+, K^+-ATPase and -ITPase, and K^+-pNPPase Activity

A.

Ions	ATPase		ITPase		pNPPase	
	(mM)	K_m (Na^+)	(mM)	K_m (Na^+)	(mM)	K_i (Na^+)
Cs^+	(40)	14 mM	(10)	6 mM	(40)	20 mM
NH_4^+	(26)	5.7	(6.4)	7		
K^+	(14)	9	(2.6)	8	(14)	16
Rb^+	(16)	12	(0.64)	8.2		
Tl^+	(1.6)	9	(0.16)	7	(1.6)	24

B.

Ions	ATPase (Na^+ 140)	ITPase (Na^+ 100)	pNPPase
Cs^+	10 mM	1.3 mM	30 mM
NH_4^+	9	1.2	4.0
K^+	1.3	0.14	2.7
Rb^+	1.2	0.12	1.4
Tl^+	0.3	0.04	0.3

In A), K_m or K_i for Na^+ was determined in the presence of one of the "K^+-series" ions at the concentration indicated in parentheses. In B), K_m for each one of the "K^+-series" ions was determined in the presence of Na^+ (concentration shown in parentheses) or without Na^+ (in the case of pNPPase).

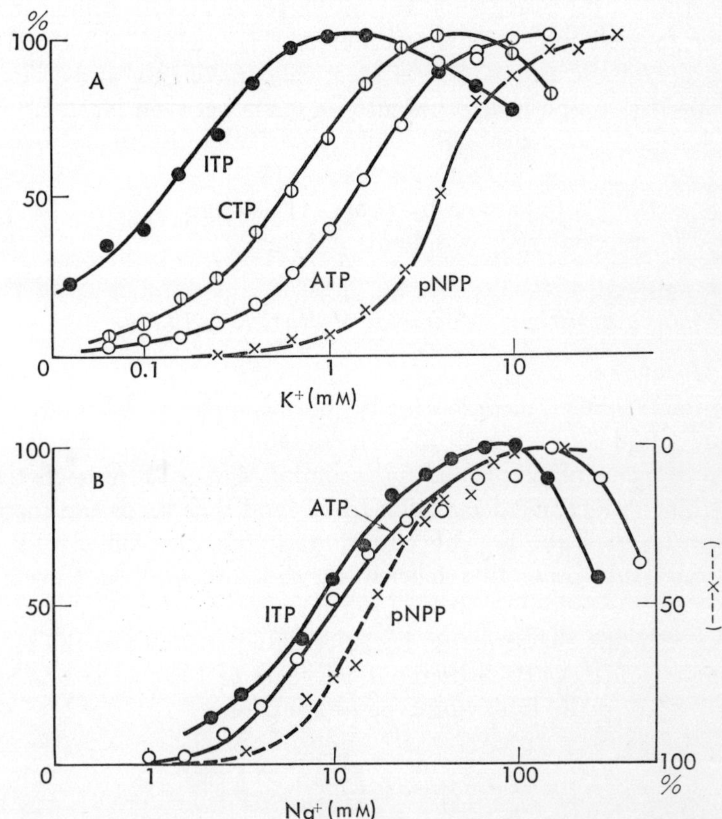

FIG. 7. K_m's of Na^+, K^+-ATPase preparation for various substrates. A: Na^+=100 mM (ITP), 140 mM (ATP and CTP), or without Na^+ (pNPP). B: K^+=2.6 mM (ITP), 14 mM (ATP), or 18 mM (pNPP).

Na^+, K^+-CTPase and -ITPase activities amount to about 20 and 8–10% of that of Na^+, K^+-ATPase, respectively. They share the characteristics of requiring Na^+ and K^+ (Fig. 6) and of being inhibited by ouabain. The $E^{32}P$ from $AT^{32}P$ and that from $IT^{32}P$ are found to be identical (*32*).

The K_m's for Na^+ in the presence of a sufficient amount of K^+, Rb^+, Cs^+, NH_4^+ or Tl ("K^+-series" ions) are roughly the same among these different ions and also between different substrates (ATP or ITP, Table III-A).

On the contrary, K_m's for K^+ in the presence of a sufficient amount of Na^+ varied widely among different "K^+-series" ions. However, the order of the five K_m's was identical for ATPase and ITPase.

A similar comparison could also be made between Na^+, K^+-ATPase and K^+-pNPPase, although comparison had to be made between K_m (ATPase) and K_i(pNPPase) for Na^+.

The results summarized in Table III strongly suggest that all these activities are catalyzed by a single enzyme.

ATP as the Sensitivity Modulator of Na^+, K^+-ATPase

1. Modulation of K_m for K^+

Both the maximal reaction velocity and the apparent K_m for K^+ differed about tenfold between Na^+, K^+-ATPase and -ITPase. This discrepancy could be plausibly explained by assuming that ATP protects the Na^+-site from competitive occupation by K^+ and thus keeps the enzyme uninhibited, unless the K^+ concentration is too high. When ITP was the substrate, this protection mechanism did not work and even a relatively low concentration of K^+ behaved as a competitive inhibitor for the Na^+-site, while the same ion is the essential activator at the K^+-site.

The effect of ATP, or the ineffectiveness of ITP, corresponds to different K'_b's in the following type of equation.

$$v=\frac{V}{\left[1+\frac{K_a}{a}\left(1+\frac{b}{K'_a}\right)\right]\left[1+\frac{K_b}{b}\left(1+\frac{a}{K'_b}\right)\right]}, \tag{4}$$

where a and b represent concentrations of Na^+ and K^+, and K_a, K'_a, K_b, and K'_b represent dissociation constants and inhibition constants (with primes) at the Na^+-site (suffix a) and K^+-site (suffix b), respectively (*33*). Quantitative treatment may be more complicated,* but a rough idea may be evident from Fig. 8.

* A more accurate formulation may be

$$v=\frac{V}{\left[1+\frac{K_a}{a}\left(1+\frac{b}{K'_a}\right)\right]^n\left[1+\frac{K_b}{b}\left(1+\frac{a}{K'_b}\right)\right]^m}. \tag{5}$$

Two or three sites may be involved for both Na^+ and K^+ (*34, 35*). Estimation of m and n shows contradictions among investigators and experimental designs. Some authors assume allosteric interactions between the sites (*28, 35, 36*).

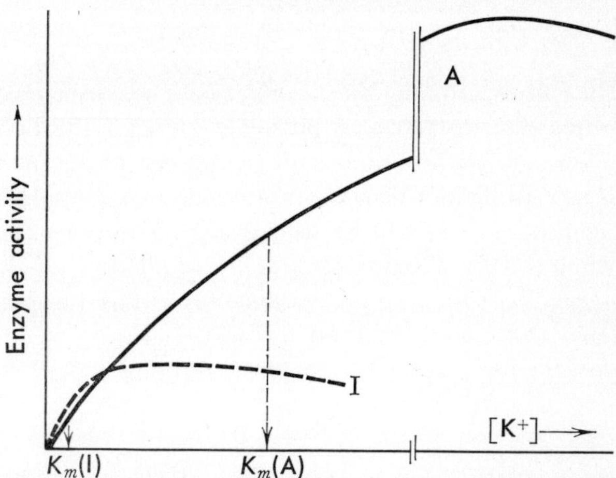

FIG. 8. Schematic illustration of the effect of different K'_b. For details see text and Reaction 1.

The effector site for ATP on the enzyme may be different from the substrate site for hydrolysis. If this is the case, K_m of Na^+, K^+-ITPase for K^+ will be greater in the presence of ATP, since some of the enzyme molecules will bind ATP on the effector site while ITP is bound to the substrate site to be hydrolyzed. Na^+, K^+-ITPase was assayed using $IT^{32}P$ with or without added cold ATP and counting the $^{32}P_i$ released. K_m for K^+ was about 0.12 mM and 0.45 mM in the absence and presence of 1 mM ATP, respectively.

2. K_i for ouabain

The K_i's of Na^+, K^+-ATPase and -ITPase, and K^+-pNPPase for ouabain also differ from each other (0.15, 3.0 and 8.0×10^{-6} M, respectively). The enzyme is apparently more sensitive to ouabain at low K^+ concentrations, and since ATP makes K^+ less effective (see above), the K_i's of Na^+, K^+-ITPase and K^+-pNPPase are likely to be smaller in the presence of ATP. In fact, in the presence of ATP, K_i's were 0.12 and 2.5×10^{-6} M for Na^+, K^+-ITPase and K^+-pNPPase, respectively.

The above results combined suggest that ATP is, in addition to being a physiological substrate, a potent sensitivity-modulating effector for the enzyme.

Concluding Remarks

The fundamental features of the Na^+, K^+-ATPase reaction mechanism are now established: formation of a high-energy acylphosphate intermediate, dependence of its formation and decomposition upon Na^+ and K^+, and possible participation of conformational change of the enzyme in the active transport of Na^+ and K^+ across cell membranes. Further details of the whole reaction sequence and its regulation especially by ATP itself, however, are in need of clarification. The purification of the enzyme, hoped to be accomplished in the near future, seems to be a critical prerequisite for a complete understanding.

SUMMARY

The present state of studies on Na^+, K^+-ATPase is discussed, mainly in terms of 1) its phylogenetic distribution and cellular localization; 2) its polar orientation in the cell membrane; 3) its molecular size and possible subunits; 4) separation and properties of the phosphorylated intermediate and its plausibility as the reaction intermediate; 5) relation to other Na^+, K^+-dependent activities observed with several other substrates, especially pNPP and ITP; and 6) features of ATP as the sensitivity modulator for K^+ and ouabain.

Acknowledgments

Kind advice and criticism by Dr.H, Tamiya and Dr. Y. Ogura are gratefully acknowledged.

REFERENCES

1 J. C. Skou, *Physiol. Rev.*, **45**, 596 (1965).
2 S. L. Bonting, *in* " Membranes and Ion Transport," ed. by E. E. Bittar, Wiley-Interscience, New York, Vol. I, p. 257 (1970).
3 H. N. Christensen, *in* " Membrane and Ion Transport," ed. by E. E. Bittar, Wiley-Interscience, New York, Vol. I, p. 365 (1970).
4 W. D. Stein, " The Movement of Molecules across Cell Membranes," Academic Press, New York, p. 177 (1967).
5 Y. F. Lai, and J. E. Thompson, *Biochim. Biophys. Acta*, **233**, 84 (1971).

6 H. D. Brown, S. K. Chattopadhyay and A. Patel, *Enzymologia*, **32**, 205 (1967).
7 J. C. M. Hafkenscheid and S. L. Bonting, *Biochim. Biophys. Acta*, **178**, 128 (1969).
8 M. Fujita, H. Matsui, K. Nagano and M. Nakao, *Biochim. Biophys. Acta*, **233**, 404 (1971).
9 D. Miller and R. K. Crane, *Biochim. Biophys. Acta*, **52**, 293 (1961).
10 Y. Asano, Y. Tashima, H. Matsui, K. Nagano and M. Nakao, *Biochim. Biophys. Acta*, **219**, 169 (1970).
11 R. Whittam and M. E. Ager, *Biochem. J.*, **93**, 337 (1964).
12 A. K. Sen and R. L. Post, *J. Biol. Chem.*, **239**, 345 (1964).
13 I. M. Glynn, *J. Physiol.*, **160**, 18P (1962).
14 H. Ohta, J. Matsumoto, K. Nagano, M. Fujita and M. Nakao, *Biochem. Biophys. Res. Commun.*, **42**, 1127 (1971).
15 N. Mizuno, K. Nagano, T. Nakao, Y. Tashima, M. Fujita and M. Nakao, *Biochim. Biophys. Acta*, **168**, 311 (1968).
16 A. Kahlenberg, N. C. Dulak, J. F. Dixon, P. R. Galsworthy and L. E. Hokin, *Arch. Biochem. Biophys.*, **131**, 153 (1969).
17 G. R. Kepner and R. I. Macey, *Biochim. Biophys. Acta*, **163**, 188 (1968).
18 J. Kyte, *Biochem. Biophys. Res. Commun.*, **43**, 1259 (1971).
19 M. Nakao, K. Nagano, N. Mizuno, Y. Tashima, M. Fujita, H. Maeda and H. Matsudaira, *Biochem. Biophys. Res. Commun.*, **29**, 588 (1967).
20 M. Nakao, K. Nagano, H. Matsui, N. Mizuno, T. Nakao and Y. Tashima, *in* "The Molecular Basis of Membrane Function," ed. by D. C. Tosteson, Prentice-Hall, Inc., New York, p. 539 (1969).
21 K. Nagano, T. Kanazawa, N. Mizuno, Y. Tashima, T. Nakao and M. Nakao, *Biochem. Biophys. Res. Commun.*, **19**, 759 (1965).
22 K. Nagano, N. Mizuno, M. Fujita, Y. Tashima and M. Nakao, *Biochim. Biophys. Acta*, **143**, 239 (1967).
23 A. Schwartz, K. Nagano, M. Nakao, G. E. Lindenmayer, J. C. Allen and H. Matsui, *in* "Methods in Pharmacology," Appleton-Century-Crofts, New York, Vol. I, p. 361 (1971).
24 M. Fujita, T. Nakao, Y. Tashima, N. Mizuno, K. Nagano and M. Nakao, *Biochim. Biophys. Acta*, **117**, 42 (1966).
25 H. Yoshida, F. Izumi and K. Nagai, *Biochim. Biophys. Acta*, **120**, 183 (1966).
26 J. D. Robinson, *Arch. Biochem. Biophys.*, **139**, 164 (1970).
27 H. Yoshida, K. Nagai, T. Ohashi and N. Yoshiko, *Biochim. Biophys. Acta*, **171**, 178 (1969).
28 J. D. Robinson, *Biochim. Biophys. Acta*, **212**, 509 (1970).
29 W. F. Dudding and C. G. Winter, *Biochim. Biophys. Acta*, **241**, 650 (1971).

30 R. L. Post, S. Kume, T. Tobin, B. Orcutt and A. K. Sen, *J. Gen. Physiol.*, **54**, 306S (1969).
31 R. W. Albers, *Ann. Rev. Biochem.*, **36**, 727 (1967).
32 K. Nagano, unpublished data.
33 A. L. Green and C. B. Taylor, *Biochem. Biophys. Res. Commun.*, **14**, 118 (1964).
34 K. Ahmed, J. D. Judah and P. G. Scholefield, *Biochim. Biophys. Acta*, **120**, 351 (1966).
35 B. Formby and J. Clausen, *Hoppe-Seyler's Z. Physiol. Chem.*, **350**, 973 (1969).
36 R. F. Squire, *Biochem. Biophys. Res. Commun.*, **19**, 27 (1965).

Received for publication December 7, 1971.

PEPTIDE BOND FORMATION ON RIBOSOMES

Takahisa Ohta
Department of Agricultural Chemistry, Faculty of Agriculture, University of Tokyo, Tokyo

In protein biosynthesis, the ribosome serves as a machine on which the specific nucleotide sequence of mRNA is translated into the corresponding amino acid sequence. Some protein factors work together in the reaction on the ribosome. The unique signal to determine initiation points of the peptide bond formation is encoded in messenger RNA. Initiation factors and the ribosome decode the initiation signal and form an initiation ribosomal complex with an initiator tRNA. One of two kinds of methionine tRNA, in which an α-amino group is formylated, serves as the initiator RNA in bacterial systems. The nucleotide sequence corresponding to the initiator tRNA has been found to be AUG, so that synthetic oligonucleotide, A_pU_pG, $A_pU_pG_pU_pU_pU_p \ldots$, or a similar nucleotide can be used as a template in the *in vitro* system for peptide synthesis.

The binding of aminoacyl-tRNA to ribosomes is schematically shown

The following abbreviations are used: AUG, A_pU_pG; aa-tRNA: aminoacyl-transfer RNA; fMet-tRNA, N-formylmethionyl-transfer RNA; fMet-puromycin, N-formylmethionyl-puromycin; GMP-PCP, 5'-guanylyl-methylenediphosphonate.

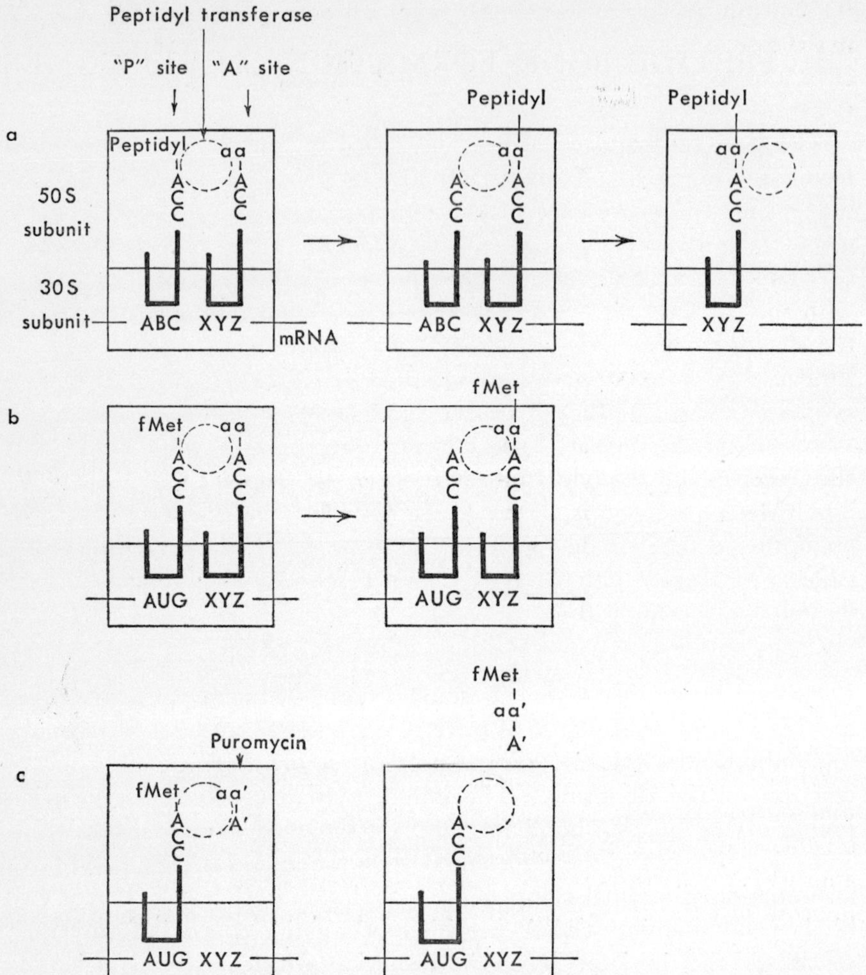

FIG. 1. Peptide bond formation on the ribosome.

in Fig. 1. It is postulated that the ribosome has two binding sites for tRNA. One is called the " peptidyl " site (P site), and the other the " aminoacyl " site (A site). A growing peptide is anchored on the P site as peptidyl-tRNA, and aminoacyl-tRNA is supplied to the A site by virtue of the T factor. Peptide is transferred from peptidyl-tRNA to the α-amino group of aa-tRNA at the A site by the peptidyl-transferase, the constitutive protein(s) of the larger ribosomal subunit (50S subunit in the case of bacteria). The G factor moves the newly formed peptidyl-tRNA from the A site to the P site, accompanying the hydrolysis of GTP.

The same mechanism has been proposed for the initiation of peptide formation (Fig. 1-b). Formylmethionyl-tRNA may be finally located on the P site in the presence of the template containing the AUG code word, the initiation factors and GTP, as shown in Fig. 1 (*1*, *2*). Amino-acyl-tRNA, corresponding to the code next to AUG, enters the A site with the T factor and GTP, and then the formylmethionyl residue is transferred to the aminoacyl-tRNA to form fMet-aa-tRNA (*3*). A similar reaction has been reported in the case of the inhibition of peptide synthesis by an antibiotic, puromycin. The antibiotic, which has a structure similar to that of the terminal part of aa-tRNA, enters the A site, receives the formylmethionyl residue, and forms fMet-puromycin. The fMet-puromycin is released from the ribosome resulting in the interruption of the peptide synthesis (Fig. 1-c). After the formation of the ribosomal complex with fMet-tRNA, the reaction with puromycin occurs within 1 min even at 0°C.

Peptide Transfer Stimulated by Solvents

The formation of the puromycin derivatives requires the binding of fMet-tRNA or peptidyl-tRNA at the P site on the ribosome with the factors and GTP. Monro found that the reaction of the peptidyl transfer on the ribosome is stimulated by alcohol (*4*). In the presence of an appropriate concentration of alcohol, fMet-puromycin is formed in a mixture of fMet-tRNA, puromycin, ribosomes, Mg^{2+} and monovalent cation, such as K^+ or NH_4^+. The factors, GTP and a template are not required in the reaction. Further studies revealed that the reaction stimulated by alcohol does not require the whole structures of the ribosome and fMet-tRNA, but only the 50S ribosomal subunit and the fragment of 3′ terminal part of fMet-tRNA (*5*).

In order to elucidate the mechanism of the reaction of peptide transfer, we examined at which step alcohol stimulates the reaction. Ribosomes were preincubated with some components of the reaction mixture, and then the reaction was started by the addition of the remainder. Both the preincubation and the reaction were carried out at 0°C. Aliquots of the

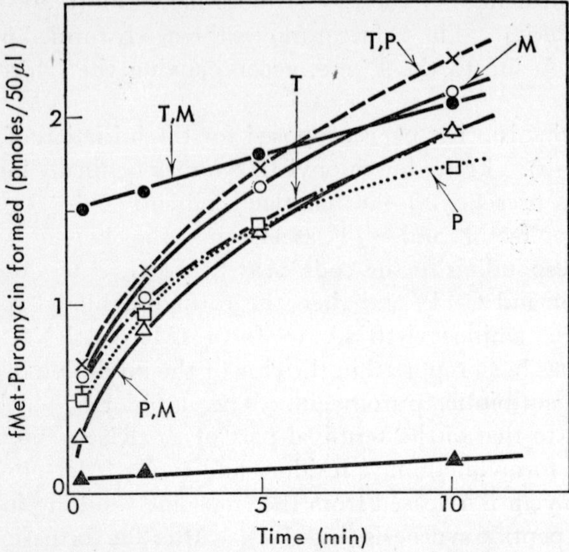

FIG. 2. Effects of preincubation on the stimulation of peptidyl transferase activity. Ribosomes were preincubated in the presence of buffer and ions for 15 min with tRNA (T) and methanol (M)(●), with tRNA and puromycin (P)(×), with tRNA(+), with puromycin (□), with puromycin and methanol (△). Reactions were started by the addition of the remaining components, with methanol (○) or without additives (▲). Preincubation and reaction were carried out at 0°C. The final reaction mixture contained in a volume of 0.1 ml: Tris-HCl, pH 7.4, 50 mM; KCl 400 mM; $Mg(OAc)_2$ 20 mM; ribosomes 1.8 A_{260} unit; puromycin 50 μg; charged tRNA; 5 pmoles as fMet-tRNA, and methanol 20 μl.

reaction mixture were taken out at appropriate intervals and added to a solution of beryllium chloride to stop the reaction. Formylmethionyl-puromycin was determined by the method used by Leder and Bursztyn (*6*).

Only one case, in which ribosomes and fMet-tRNA were incubated in the presence of methanol prior to the addition of puromycin, showed the formation of a considerable amount of fMet-puromycin immediately after the start of the reaction (Fig. 2). In all other cases, extrapolation of the amount of fMet-puromycin to zero time gave extremely small values. Formylmethionyl-tRNA bound enzymatically to ribosomes reacted very rapidly with puromycin, as mentioned previously. Therefore, the early formation of fMet-puromycin indicates the presence of a similar complex between fMet-tRNA and ribosomes. Detection of the complex by its retention on a membrane filter was not successful. Methanol probably stimulates the interaction between fMet-tRNA and ribosomes to form a weak complex between them. Specific interaction between ribosomes and fMet-tRNA promoted by alcohol was also reported by Monro's group (*6*).

Specificity of solvents for the reaction has been studied by Monro's group (*7*). Their experiments, using a 50S ribosomal subunit and tRNA fragments, showed that methanol, ethanol, isopropanol and acetone were active in the stimulation of peptide transfer. Other solvents, including ethylene glycol, dimethylformamide did not promote the reaction. They concluded that the presence of a hydroxyl group was important and that the activity of acetone is probably a result of tautomerism forming the enol form.

Similar experiments were attempted using fMet-tRNA and 70S ribosomes. As expected, methanol, ethanol, isopropanol and acetone were active in these experiments (Fig. 3-a). However, *n*-propanol and dimethylformamide, which have been reported as inactive solvents, also stimulated the formation of fMet-puromycin (Figs. 3-a and-b). Although ethylene glycol (20%) was inactive, the replacement of its hydroxyl group by a methoxy group turned it into an active solvent (Fig. 3-a).

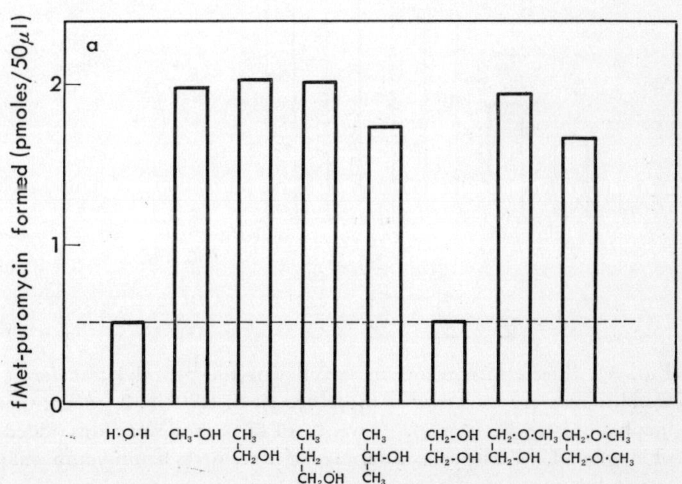

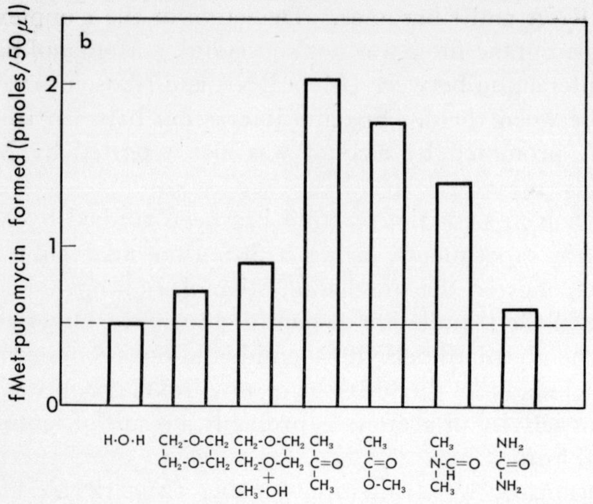

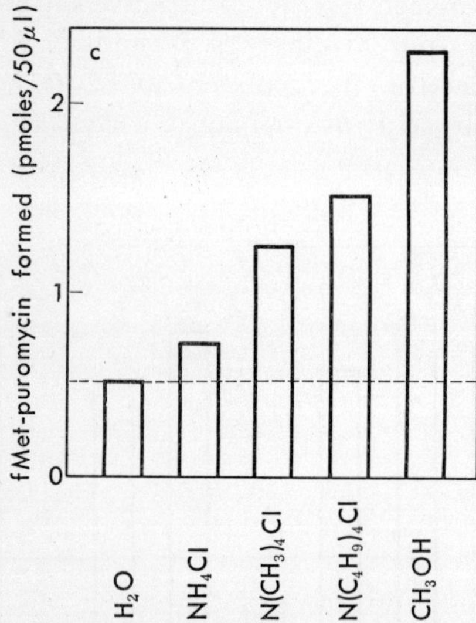

FIG. 3. Effects of solvents in stimulating the peptidyl transferase activity. Conditions were the same as those described for Fig. 2, except that no pre-incubation was performed. Twenty μl of each solvent was added in place of methanol. Final concentrations of tetramethylammonium chloride and tetrabutylammonium chloride were 1M and 2M, respectively.

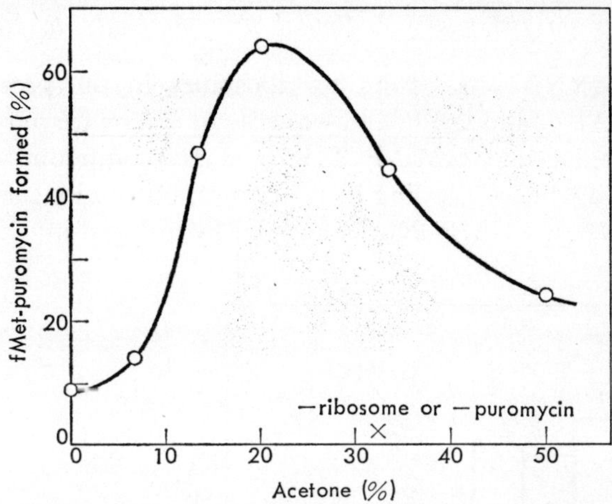

FIG. 4. Effect of concentration of acetone on the peptidyl transferase activity. The reaction mixture contained in a volume of 0.15 ml: Tris-HCl, pH 7.4, 33 mM; KCl 270 mM; $Mg(OAc)_2$ 13 mM; ribosomes $2A_{260}$ units; charged tRNA 5 pmoles as fMet-tRNA; puromycin 50 μg; and acetone as indicated in the figure. No correction was made for the volume change caused by mixing. Temperature: 0°C.

Significant activity of ethylene gylcol dimethylether indicates that the presence of a hydroxyl group is not always necessary for the stimulation of the reaction. Methylacetate and dimethylformamide do not contain the hydroxyl group either. An important requirement for the stimulating activity of solvents seemed to be the presence of an alkyl group rather than a hydroxyl group. Thus, the effects of tetraalkylammonium salts were investigated. Although their stimulating activities were not so high, tetramethylammonium and tetrabutylammonium chlorides were active while ammonium chloride at similar concentrations was not (Fig. 3-c). The effect of concentrations of acetone is shown in Fig. 4. At the optimal concentration, the stimulating activity of acetone was as high as that of methanol. Dioxane was inactive and rather inhibitory for stimulation by methanol, probably by its specific interaction at the active site. The discrepancy between our results and those reported previously cannot be explained at the present time. Further experiments using the 50S subunits must be carried out.

Effects of Solvents on fMet-tRNA Bound Enzymatically to Ribosomes

Formylmethionyl-tRNA was bound to ribosomes in the presence of initiation factors, AUG and GTP, and was able to react with puromycin. When GTP was replaced by GMP-PCP, an analogue compound of GTP, fMet-tRNA was capable of binding to ribosomes, but no longer reacted with puromycin (*1–3*). GMP-PCP cannot be hydrolyzed into GDP by

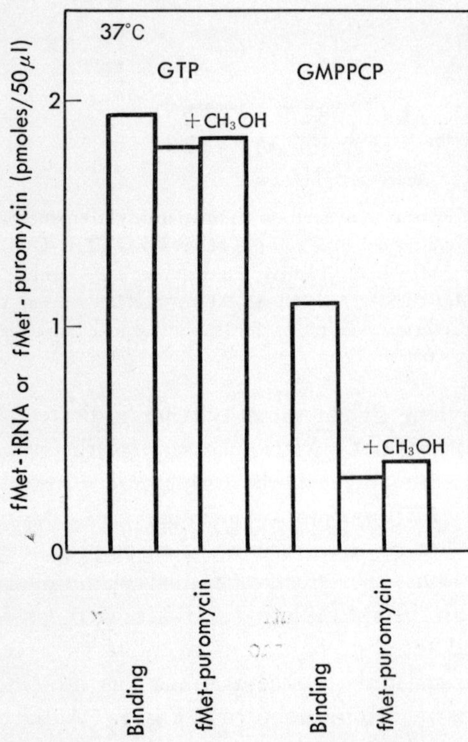

FIG. 5. Effect of methanol on the formation of fMet-puromycin from fMet-tRNA bound enzymatically to ribosomes. The binding of fMet-tRNA to ribosomes was carried out at 37°C for 3 min. The reaction mixture contained in a volume of 0.05 ml: Tris-HCl, pH 7.4, 50mM; NH_4Cl 300 mM; $Mg(OAc)_2$ 6 mM; ribosomes $2A_{260}$ units; charged tRNA 20.5 pmoles as fMet-tRNA; AUG 3 μg; GTP or GMP-PCP 1 mM. Thirty μg of puromycin was added in the presence or absence of methanol at 0°C to the reaction mixture after the binding reaction. Incubation for the puromycin reaction was carried out at 0°C for 5 min.

the replacement of oxygen between β and γ phosphorus with a methylene group. Aminoacyl-tRNA bound enzymatically to the ribosomes with the T factor in the presence of GMP-PCP was also inactive in peptide bond formation (*8*). Therefore, the reactivity with peptidyl transferase seems to be produced by the hydrolysis of GTP.

If fMet-tRNA in the presence of GMP-PCP was locked at an entry site or at a site where the tRNA could not react with peptidyl transferase, free fMet-tRNA may be able to react with peptidyl transferase on the addition of the solvent. Enzymatic binding of fMet-tRNA in the presence of GTP or GMP-PCP was carried out at 37°C, and aliquots were taken out for the binding assay and for the puromycin reaction at 0°C in the presence and absence of methanol. The results shown in Fig. 5 indicate that fMet-tRNA probably occupied a site near peptidyl transferase even in the presence of GMP-PCP, and the bound fMet-tRNA as well as free fMet-tRNA was unable to form a puromycin derivative on the addition of methanol.

These experiments may suggest the following mechanism. In the presence of initiation factors, GTP and the template, fMet-tRNA binds to the site near the peptidyl transferase. The hydrolysis of GTP turns fMet-tRNA to a state reactive with puromycin. Some solvent containing an alkyl group may alter the structure of the ribosome (and probably of tRNA) resulting in the formation of a reactive complex similar to that formed enzymatically.

Summary

The activity of peptidyl transferase on ribosomes of *Escherichia coli* was measured by determining the formation of formylmethionyl-puromycin from formylmethionyl-tRNA and puromycin. Several solvents as well as tetraalkylammonium ions stimulated the activity. Preincubation experiments showed that the stimulation occurred at the interaction step of fMet-tRNA with ribosomes. Methanol did not stimulate the fMet-puromycin formation from fMet-tRNA bound enzymatically to ribosomes in the presence of GMP-PCP.

REFERENCES

1 T. Ohta, S. Sarker and R. E. Thach, *Proc. Natl. Acad. Sci. U.S.*, **58**, 1638 (1967).
2 D. Kolakofsky, T. Ohta and R. E. Thach, *Nature*, **220**, 244 (1968).
3 T. Ohta and R. E. Thach, *Nature*, **219**, 238 (1968).
4 R. E. Monro, *J. Mol. Biol.*, **26**, 147 (1967).
5 R. E. Monro, T. Staehelin, M. L. Celma and D. Vazquez, *Cold Spring Harbor Symp. Quant. Biol.*, **34**, 357 (1969).
6 P. Leder and H. Bursztyn, *Biochem. Biophys. Res. Commun.*, **25**, 233 (1966).
7 M. L. Celma, R. E. Monro and D. Vazquez, *Fed. Eur. Biochem. Soc. Letters*, **6**, 273 (1970).
8 Y. Ono, A. Skoultchi, J. Waterson and P. Lengyel, *Nature*, **222**, 645 (1969).

Received for publication December 20, 1971.